Unified Theoretical Analysis of Nonlinear Multicarrier Schemes

This book provides the analytical tools to characterize nonlinear distorted multicarrier signals and optimal/sub-optimal receivers employed in high data rate communication systems.

Unified Theoretical Analysis of Nonlinear Multicarrier Schemes introduces new optimal and sub-optimal receivers for nonlinear distorted signals that can use nonlinear distortion to improve performance when compared with common receivers. It addresses the analysis of nonlinear systems with stochastic inputs and establishes new receivers designs for multi-carrier communication systems with nonlinearities. The authors also include the characterization and definition of optimum and sub-optimum receivers for nonlinear distorted signals that may use the nonlinear distortion to improve receiver performance over the existing ones. The book also includes a set of applications where the analytical unified method for characterization of nonlinear distortion can be applied. The two final chapters of the book include systems like MIMO OFDM and the extension to optical systems. These techniques are the base of 5G, Wi-fi and future 6G mobile networks.

The book will be a valuable resource for design engineers, industrial engineers, applications engineers and researchers working on multi-carrier systems, power amplifiers modelling and design.

Paulo Montezuma received his Ph.D. degree from Faculdade de Ciências e Tecnologia - Universidade Nova de Lisboa, Portugal. Currently, he is a professor at FCT-UNL. Since 2013, he has been a researcher at Instituto de Telecomunicações. He has been actively involved in several international research projects in the wireless communications area. His main research activities are on transmitter design, coding, and receiver design, namely for MIMO systems, OFDM, and SC-FDE modulations.

João Guerreiro received his Ph.D. in Electrical and Computer Engineering from Faculdade de Ciências e Tecnologias da Universidade Nova de Lisboa (FCT-UNL) in 2016. He is currently an assistant professor at FCT-UNL and a researcher at Instituto de Telecomunicações (IT). Since 2013, he has been actively involved in scientific research in wireless communications and has participated in several national and international research projects. His main research interests are signal processing, transceiver design, and channel modeling for single- and multi-antenna wireless communications.

Rui Dinis received his Ph.D. degree from Instituto Superior Técnico, Portugal. From 2001 to 2008, he was a Professor at IST. Currently, he is an associate professor at FCT-UNL. Since 2009, he has been a researcher at Instituto de Telecomunicações. He has been involved in several international research projects in the broadband wireless communications area. His main research activities are on modulation and transmitter design, nonlinear effects on digital communications and receiver design, with emphasis on MIMO systems, OFDM and SC-FDE.

Unified Theoretical Analysis of Nonlinear Multicarrier Schemes

Paulo Montezuma, João Guerreiro,
and Rui Dinis

CRC Press
Taylor & Francis Group
Boca Raton London New York

CRC Press is an imprint of the
Taylor & Francis Group, an **informa** business

Designed cover image: Shutterstock

First edition published 2024
by CRC Press
2385 NW Executive Center Drive, Suite 320, Boca Raton FL 33431

and by CRC Press
4 Park Square, Milton Park, Abingdon, Oxon, OX14 4RN

CRC Press is an imprint of Taylor & Francis Group, LLC

ISBN: 978-1-032-70872-0 (hbk)
ISBN: 978-1-032-70876-8 (pbk)
ISBN: 978-1-032-70874-4 (ebk)

DOI: 10.1201/9781032708744

Typeset in Nimbus Roman
by KnowledgeWorks Global Ltd.

To our Families

Contents

Preface

It is widely recognized that multicarrier systems such as orthogonal frequency division multiplexing (OFDM) are suitable for severely time-dispersive channels. However, multicarrier signals have high envelope fluctuations which make them especially sensitive to nonlinear distortion effects. In fact, it is almost unavoidable to have nonlinear distortion effects in the transmission chain. For this reason, it is essential to have a theoretical, accurate characterization of nonlinearly distorted signals not only to evaluate the corresponding impact of these distortion effects on the system's performance, but also to develop mechanisms to combat them. One of the goals of this book is to address these challenges and present a theoretical characterization of nonlinearly distorted multicarrier signals in a simple, accurate way. This analytical characterization is then particularized for several multicarrier schemes considering both single-antenna and multi-antenna scenarios.

Conventionally, nonlinear distortion is seen as a noise term that degrades the system's performance, leading even to irreducible error floors. Even receivers that try to estimate and cancel it have a poor performance, comparatively to the performance associated with a linear transmission. Another goal of this book is to present a study on the optimum detection of nonlinearly distorted, multicarrier signals. Throughout the book, it is shown that the nonlinear distortion should not be considered as a noise term, but instead as something that contains useful information for detection purposes. The adequate receiver to take advantage of this information is the optimum receiver, since it makes a block-by-block detection, allowing us to exploit the nonlinear distortion which is spread along the signal's band. Although the optimum receiver for nonlinear multicarrier schemes is too complex, due to its necessity to compare the received signal with all possible transmitted sequences, it is important to study its potential performance gains. It is demonstrated that the optimum receiver outperforms the conventional detection, presenting gains not only relatively to conventional receivers that deal with nonlinear multicarrier signals, but also relatively to conventional receivers that deal with linear, multicarrier signals. Sub-optimum receivers able to approach the performance gains associated with the optimum detection and that can even outperform the conventional linear, multicarrier schemes are also presented.

Authors

Paulo Montezuma Carvalho is an associate professor at Faculdade de Ciências e Tecnologia (FCT), Universidade Nova de Lisboa (UNL) (he has a Ph.D and a Master degree in Telecommunications). He was a researcher at CAPS (Centro de Análise e Processamento de Sinal), IST, from 1997 to 2005 and since 2013, he has been a researcher at IT (Instituto de Telecomunicacões). He has been actively involved in several international research projects in the broadband wireless communications area and many national industrial and research projects, most of them as nuclear researcher and/or in charge of his research centre. He published over 44 journal papers and four book chapters, one book and more than 120 conference papers (1 best paper award), has 16 patents assigned and three pending and is editor in several Journals. His main research activities are on modulation and transmitter design, coding, nonlinear effects on digital communications and receiver design, with emphasis on frequency-domain implementations, namely for MIMO systems and/or OFDM and SC-FDE modulations and energy efficient transmission techniques.

João Guerreiro received the MSc and Ph.D. degrees in Electrical and Computer Engineering from Faculdade de Ciências e Tecnologia (FCT), Universidade Nova de Lisboa (UNL) in 2012 and 2016, respectively. Since 2013, he has been actively involved in scientific research in the field of wireless communications, including signal processing, transceiver design, and channel modeling. He published more than 30 journal papers, 50 conference papers (1 best paper award at IEEE Vehicular Technology Conference'15 (Boston)), three patents (granted or pending), and two book chapters. He has participated in several national and international research projects in the last few years. He is currently an assistant professor at FCT-UNL and a researcher at IT (Instituto de Telecomunicacões). He is also editor at IEEE Communications Letters.

Rui Dinis received the Ph.D. degree from Instituto Superior Técnico (IST), Technical University of Lisbon, Portugal, in 2001 and the Habilitation in Telecommunications from Faculdade de Ciências e Tecnologia (FCT), Universidade Nova de Lisboa (UNL), in 2010. From 2001 to 2008 he was a Professor at IST. Currently he is full professor at FCT-UNL. During 2003 he was an invited professor at Carleton University, Ottawa, Canada. He was a researcher at CAPS (Centro de Análise e Processamento de Sinal), IST, from 1992 to 2005 and a researcher at ISR (Instituto de Sistemas e Robótica) from 2005 to 2008. Since 2009, he has been a researcher at IT (Instituto de Telecomunicacões). He is an IEEE Senior Member and editor at IEEE Transactions on Wireless Communications, IEEE Transactions on Communications, IEEE Transactions on Vehicular Technology and Elsevier Physical Communication. He is or was of the Organizing Committee of several international conferences (such as the IEEE conferences ICT'2014, VTC'2017-Fall, VTC'2018-Spring and ISWCS'2018). He is also president of VTS Portugal Chapter since 2016 and a member of several technical committees of IEEE Communications Society (SPCE, RCC, WC and CT). He has been actively involved in several international research projects in the broadband wireless communications area (RACE project MBS, ACTS project SAMBA and IST projects B-BONE and C-MOBILE) and many national projects, most of them as nuclear researcher and/or in charge of his research center in multi-institutional projects. He published six books, over 100 journal papers and book chapters and over 300 conference papers (of which five received best papers' awards), and over 16 patents (attributed or pending). He was involved in pioneer projects on the use of mm-waves for broadband wireless communications (international projects MBS and SAMBA) and his main research activities are on modulation and transmitter design, nonlinear effects on digital communications and receiver design (detection, equalization, channel estimation and carrier synchronization), with emphasis on frequency-domain implementations, namely for MIMO systems and/or OFDM and SC-FDE modulations. He is also working on cross-layer design and optimization involving PHY, MAC and LLC issues, as well as indoor positioning techniques.

List of Acronyms

OWC optical wireless communications
NLPN nonlinear phase noise
CO coherent optical
CE constant envelope
EHF extremely high frequency
ESNR equivalent signal-to-noise plus self-interference ratio
DF decode and forward
AF amplify and forward
COFDM coded OFDM
CEPB constant envelope paired burst
SIR signal-to-interference ratio
ACO asymmetrical clipping optical
DCO DC-biased optical
LINC linear amplification with nonlinear components
MAP maximum a posteriori
ML maximum likelihood
PEP pairwise error probability
BNC Bussgang noise cancellation
SER symbol error rate
IMP intermodulation product
AM/AM amplitude modulation/amplitude modulation
AM/PM amplitude modulation/phase modulation
SSPA solid state power amplifier
TWTA traveling wave tube amplifier
PTS partial transmit sequences
OBO output back-off
FDF frequency-domain filter
CDF cumulative density function
SISO single-input, single-output
IEEE institute of electrical and electronic engineers
SVD singular value decomposition
BLAST Bell Laboratories layered space time
STBC space time block codes
MIMO multiple-input, multiple-output
QAM quadrature amplitude modulation
TDM time division multiplexing
QoS quality of service
DMT discrete multi tone
ADSL asymmetrical digital subscriber line

ACI adjacent channel interference
AWGN additive white Gaussian noise
IBI inter-block interference
PAE power added efficiency
PA power amplifier
CP cyclic prefix
CIR channel impulsive response
LTE long term evolution
ACPR adjacent channel power ratio
DE drain efficiency
PAPR peak-to-average power ratio
RF radio frequency
BS base station
FFT fast Fourier transform
PAN personal area network
WAN wide area network
LAN local area network
MAN metropolitan area network
UWB ultra-wide band
DAB digital audio broadcasting
DVB digital video broadcasting
FDM frequency division multiplexing
QPSK quadrature phase shift keying
IDFT inverse discrete Fourier transform
DFT discrete Fourier transform
DAC digital-to-analogue converter
ADC analogue to digital converter
PHY physical layer
ISI inter-symbol interference
IBO input back-off
SRRC square-root raised-cosine
MC multi carrier
SC-FDE single carrier with frequency domain equalization
SNR signal-to-noise ratio
SC single carrier
WiMAX worldwide interoperability for microwave access
PDP power delay profile
CLT central limit theorem
ICI inter-channel interference
SPC serial-to-parallel converter
BER bit error rate
ZP zero padded
FIR finite impulse response
PSD power spectral density

HPA high power amplifier
ZF zero forcing
MMSE minimum mean square error
PDF probability density function
CCDF complementary cumulative density function

1 Introduction

1.1 SCOPE AND MOTIVATION

The evolution of wireless communications has been constant in the past decades, which makes it possible to achieve very high data rates, low latencies and considerable improvements in the quality of service (QoS) experienced by each network user. However, a large number of users have started to use connected devices and their needs may include very demanding multimedia applications such as transmission of ultra high definition (4K) videos. This means that a single user may be served with several tens of Mbit/s. In addition, concepts such as internet of things [1], where everything is connected everywhere at anytime, have become a part of the people's life. Undoubtedly, this requires large efforts to increase the data rates, reliability, security, spectral efficiency and energy efficiency of wireless communications.

As widely known, the transmission of very large data rates in wireless channels is a challenging task, not only due to the severe inter-symbol interference (ISI) generated by the multipath propagation, which includes signal copies whose corresponding delays can be many times larger than the symbol's duration, but also due to the increasing user's mobility [2]. The multicarrier MC systems, such as orthogonal frequency division multiplexing (OFDM) [3], appeared as systems that are resilient to the ISI caused by the multipath propagation phenomena. In fact, multi carrier (MC) systems are widely used in wireless and wireline communication systems. This choice is justified by their capability to mitigate the ISI by converting a frequency-selective channel into a set of flat-fading channels, which can allow an ISI-free transmission. In addition, MC systems such as OFDM are spectrally efficient due to the orthogonality of the subcarriers and easy to implement digitally through fast Fourier transform (FFT) algorithms [4]. Nevertheless, multicarrier signals have large envelope fluctuations and a high peak-to-average power ratio (PAPR), which lead to amplification difficulties. In fact, it is very hard to perform a linear amplification of a multicarrier signal, since it requires costly and inefficient power amplifiers – two unwanted things in a hand-held device – that must operate with considerable back-offs. This is also extremely inefficient for the base stations (BSs), where one of the main energy consumers is the power amplifier. Therefore, the solution is to employ nonlinear power amplifiers that are cheaper and more efficient [5]. However, such amplifiers can only be employed if the signal's envelope fluctuations are reduced to avoid significant nonlinear distortion levels. Among the so-called PAPR-reducing techniques proposed in the last years [6], it is consensual that the simpler and more flexible ones are those involving a clipping operation [7]. As a nonlinear operation, the clipping introduces nonlinear distortion effects that may not only distort the transmitted signal, but also affect communication systems operating in adjacent bands, reducing the performance and the spectral efficiency, respectively. For this reason, it becomes clear that the amplification of multicarrier signals is a

DOI: 10.1201/9781032708744-1

demanding task, which includes a trade-off between energy efficiency and spectral efficiency. Moreover, nonlinear distortion effects resulting from low-resolution quantization processes [8] or nonlinear phase noise (NLPN) [9] can take place. For this reason, it is very likely that multicarrier signals will be nonlinearly distorted. Therefore, it is important to obtain the impact of the nonlinear distortion effects on the performance of MC systems. This is usually done by taking advantage of the Gaussian nature of multicarrier signals with a large number of subcarriers, which allows to decompose the nonlinearly distorted signal in uncorrelated useful and distortion components [10], and the spectral characterization of the distortion component usually resorts to intermodulation product (IMP) tools [11, 12]. The complexity and accuracy of these IMP tools are intimately related to the severeness of nonlinearity. In fact, they may present not only complexity problems but also convergence problems, especially for nonsmooth nonlinearities such as the ones associated with quantization operations.

The analytical characterization and the optimum detection of nonlinearly distorted multicarrier signals are the two main goals of this book. Regarding the former, it presents the analytical characterization of several nonlinear multicarrier schemes that can be encountered in actual communication systems. This analytical characterization is based on IMP tools, directly by employing the truncated IMP approach or through the novel concept of equivalent nonlinearities that is suitable even for severe nonlinearities, where the truncated IMP approach is not adequate due to accuracy and convergence problems. This analytical characterization is suitable for systems impaired by different types of nonlinearity such as bandpass, Cartesian or real-valued nonlinearities. We propose an efficient and simple theoretical method for obtaining an accurate spectral characterization of nonlinearly distorted multicarrier signals that is suitable even for severe nonlinearities. This method involves the creation of an equivalent nonlinearity that can be used to substitute the conventional nonlinear characteristic, but that leads to the same spectral characterization of the signals distorted by the original nonlinearity.

Another subject analyzed in this book is the optimum detection of nonlinearly distorted multicarrier signals. Conventionally regarded as a noise term, the nonlinear distortion component can severely prejudice the performance leading to a substantial bit error rate (BER) increase and even error floors. To mitigate this performance penalty, receivers that try to estimate and cancel the nonlinear distortion have been proposed [13]. However, due to the difficulty of estimating the nonlinear distortion, even those receivers show a poor performance in comparison to the performance of linear, multicarrier systems. Additionally, it is shown that the nonlinear distortion component can be seen as useful information and used to improve the performance of multicarrier systems, provided that an optimum receiver is considered [14, 15]. The analysis includes a theoretical study on the optimum detection of nonlinearly distorted, multicarrier schemes. It is demonstrated, for several multicarrier systems, impaired by different nonlinearities and under different channel types, that the optimum detection of nonlinearly distorted multicarrier signals presents better performance than the conventional receivers and can even outperform the performance associated to linear, multicarrier systems. These results are based on the optimum asymptotic performance that is obtained theoretically through the squared Euclidean distance between nonlinearly distorted, multicarrier signals. It is also shown that even when a sub-optimum receiver is considered, substantial performance improvements can be obtained. This means that besides the large complexity of optimum detection, low-complex, sub-optimum receivers can be practically implemented to explore the potential gains associated by the optimum detection. This book also includes a set of applications for which both the analytical characterization and the optimum performance are studied. Moreover, these analytical tools are also employed for the characterization of nonlinear distortion effects associated not only with single antenna systems but also with multiple-input, multiple-output (MIMO)/massive MIMO multicarrier systems.

1.2 BOOK STRUCTURE

Chapter 2 is concerned with the characterization of multicarrier systems. It includes an introduction to multicarrier schemes in general and a statistical characterization of the signals at the receiver and at the transmitter for the particular case of OFDM. The main drawback of multicarrier schemes – the large envelope fluctuations and the corresponding amplification problems – is also analyzed in this chapter. It is pointed out that the amplification process involves trade-offs between the spectral and energy efficiency and that usually PAPR-reducing techniques are employed prior to that process. In addition, the problems associated with transmission in frequency-selective channels and the amplification problems associated to the multicarrier signals are analyzed. The chapter ends with a brief characterization of the most common PAPR-reducing techniques employed to mitigate the amplification issues of multicarrier signals.

Chapter 3 presents the theory behind the analytical characterization and the optimum detection of nonlinearly distorted multicarrier signals, which is applied in subsequent chapters. Different nonlinearities that can have either a baseband or bandpass nature are identified and characterized. The conventional spectral characterization made through IMP tools is considered to introduce and motivate the concept of equivalent nonlinearities. These equivalent nonlinearities are polynomial nonlinearities that can substitute the conventional ones. However, although being smoother and suitable to be employed in an IMP analysis, they give rise to signals with the same spectral characterization than that of the ones distorted by the original nonlinearities. This means that they can be employed to obtain a simple, accurate spectral characterization of nonlinearly distorted multicarrier signals.

Chapter 4 extends the results of previous chapter to the characterization of optimum receiver's performance. It starts with the definition of the optimum receiver. Further, the optimum performance regarding linear OFDM schemes is also derived. It is shown that an insight into the optimum performance can be computed through the squared Euclidean distance between the signals. In fact, when the transmission is linear, this distance is equal regardless of the subcarrier that is in error. However, when the transmission is nonlinear, the squared Euclidean distance depends on the information spread along the signal band. For this reason, and due to the large number of possible transmitted sequences, the distribution of the squared Euclidean distance (and of the corresponding asymptotic gain) was obtained. It shows that when the number of subcarriers is large, the variance of this distribution decreases, meaning that for a very large number of subcarriers, the squared Euclidean distance between two nonlinearly distorted signals assumes a unique value. This allows us to obtain the asymptotic gain for different types of nonlinearities. For the case of frequency-selective channels, the squared Euclidean distance between the nonlinearly distorted signals is random even if the number of subcarriers is very large, since it is dependent on the channel's frequency responses that present a Rayleigh distribution. For this reason, it also includes a statistical characterization of the channel's frequency responses, which can be used to obtain the distribution of asymptotic gain associated with the optimum detection.

Chapter 5 is focused on the different applications where the theory presented in Chapter 3 can be employed. It is concerned with the applications for which the analytical characterization and the asymptotic optimum performance presented in Chapter 3 can be employed. It starts with the conventional OFDM impaired by either an envelope clipping or nonlinear power amplifiers. For that system, it is shown that the equivalent nonlinearities may be used to obtain an accurate spectral characterization of the corresponding nonlinearly distorted signals and that the optimum performance presents considerable gains when compared to the traditional detection of nonlinear and even of linear, OFDM schemes. In addition, are included performance results regarding a sub-optimum receiver considering the effect of receive diversity, clipping and filtering techniques and M-quadrature amplitude modulation (QAM) constellations, showing that even when a sub-optimum detection is considered, the potential performance gains associated to the optimum detection might be obtained. The linear amplification with nonlinear components (LINC) techniques are also analyzed since they allow an efficient amplification through power amplifiers of class D or E, despite the nonlinear distortion effects introduced in the transmitted signals. For this reason, both the analytical characterization and the optimum detection of the corresponding nonlinearly distorted signals are characterized. Another alternative for the conventional OFDM that allows the use of highly efficient amplifiers is the constant envelope (CE)-OFDM. It is noted that the severe nonlinear distortion effects created by the phase modulation process of CE-OFDM schemes can be accurately characterized using IMP tools and that the optimum detection outperforms the conventional detection based on a phase demodulator, especially in scenarios where the power efficiency is crucial. Closing the chapter, it is presented the analytical spectral characterization for the signals along the transmission chain of an OFDM-based amplify and forward (AF) relay system, i.e., in a scenario where there are nonlinearities both at the transmitter and at the relay.

An extension to OFDM–MIMO systems is presented in Chapter 6. It focuses on the analytical characterization of nonlinear, MIMO-OFDM systems with an envelope clipping operation employing a singular value decomposition (SVD) decomposition and massive MIMO-OFDM systems with maximum ratio transmission (MRT) approach, employing both Cartesian, 1-bit quantizers and nonlinear amplifiers. For both systems, it is demonstrated that the nonlinear distortion at the subcarrier level decreases when the number of transmit antennas increases, provided that the number of streams is fixed. This means that when we have much more transmit antennas than receive antennas, the nonlinear distortion can be substantially reduced. Therefore, massive MIMO-OFDM may have nonlinear amplification processes and/or employ low-complex, low-resolution quantizers that can meet their need to have simple and low-cost transmitting branches, since the corresponding performance penalty associated with the use of such nonlinear devices can be tolerated. Baseband, nonlinear discrete multi-tone schemes that have a quantization operation on the transmission chain are also analyzed. Due to the severity of the nonlinearity associated with the quantization operation, especially when the resolution is low, equivalent nonlinearities are used to provide an accurate spectral characterization of quantized, multicarrier signals. Additionally, the optimum asymptotic performance is derived and its performance gains over the linear discrete multi tone (DMT) schemes are quantified through a set of performance results. This chapter ends with the analytical characterization and a study on the optimum detection of optical OFDM signals and includes the characterization of nonlinear distortion effects on fiber optical systems, i.e., the NLPN, as well as the characterization of the distortion effects related to optical wireless OFDM systems, such as DC-biased optical (DCO)-OFDM systems, where asymmetrical clipping operations take place.

Besides these particular applications, the theory associated with the analytical characterization of nonlinearly distorted multicarrier signals presented here can be employed more generally in sampled, Gaussian signals, regardless of their multicarrier or single carrier nature, due to the accuracy associated with the use of equivalent nonlinearities to characterize the nonlinear effects.

2 Multicarrier Systems

2.1 HISTORICAL PERSPECTIVE

MC systems result from the evolution of frequency division multiplexing (FDM) systems, which were initially proposed for the transmission of multiple telegraph channels and for analog voice in the first decades of the 20th century, becoming the main multiplexing technique for telephony [16, 17]. In the 1970s, with the advance of digital communications, the FDM systems began to be replaced by time division multiplexing (TDM) systems. However, as the speed of TDM lines increased, the corresponding larger channel bandwidths brought problems such as severe ISI and frequency-selective fading. The solution to these problems was to come back to FDM systems, where the subchannels with better conditions were allocated with more data, i.e., the loading was adapted to the channel characteristics. Nevertheless, FDM systems also had disadvantages such as the large implementation complexity, which increases with the number of subchannels, and the low bandwidth efficiency, due to the existence of guard bands between the subchannels. MC systems such as OFDM were proposed to alleviate these problems.

The main idea behind MC systems, the parallel data transmission over several sub-channels, was firstly introduced in the mid-1960s [18–20]. However, this idea was only further developed and investigated in the late 1980s [3], when the MC systems were solidly stated. The cause for the time-span between the introduction of the MC systems and their effective development and implementation resided on the very high complexity associated with their analog transceivers, which could not be handled by the existing technology in 1960s, making them not suitable to compete with the single carrier (SC) systems of that time. However, the advance and improvement of signal processing techniques and electronics in 1990s changed this scenario, allowing the implementation of MC systems, such as OFDM systems, without the necessity of having independent modulators for each subcarrier, which drastically decreased its implementation complexity. In fact, OFDM systems had a big growth in this decade [17]. This growth is explained by the introduction of digital signal processing techniques that allows ease implementation of efficient FFT algorithms such as the Cooley-Tukey algorithm [4] by computers or by low-cost, massively produced electronic chips. This advancement significantly increased the role of multicarrier schemes in both wireless and wireline systems.

The first major OFDM application with big commercial success was the DMT, that was proposed for asymmetrical digital subscriber line (ADSL) systems in 1990 [21]. In fact, an immense growth of OFDM continued over this decade and, from the late 1990's, OFDM and OFDM-based systems started to be employed in the physical layer (PHY) of several wireless standards with application in different types of networks. For instance, these systems were considered for ultra-wide band (UWB) [22], in personal area networks (PANs) and for several variants of the standard institute

DOI: 10.1201/9781032708744-2

of electrical and electronic engineers (IEEE) 802.11 [23], in local area networks (LANs). They were also considered in wide area networks (WANs) for worldwide interoperability for microwave access (WiMAX) 802.16 [24], and in metropolitan area networks (MANs), for long-term evolution [LTE] [25], digital video broadcasting (DVB) [26] and digital audio broadcasting (DAB) [27]. For this reason, MC systems such as OFDM systems have been extensively studied over the past decade and many research have been done toward their continuous evolution.

Nowadays, OFDM and single carrier with frequency domain equalization (SC-FDE) systems are the two main contenders for wireless broadband systems. In fact, they do not have considerable differences in achievable data-rate or bandwidth efficiency, and it can be shown that the complexity of generating and detecting an OFDM signal is similar to the complexity associated with the equalization of the SC-FDE systems [28]. The choice of OFDM for some applications is justified by its larger flexibility, i.e., by its high capability to adapt the loading by taking into account the signal-to-noise ratio (SNR) of each subcarrier.

2.2 OFDM

In this section, OFDM systems are characterized both in time and frequency domain and the motivation behind their digital implementation is given. The main reasons that explain the high resilience of OFDM systems to multipath propagation are also presented.

2.2.1 TIME AND FREQUENCY-DOMAIN CHARACTERIZATION

In contrast with SC systems, where a single carrier is modulated and transmitted within a given symbol time duration, in MC systems such as OFDM, several carriers are used for data transmission during the symbol period. Basically, the total available bandwidth B_s is divided into N subchannels, on which N low-rate data streams are transmitted in parallel. The center frequencies of these sub-channels, the so-called subcarriers, are spaced by $\Delta f = B_s/N$ and are mutually orthogonal within the symbol duration T_u.

The complex envelope of an OFDM signal is composed of consecutive OFDM "blocks" spaced by T_u seconds. Let $s_{bb}(t)$ be the complex envelope of a given OFDM signal composed by N subcarriers, i.e.,

$$s_{bb}(t) = \sum_{m=-\infty}^{+\infty} s^{(m)}(t - mT_u), \tag{2.1}$$

with $s^{(m)}(t)$ denoting the waveform of the mth OFDM block. Without loss of generality, let us consider the OFDM signal of the 0th block ($m = 0$)[1], that is given

---

[1]For the sake of notation simplicity, the subscript m is omitted when there is no risk of ambiguity.

by

$$s(t) = \sum_{k=-\frac{N}{2}}^{\frac{N}{2}-1} S_k \exp\left(j2\pi k\frac{t}{T_u}\right) w(t), \tag{2.2}$$

where S_k is the data symbol transmitted on the kth subcarrier and

$$w(t) = \text{rect}(t/T_u), \tag{2.3}$$

is a rectangular window function with duration T_u. Each data symbol S_k is selected from a given QAM constellation with M points. For instance, when quadrature phase shift keying (QPSK) constellations are employed, we have $M = 4$ and $S_k = \pm\sigma_S \pm j\sigma_S$, where σ_S is the amplitude of the real and imaginary parts of each data symbol. Thorough this work, it is assumed zero mean (i.e., $\mathbb{E}[S_k] = 0$) and uncorrelated data symbols, since

$$\mathbb{E}\left[S_k S_{k'}^*\right] = \begin{cases} \mathbb{E}\left[|S_k|^2\right] = 2\sigma_S^2, & k = k' \\ 0, & k \neq k', \end{cases} \tag{2.4}$$

which means that power is uniformly distributed along the N subcarriers and equal to $2\sigma_S^2$.

As aforementioned, one of the most important particularities of OFDM relatively to conventional FDM schemes is the orthogonality of the subcarriers. Indeed, by defining the waveform of the kth subcarrier as

$$\phi_k(t) = \exp\left(j2\pi\frac{kt}{T_u}\right), \tag{2.5}$$

one can note that the subcarriers are mutually orthogonal during the block duration T_u, since

$$\begin{aligned} \langle\phi_k(t), \phi_{k'}(t)\rangle &= \frac{1}{T_u}\int_0^{T_u} \phi_k(t)\phi_{k'}(t)\mathrm{d}t \\ &= \frac{1}{T}\int_0^{T_u} \exp\left(j2\pi k\frac{t}{T_u}\right)\exp\left(-j2\pi k'\frac{t}{T_u}\right)\mathrm{d}t \\ &= \frac{1}{T}\int_0^{T_u} \exp\left(j2\pi\frac{(k-k')t}{T_u}\right)\mathrm{d}t \\ &= \begin{cases} 1, & k = k' \\ 0, & k \neq k', \end{cases} \end{aligned} \tag{2.6}$$

where $\langle\cdot,\cdot\rangle$ represents the dot product. Let us consider the signal $s^P(t)$, which is a periodic signal composed by the subcarriers associated with a given OFDM signal

$$s^P(t) = \sum_{k=-\frac{N}{2}}^{\frac{N}{2}-1} S_k \exp\left(j2\pi k\frac{t}{T_u}\right). \tag{2.7}$$

Indeed, equation (2.2) can be rewritten as the multiplication between the periodic signal $s^P(t)$ and the window function represented in (2.3), i.e.,

$$s(t) = s^P(t)w(t). \tag{2.8}$$

The frequency-domain version of the signal represented in (2.7) is

$$S^P(f) = \mathscr{F}\{s^P(t)\} = \sum_{k=-\frac{N}{2}}^{\frac{N}{2}-1} S_k \delta\left(f - \frac{k}{T_u}\right), \tag{2.9}$$

where $\mathscr{F}\{\cdot\}$ represents the Fourier transform and $\delta(\cdot)$ is the Dirac delta function. The corresponding power spectral density (PSD) associated with (2.7) can be obtained as

$$\begin{aligned} G_{s^P}(f) &= \frac{1}{T_u}\mathbb{E}\left[\left|S^P(f)\right|^2\right] \\ &= \frac{1}{T_u}\sum_{k=-\frac{N}{2}}^{\frac{N}{2}-1} \mathbb{E}\left[|S_k|^2\right]\left|\delta\left(f - \frac{k}{T_u}\right)\right|^2 \\ &= \frac{2\sigma_S^2}{T_u}\sum_{k=-\frac{N}{2}}^{\frac{N}{2}-1}\left|\delta\left(f - \frac{k}{T_u}\right)\right|^2. \end{aligned} \tag{2.10}$$

Fig. 2.2 shows the normalized PSD associated with $s^P(t)$, considering $N = 16$ subcarriers. In fact, from (2.9), it can be noted that S_k are the complex Fourier coefficients associated with the Fourier series of (2.7). Let us now focus on the frequency-domain version of (2.2), that is

$$S(f) = \mathscr{F}\{s(t)\} = \sum_{k=-\frac{N}{2}}^{\frac{N}{2}-1} S_k W\left(f - \frac{k}{T_u}\right). \tag{2.11}$$

where

$$W(f) = T_u \mathrm{sinc}(fT_u), \tag{2.12}$$

is the Fourier transform of the window function $w(t)$. Using (2.11), the PSD associated to (2.2) can be obtained as

$$\begin{aligned} G_s(f) &= \frac{1}{T_u}\mathbb{E}[|S(f)|^2] \\ &= \sum_{k=-\frac{N}{2}}^{\frac{N}{2}-1} \mathbb{E}\left[|S_k|^2\right]\left|W\left(f - \frac{k}{T_u}\right)\right|^2 \\ &= 2\sigma_S^2 \sum_{k=-\frac{N}{2}}^{\frac{N}{2}-1} |\mathrm{sinc}\,(fT_u - k)|^2. \end{aligned} \tag{2.13}$$

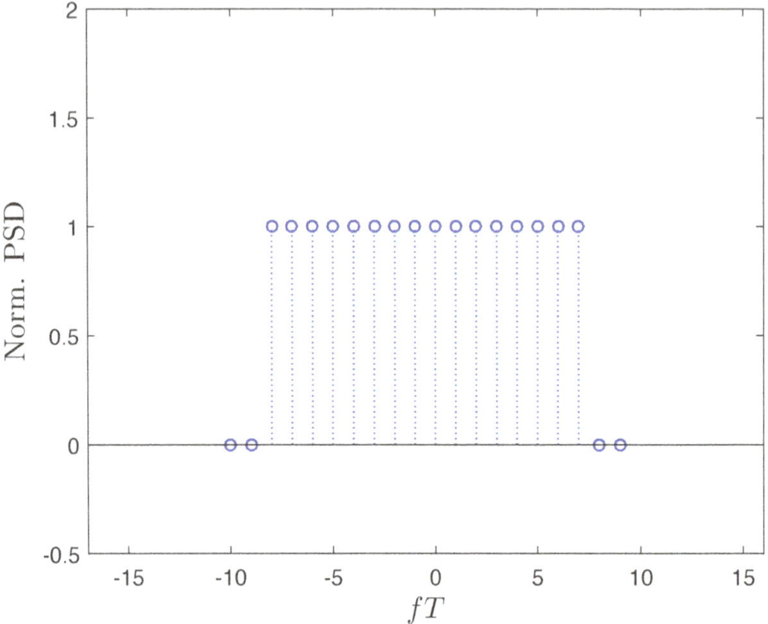

Figure 2.1 Normalized PSD of $s^P(t)$ considering $N = 16$ subcarriers.

From (2.13), it can be seen that PSD is composed by the summation of shifted "sinc-like" functions, each one corresponding to a given subcarrier. Fig. 2.2 represents the normalized PSD of a given OFDM block with $N = 8$ subcarriers. From the figure, it can be noted that even with a moderate number of subcarriers, the PSD of an OFDM signal is approximately rectangular.

As the digital implementation of OFDM is based on the discrete Fourier transform (DFT), one can make use of the periodicity of the frequency-domain samples and consider the following relation between the transmitted symbols,

$$S_k = \begin{cases} S_k, & 0 \leq k \leq \frac{N}{2} - 1 \\ S_{k-N}, & \frac{N}{2} \leq k < N - 1. \end{cases} \tag{2.14}$$

This relation allows the subcarrier indexes to run from $k = 0$ to $k = N - 1$ and, in these conditions, group the symbols to be transmitted in the block[2] $\mathbf{S} = [S_0\ S_1\ S_2\ \ldots\ S_{N-1}]^T \in \mathbb{C}^N$. Further, considering (2.14) in (2.2), we may write

$$s(t) = \sum_{k=0}^{N-1} S_k \exp\left(j2\pi k \frac{t}{T_u} \right) w(t). \tag{2.15}$$

[2]In this work the matrices and arrays start with the index 0.

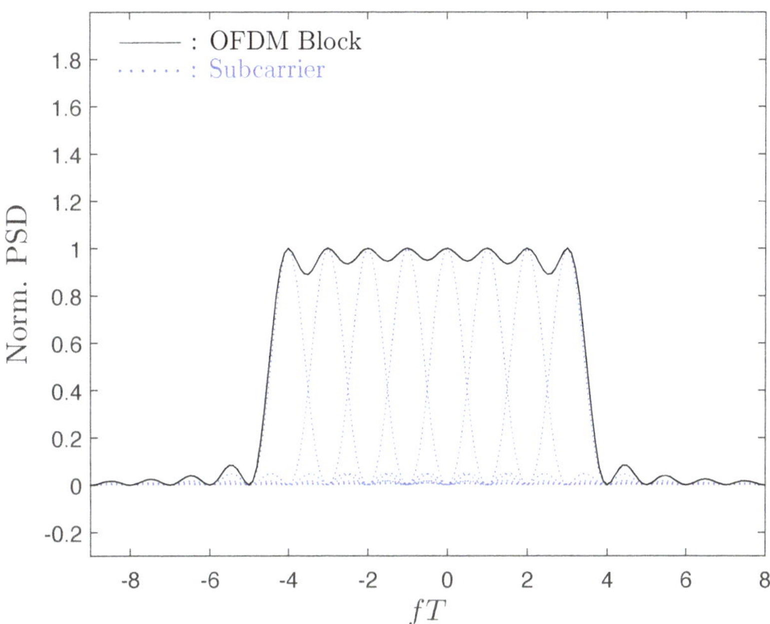

Figure 2.2 Normalized PSD associated with an OFDM signal with $N = 8$ subcarriers.

By sampling the above equation at the time instants spaced by $T_s = T_u/N$, i.e., considering a sampling rate $F_s = N/T_u = N\Delta f$, the nth time-domain sample of (2.15) is given by

$$\begin{aligned} s_n &= s\left(n\frac{T_u}{N}\right) \\ &= \sum_{k=0}^{N-1} S_k \exp\left(j2\pi\frac{kn}{N}\right) w\left(n\frac{T_u}{N}\right) \\ &= \sum_{k=0}^{N-1} S_k \exp\left(j2\pi\frac{kn}{N}\right), \end{aligned} \tag{2.16}$$

since $w\left(n\frac{T_u}{N}\right) = 1$. On the other hand, the nth output of the inverse discrete Fourier transform (IDFT) of the block $\mathbf{S} = [S_0\ S_1\ S_2\ \ldots\ S_{N-1}]^T \in \mathbb{C}^N$ is

$$s_n = \frac{1}{N}\sum_{k=0}^{N-1} S_k \exp\left(j2\pi\frac{kn}{N}\right). \tag{2.17}$$

From the comparison between (2.16) and (2.17) it can be noted that, apart from the scale factor $1/N$, equation (2.16) represents the nth output of the IDFT of

$\mathbf{S} = [S_0 \; S_1 \; S_2 \; ... \; S_{N-1}]^T \in \mathbb{C}^N$. In fact, by considering the IDFT to generate the time-domain samples of an OFDM signal, the use of a large number of modulators and oscillators (which is as high as N) implicit in (2.7) can be avoided. Instead, the complex envelope of an OFDM symbol can be generated in a much easier way by means of efficient FFT algorithms, which constitutes a very important advantage and allows to greatly simplify the transceivers. In what follows, the procedures associated with the DFT-based, digital implementation of OFDM are briefly described.

The first task is performed by the serial-to-parallel converter (SPC), which gathers a set of data symbols from a given modulator to be transmitted onto the N subcarriers. These data symbols form the block $\mathbf{S} = [S_0 \; S_1 \; S_2 \; ... \; S_{N-1}]^T \in \mathbb{C}^N$. In this work, the DFT of the block $\mathbf{a} = [a_0 \; a_1 \; a_2 \; ... \; a_{N-1}]^T \in \mathbb{C}^N$ is represented as $\mathbf{A} = \mathbf{Fa} = [A_0 \, A_1 \, A_2 \, ... \, A_{N-1}]^T \in \mathbb{C}^N$, where $\mathbf{F}$ denotes the DFT matrix, that is an $N \times N$ matrix defined by[3]

$$\mathbf{F} = \begin{bmatrix} 1 & 1 & 1 & 1 & \cdots & 1 \\ 1 & \omega & \omega^2 & \omega^3 & \cdots & \omega^{N-1} \\ 1 & \omega^2 & \omega^4 & \omega^6 & \cdots & \omega^{2(N-1)} \\ 1 & \omega^3 & \omega^6 & \omega^9 & \cdots & \omega^{3(N-1)} \\ \vdots & \vdots & \vdots & \vdots & \ddots & \vdots \\ 1 & \omega^{N-1} & \omega^{2(N-1)} & \omega^{3(N-1)} & \cdots & \omega^{(N-1)^2}, \end{bmatrix}, \tag{2.18}$$

where $\omega = \exp(-j2\pi/N)$. Therefore, the time-domain samples of a given OFDM block $\mathbf{s} = [s_0 \; s_1 \; s_2 \; ... \; s_{N-1}]^T \in \mathbb{C}^N$ are obtained as

$$\mathbf{s} = \mathbf{F}^{-1}\mathbf{S}, \tag{2.19}$$

where

$$\mathbf{F}^{-1} = \frac{1}{N}\mathbf{F}^*, \tag{2.20}$$

denotes the inverse DFT matrix and $(\cdot)^*$ is the complex conjugate operator.

It is important to point out that if the IDFT is applied directly to the block of data symbols to be transmitted, it yields time-domain samples obtained at $F_s = N/T_u$. In fact, this sampling rate is adequate to perfectly recover (2.7) from its samples, since it is its "Nyquist" rate. However, it is not sufficient to perfectly recover the signal in (2.8), since this signal is not bandlimited. For this reason, even when $F_s > N/T_u$, there is always some aliasing effects since the sampling theorem does not hold. However, to mitigate such aliasing effects and reduce the order of the reconstruction filter after the digital-to-analog converter (DAC), it is common to consider an oversampling operation. To obtain an oversampling factor of O, $N_g = N_u(O-1)$ out of $N = ON_u$ subcarriers are left idle, i.e., with $S_k = 0$. Indeed, as N_u/T_u is the highest frequency component, this is equivalent of having an increased sampling rate given by $F_s = N/T_u = ON_u/T_u$, where N_u is the number of subcarriers that are effectively used

[3] In this work, the conventional, non-normalized DFT definition is adopted.

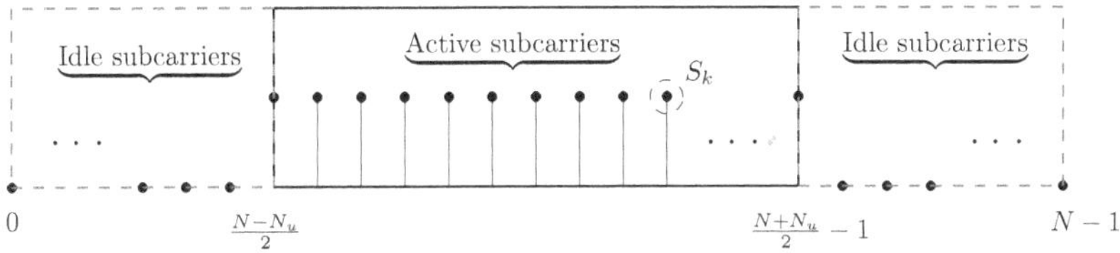

Figure 2.3 Format of a general OFDM block with idle subcarriers for oversampling purposes.

to transmit data[4]. In these conditions, if QPSK constellations are considered, the subcarriers are mapped in the following way

$$S_k = \begin{cases} 0, & 0 \leq k \leq \frac{N-N_u}{2} - 1 \\ \pm\sigma_S \pm j\sigma_S, & \frac{N-N_u}{2} \leq k \leq \frac{N+N_u}{2} - 1 \\ 0, & \frac{N+N_u}{2} \leq k \leq N-1, \end{cases} \tag{2.21}$$

which means that the IDFT of $\mathbf{S} = [S_0\ S_1\ S_2\ ...\ S_{N-1}]^T \in \mathbb{C}^N$ yields oversampled samples of the underlying continuous-time OFDM signal. The nth IDFT output sample can be written as

$$s_n = \sum_{k=0}^{N-1} S_k \exp\left(j2\pi \frac{kn}{N}\right) = \sum_{k=\frac{N-N_u}{2}}^{\frac{N+N_u}{2}-1} S_k \exp\left(j2\pi \frac{kn}{N}\right). \tag{2.22}$$

Fig. 2.3 illustrates the arrangement of the OFDM block when $N_u(O-1)$ subcarriers are left idle for oversampling purposes (see (2.21)). It should also be noted that due to the periodicity of the DFT, the time-domain samples $\mathbf{s} = [s_0\ s_1\ s_2\ ...\ s_{N-1}]^T \in \mathbb{C}^N$ are periodic with period N, since

$$\begin{aligned} s_{n+N} &= \sum_{k=\frac{N-N_u}{2}}^{\frac{N+N_u}{2}-1} S_k \exp\left(j2\pi \frac{kn}{N}\right) \exp(j2\pi k) \\ &= s_n. \end{aligned} \tag{2.23}$$

Fig. 2.4 shows the real parts of time-domain samples of an OFDM signal with $N_u = 64$ useful subcarriers and oversampling factor $O = 4$. Clearly, the amplitude of the samples varies considerably along the block. In fact, even for relatively low values of N_u, these samples present an approximate Gaussian distribution. This is a consequence of the central limit theorem (CLT), since each sample is composed by the sum of several independent complex symbols [29]. Fig. 2.5 shows the probability density function (PDF) of the real part of time-domain samples of an OFDM signal with oversampling factor $O = 4$ and different values of N_u. From the figure one can

[4]For instance, in IEEE 802.11a wireless LAN standard, we have $B = 20$ MHz and $N = 64$. However, only $N_u = 52$ out of $N = 64$ subcarriers are used for data transmission.

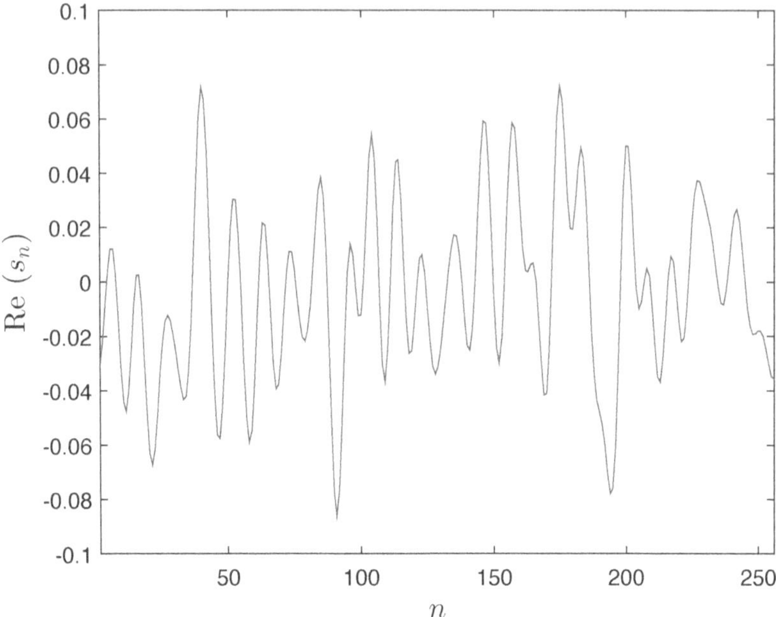

Figure 2.4 Real part of time-domain samples of an OFDM signal with $N_u = 64$ useful subcarriers and oversampling factor $O = 4$.

note that the real part of the samples of an OFDM signal has Gaussian nature (the same conclusion applies for the imaginary part). This Gaussian approximation is tight even for a moderate number of subcarriers, i.e., when $N_u = 32$, the theoretical results are almost equal to the simulated results. For $N_u = 64$, the differences between the simulation and the theory are negligible. Let us decompose s_n as

$$s_n = s_{n,I} + js_{n,Q}, \tag{2.24}$$

where $s_{n,I}$ and $s_{n,Q}$ denote the real and imaginary parts of the nth time-domain sample of a given OFDM signal, respectively. In fact, $s_{n,I}$ and $s_{n,Q}$ can be modeled by the random variable $s \sim \mathcal{N}(0, \sigma^2)$. According to the DFT definition presented in (2.18), the variance of this Gaussian random variable can be defined as

$$\sigma^2 = \frac{\sigma_S^2}{N_u O^2}. \tag{2.25}$$

Therefore, the theoretical PDF of the real (and imaginary) part of s_n is

$$p(s) = \frac{1}{\sqrt{2\pi\sigma^2}} \exp\left(-\frac{s^2}{2\sigma^2}\right). \tag{2.26}$$

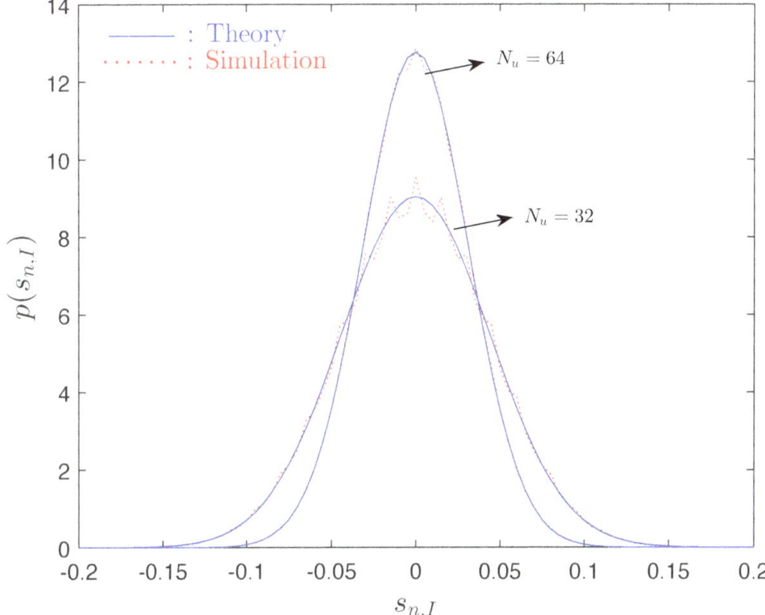

Figure 2.5 Distribution of the real part of time-domain samples of OFDM signals with $O = 4$ and different values of N_u.

The continuous-time envelope of an OFDM signal is constructed from the time-domain samples $\mathbf{s} = [s_0\ s_1\ s_2\ \ldots\ s_{N-1}]^T \in \mathbb{C}^N$ generated by the IDFT using a digital-to-analogue converter (DAC) followed by a low-pass reconstruction filter, that is characterized by the impulse response $h_{\text{rec}}(t)$. The OFDM transmitter structure associated with the digital implementation of OFDM is depicted in Fig. 2.6. Note that

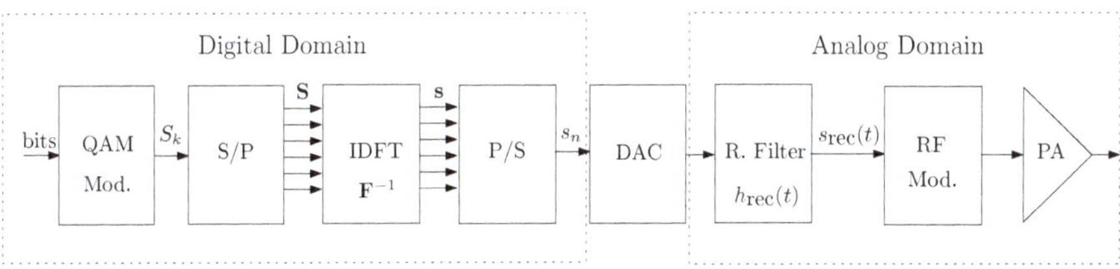

Figure 2.6 OFDM transmitter structure.

the output of the reconstruction filter can be written as

$$s_{\text{rec}}(t) = \left(\left(\underbrace{\sum_{n=-\infty}^{+\infty} s_n \delta\left(t - \frac{nT_u}{N}\right)}_{s_{\delta,n}^P(t)} \right) w(t) \right) * h_{\text{rec}}(t)$$

$$= \left(s_{\delta,n}^P(t) w(t) \right) * h_{\text{rec}}(t), \tag{2.27}$$

where $*$ denotes the convolution operator and $s_{\delta,n}^P(t)$ is a periodic signal formed by the periodic samples s_n that has period T_u, i.e.,

$$s_{\delta,n}^P(t) = \sum_{n=-\infty}^{+\infty} s_n \delta\left(t - n\frac{T_u}{N}\right) = \sum_{l=-\infty}^{+\infty} s_{\delta,n}(t - lT_u), \tag{2.28}$$

where

$$s_{\delta,n}(t) = \sum_{n=0}^{N-1} s_n \delta\left(t - n\frac{T_u}{N}\right). \tag{2.29}$$

Naturally, even when an oversampling operation is considered, the continuous-time signal $s_{\text{rec}}(t)$ does not constitute an exact representation of the "ideal" OFDM signal of (2.15). As mentioned before, this is due to the fact that $s(t)$ is not a bandlimited signal, which means that the sampling theorem does not hold and the reconstructed signal exhibits aliasing effects, whatever the sampling frequency. In the following, through an analysis regarding the frequency-domain, it is shown what is the impact of these aliasing effects on the reconstructed signal. The PSD of the reconstructed OFDM signals can be obtained as

$$G_{s_{\text{rec}}}(f) = \frac{\mathbb{E}\left[|S_{\text{rec}}(f)|^2\right]}{T_{\text{symb}}}, \tag{2.30}$$

where T_{symb} is the total duration of a given OFDM signal (including the guard interval that will be discussed later) and $S_{\text{rec}}(f)$, the Fourier transform of $s_{\text{rec}}(t)$ represented in (2.27), is defined as

$$S_{\text{rec}}(f) = \frac{1}{T_u}\left(S_{\delta,n}^P(f) * W(f) \right) H_{\text{rec}}(f), \tag{2.31}$$

where $S_{\delta,n}(f)$ and $H_{\text{rec}}(f)$ are the Fourier transform of (2.28) and of $h_{\text{rec}}(t)$, respectively. In fact, $S_{\delta,n}(f)$ can be written as

$$S_{\delta,n}^P(f) = \frac{1}{T_u} \sum_{k=-\infty}^{+\infty} S_{\delta,n}\left(\frac{k}{T_u}\right) \delta\left(f - \frac{k}{T_u}\right). \tag{2.32}$$

As

$$S_{\delta,n}(f) = \sum_{n=0}^{N-1} s_n \exp\left(j2\pi f n\frac{T_u}{N}\right), \tag{2.33}$$

we have that

$$S_{\delta,n}\left(\frac{k}{T_u}\right) = \sum_{n=0}^{N-1} s_n \exp\left(j2\pi n\frac{k}{N}\right)$$
$$= S_k, \tag{2.34}$$

and

$$S_{\text{rec}}(f) = \frac{1}{T_u}\left(\left(\sum_{k=-\infty}^{+\infty} S_k \delta\left(f - \frac{k}{T_u}\right)\right) * W(f)\right) H_{\text{rec}}(f)$$
$$= \frac{1}{T_u}\left(\sum_{k=-\infty}^{+\infty} S_k W\left(f - \frac{k}{T_u}\right)\right) H_{\text{rec}}(f)$$
$$= \frac{1}{T_u}\left(\sum_{k=-N/2}^{N/2-1} S_k W^{\text{a}}\left(f - \frac{k}{T_u}\right)\right) H_{\text{rec}}(f), \tag{2.35}$$

where

$$W^{\text{a}}(f) = \sum_{l=-\infty}^{+\infty} W\left(f - \frac{lN}{T_u}\right)$$
$$= \sum_{l=-\infty}^{+\infty} W(f - lF_s), \tag{2.36}$$

is the Fourier transform of an "equivalent window function". In fact, $W^{\text{a}}(f)$ includes the aliasing effects inherent to the digital generation of the OFDM signal, since it accounts for the impact of the replicas of $W(f)$ that are centered at multiples of F_s (see (2.36)). Note also that by replacing (2.35) in (2.30) and considering (2.4), we obtain

$$G_{s_{\text{rec}}}(f) = \frac{\mathbb{E}\left[S_{\text{rec}}(f)S^*_{\text{rec}}(f)\right]}{T_{\text{symb}}} = \frac{|H_{\text{rec}}(f)|^2}{T_u^2 T_{\text{symb}}}$$
$$.\mathbb{E}\left[\left(\sum_{k=-N/2}^{N/2-1} S_k W^{\text{a}}\left(f - \frac{k}{T_u}\right)\right)\left(\sum_{k=-N/2}^{N/2-1} S^*_{k'} W^{\text{a}*}\left(f - \frac{k}{T_u}\right)\right)\right]. \tag{2.37}$$

In addition, taking into account (2.4), we have

$$G_{s_{\text{rec}}}(f) = \frac{|H_{\text{rec}}(f)|^2}{T_u^2 T_{\text{symb}}} \sum_{k=-N/2}^{N/2-1} \mathbb{E}\left[|S_k|^2\right] \left|W^{\text{a}}\left(f - \frac{k}{T_u}\right)\right|^2$$
$$= \frac{2\sigma_S^2 |H_{\text{rec}}(f)|^2}{T_u^2 T_{\text{symb}}} \sum_{k=-N_u/2}^{N_u/2-1} \left|W^{\text{a}}\left(f - \frac{k}{T_u}\right)\right|^2. \tag{2.38}$$

Naturally, the aliasing effects decrease when the oversampling factor increases. Indeed, when N is large and/or $O \geq 2$, it can be shown that [30]

$$W^{\text{a}}\left(f - \frac{k}{T_u}\right) \approx W\left(f - \frac{k}{T_u}\right). \tag{2.39}$$

In these conditions, the aliasing effects can be considered negligible and the differences between $s_{\text{rec}}(t)$ and $s(t)$ are small. Moreover, if the frequency response of the reconstruction filter is flat over the signal band (i.e., $H_{\text{rec}}(f) = 1$), we have that $|H_{\text{rec}}(f)|^2 = 1$ and

$$G_{s_{\text{rec}}}(f) \approx \frac{2\sigma_S^2}{T_u^2 T_{\text{symb}}} \sum_{k=-N_u/2}^{N_u/2-1} \left| \text{sinc}\left(f - \frac{k}{T_u}\right) \right|^2 . \tag{2.40}$$

From (2.40), one can conclude that the PSD of reconstructed OFDM signals represented in (2.27) is approximately proportional to the PSD in (2.13), i.e., to the PSD of the continuous-time "perfect" OFDM signal $s(t)$.

Even with a quasi-perfect reconstruction of OFDM signals through the samples generated by the IDFT block, it should be noted that the use of rectangular windows lead to large side lobes outside the "useful band", which may compromise the commitment to the spectral mask of the system, leading to high adjacent channel interference (ACI) levels that prejudice the communication systems that operate in adjacent bands. In order to reduce this out-of-band radiation, different windows should be considered. A common window employed in several OFDM systems is the square-root raised-cosine (SRRC) window [31], since it can greatly compact the PSD of OFDM signals and reduce the ACI [30, 32].

After the DAC, the baseband OFDM signal $s(t)$ passes through a radio frequency (RF) modulator to be up converted and translated to a given carrier frequency f_c[5]. The resultant bandpass OFDM is

$$s_{bp}(t) = \text{Re}\left(s(t) \exp\left(j2\pi f_c t\right)\right). \tag{2.41}$$

After the up conversion to the RF band, the signal is amplified and transmitted through the channel.

2.2.2 TRANSMISSION OVER FREQUENCY-SELECTIVE CHANNELS

Due to the multipath phenomena in wireless propagation scenarios, at the receiver may arrive not only the transmitted signal, but also several copies of it. These copies may have different gains and delays and add destructively or constructively, causing fading and ISI, which can severely degrade the system's performance.

Regarding the channel characteristics, the fading can be classified as [2]: (i) slow or fast, according to the time-variation of the channel and (ii) frequency-selective or frequency-nonselective (flat), according to the channel delay spread. In this book, time-invariant (slow-fading), frequency-selective channels are considered. These channels can be generically characterized by a linear, time-invariant filter. The impulse response of this filter is conventionally termed channel impulsive response (CIR) and can be expressed as

$$h(t) = \sum_{l=0}^{L-1} a_l \delta(t - \tau_l), \tag{2.42}$$

[5] As an example, the carrier frequency in the IEEE 802.11ac wireless LAN standard is $f_c = 5$ GHz.

where L is the number of multipath components and α_l and delay τ_l are the complex amplitude and the delay of the lth ray, respectively. In the following, it is assumed that the delays of the multipath components are equally-spaced by N/T_u. Therefore, (2.42) can be rewritten as

$$h(t) = \sum_{l=0}^{L-1} a_l \delta\left(t - l\frac{N}{T_u}\right). \tag{2.43}$$

At the frequency-selective channel output, we have

$$z(t) = s(t) * h(t). \tag{2.44}$$

Naturally, the complex amplitude and the delay of the lth multipath component are random quantities due to the constant changes in the multipath environment (a consequence of the users' mobility). The power delay profile (PDP) of the CIR represented in (2.43) is defined as

$$v(t) = \sum_{l=0}^{L-1} |\alpha_l|^2 \delta\left(t - l\frac{N}{T_u}\right). \tag{2.45}$$

On the other hand, regarding the frequency-domain, the Fourier transform of $h(t)$ is

$$H(f) = \sum_{l=0}^{L-1} \alpha_l \exp(j2\pi f \tau_l). \tag{2.46}$$

As the performance analysis of OFDM systems presented in this section is based on discrete-time and discrete-frequency samples, we considered the discrete version of the frequency-selective channel. Therefore, it is firstly shown what information is contained in the channel samples regarding both the time and the frequency domains. Let us start by considering a frequency-domain signal composed of periodic repetitions of the channel frequency response $H(f)$. If these repetitions are spaced by $F_s = 1/T_s$, we may define the periodic frequency-domain signal

$$H^P(f) = \sum_{m=-\infty}^{+\infty} H(f - mF_s). \tag{2.47}$$

Note that taking into account its periodicity, we can expand $H^P(f)$ in a Fourier series, namely

$$H^P(f) = \sum_{n=-\infty}^{+\infty} \beta_n \exp(j2\pi f n T_s), \tag{2.48}$$

where β_n represents the Fourier coefficient of the nth complex sinusoid. This coefficient can be obtained as

$$\beta_n = \frac{1}{F_s} \int_{-F_s/2}^{+F_s/2} H^P(f) \exp(-j2\pi n f T_s)\, df$$
$$\approx \frac{1}{F_s} \int_{-F_s/2}^{+F_s/2} H(f) \exp(-j2\pi n f T_s)\, df, \tag{2.49}$$

where the previous approximation is made under the condition of negligible aliasing effects (the separation between the different replicas, given by F_s, is large enough to neglect their overlap). In fact, it should be noted that this definition of the Fourier coefficients is strictly related with the inverse Fourier transform of $H(f)$ (see (2.43)), that is

$$h(t) = \int_{-F_s/2}^{+F_s/2} H(f) \exp(j2\pi f t)\, df. \tag{2.50}$$

In fact, β_n is given by

$$\beta_n = \frac{1}{F_s} h(-nT_s) = T_s h(-nT_s). \tag{2.51}$$

Therefore, (2.48) can be rewritten as

$$H^P(f) = \sum_{n=-\infty}^{+\infty} T_s h(nT_s) \exp(-j2\pi f n T_s). \tag{2.52}$$

Let us now define

$$H_k \triangleq F_s H^P(k/T_u). \tag{2.53}$$

As $h(nT_s)$ is only defined for $0 \leq n \leq N-1$, we have

$$H_k = \sum_{n=0}^{N-1} h(nT_s) \exp\left(-j2\pi \frac{k}{T_u} n T_s\right)$$
$$= \sum_{n=0}^{N-1} h(nT_s) \exp\left(-j2\pi \frac{k}{N} n\right). \tag{2.54}$$

Moreover, by defining

$$h_n \triangleq h^P(nT_s), \tag{2.55}$$

where

$$h^P(t) = \sum_{i=-\infty}^{+\infty} h(t - iT_u), \tag{2.56}$$

consists in periodic repetitions of $h(t)$ spaced by T_u, it can be noted that

$$\mathbf{h} = \mathbf{F}^{-1}\mathbf{H}, \tag{2.57}$$

i.e., the block of time-domain channel samples $\mathbf{h} = [h_0\ h_1\ h_2\ \ldots\ h_{N-1}]^T \in \mathbb{C}^N$ is the IDFT of $\mathbf{H} = [H_0\ H_1\ H_2\ \ldots\ H_{N-1}]^T \in \mathbb{C}^N$. Moreover, if the different replicas of $H(f)$ for $m \neq 0$ in $H^P(f)$ are removed through filtering operations, then

$$H_k \approx F_s H(k/T_u), \tag{2.58}$$

which means that apart from a scale factor F_s, H_k are samples of the channel frequency response represented in (2.46). Clearly, the kth sample of the channel frequency response, H_k, is random, since it depends on the random amplitude of the different multipath components. In this work, the channel frequency responses are characterized through a statistical model. More concretely, we consider Rayleigh multipath fading, which is a special case of the Nakagami-m fading [33], where there is not a line of sight component between the transmitter and the receiver. In these conditions, the complex amplitude of the lth multipath component of the channel CIR has a complex Gaussian distribution, i.e.,

$$\alpha_l \sim \mathscr{CN}(0, \sigma_l^2). \tag{2.59}$$

As in average we want that each subcarrier experiences a unitary "channel gain", i.e., we aim that $\mathbb{E}[|H_k|^2] = 1$, the variance associated with the lth multipath component, σ_l^2, is defined as

$$\sigma_l^2 = \frac{1}{L}, \qquad 0 \leq l \leq L-1, \tag{2.60}$$

which means that the average power of the different multipath components is equal. Therefore, considering (2.54), one can note that the real and imaginary parts of H_k have approximately a Gaussian distribution with zero mean and variance 1/2, i.e., $\mathrm{Re}(H_k) \sim \mathscr{N}(0, 1/2)$ and $\mathrm{Im}(H_k) \sim \mathscr{N}(0, 1/2)$. On the other hand, the absolute value of the channel frequency responses has a Rayleigh distribution, i.e., $|H_k| \sim \mathrm{Rayleigh}(1/\sqrt{2})$, which means that $|H_k|^2 \sim \Gamma(1, 1)$, i.e., the squared absolute value of the channel frequency responses has a Gamma distribution with unitary shape and scale parameters. Fig. 2.7 shows the simulated and theoretical distributions associated with the real part, absolute value and squared absolute value of H_k, for frequency-selective channels with $L = 64$ uncorrelated multipath components. From the results depicted in the figure, it can be seen that the Gaussian approximation for the channel frequency responses is very tight and, even with a moderate number of L, it can be verified that the squared absolute value of the channel frequency responses has indeed a Gamma distribution. Therefore, by defining the average power of the multipath rays as in (2.60), we do have a unitary channel gain for each subcarrier, i.e., $\mathbb{E}[|H_k|^2] = 1$, which means that the average power of the received signals is the same of the transmitted signals. Fig. 2.8 shows the evolution of the squared absolute value of the channel frequency responses $|H_k|^2$ considering $N_u = 256$, $O = 4$ and different values of L. From the figure, it can be noted that the average value of the $|H_k|^2$

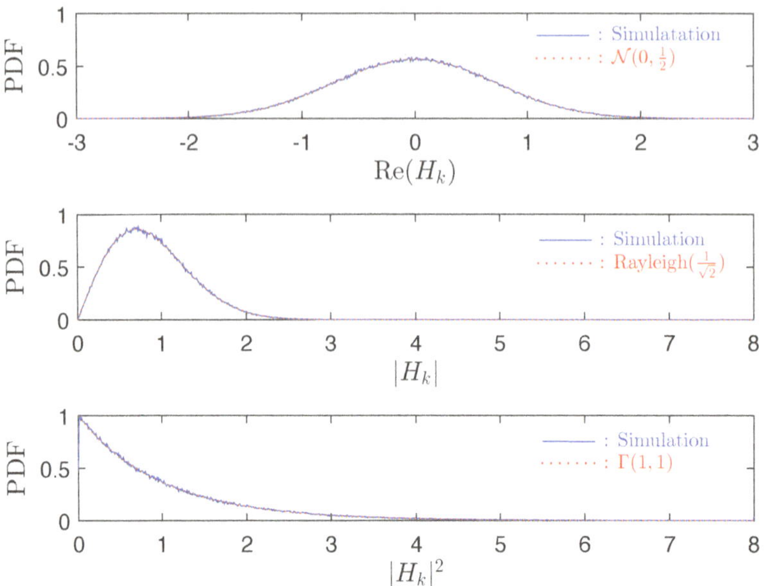

Figure 2.7 PDF of the real part, absolute value and squared absolute value of the channel frequency responses for $L = 64$ multipath components.

is unitary, regardless of L, which is expected since the distribution of $|H_k|^2$ does not depend on L. However, as L increases, the fluctuations also increase, i.e., the channel presents a more dynamic behavior.

In time-dispersive channels, the data detection can be severely affected, leading to poor performances and considerable BER degradations. This is specially important in SC systems, where the symbol time is very small and the ISI is severe. Thus, to compensate the large ISI in such systems, very high complex equalization techniques have to be employed. In OFDM, although, the higher duration of the symbols greatly reduces the ISI comparably to that of SC systems and this is one of its main advantages. However, even with a symbol duration that is many times larger, there is always some residual ISI between two consecutive OFDM blocks. In fact, the ISI in OFDM is typically constrained to one OFDM block, i.e., it does not spread over several symbols as typically occurs in SC systems, where the symbol time is smaller than the delay spread of the channel. For this reason, it is also known as inter-block interference (IBI).

Another important problem of frequency-selective channels is that delayed replicas of the transmitted signal may destroy the orthogonality between the subcarriers and cause inter-channel interference (ICI), which can severely degrade the BER of

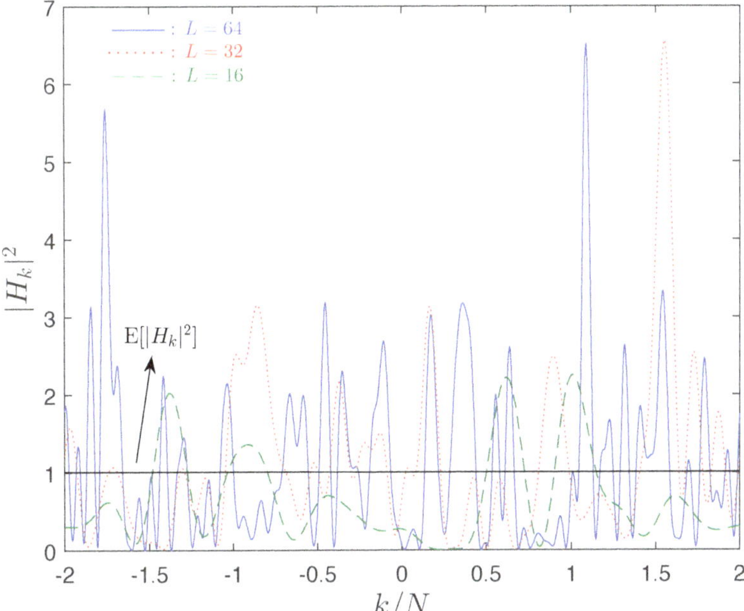

Figure 2.8 Evolution of $|H_k|^2$ considering channels with a different number of multipath components.

the OFDM system. For these reasons, it is common to consider a guard interval between OFDM blocks to guarantee both an IBI and an ICI-free transmission in a multipath propagation environment.

In the so-called cyclic prefix (CP)-OFDM [34], the guard interval is composed by a repetition of the last part of the OFDM block, i.e., it consists of a cyclic extension. The CP has mainly two purposes: (i) increase the resilience/robustness to IBI by extending the OFDM block in such a way that the interference caused by the previous block falls into the CP and (ii) reduce the complexity of the equalization process, allowing the use of simple one-tap equalizers that simplify the detection. While the former goal is achieved by discarding the CP samples at the receiver, the latter is attained through the cyclic extension of the subcarriers. With that cyclic extension, their delayed versions have an integer number of cycles during the detection interval, which allows to maintain the orthogonality. However, the advantages of the CP come at a cost of power and spectral efficiencies. In fact, there is power spent in the transmission of redundant information as well as a loss in the symbol rate, which means that both the energy and the bandwidth efficiency decrease. Thus, when choosing the duration of the CP, the system designer faces a trade-off between the robustness to IBI and ICI and the deterioration of the power/bandwidth efficiencies. Commonly,

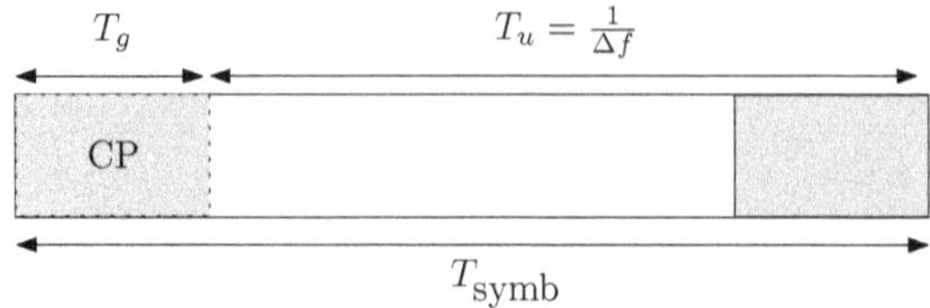

Figure 2.9 Composition of the OFDM block after the CP addition.

the number of samples that compose the CP, N_{cp}, and its corresponding duration, $T_g = N_{cp}T_s$, is dictated by the CIR length. Ideally, to completely eliminate the IBI, the CP should be, at least, equal to the CIR length. Nevertheless, in order to avoid a significant overhead, the CP duration T_g is typically chosen to be no longer than one-fourth of the symbol duration, which means that in terms of samples, we have $N_{cp} \leq N/4$[6]. After the CP addition, the duration of the OFDM symbol is

$$T_{\text{symb}} = N_t T_s = (N + N_{cp})T_s = T_g + T_u. \tag{2.61}$$

Note also that due to the CP transmission, there is a degradation in the transmission efficiency, which is given by

$$\eta_{\text{CP}} = \frac{N}{N_t} = \frac{N}{N + N_{cp}}. \tag{2.62}$$

Fig. 2.9 depicts the composition of a time-domain OFDM block after the CP addition. An alternative for CP-OFDM is the zero padded (ZP)-OFDM. In ZP-OFDM schemes, no power is wasted to transmit the CP samples since the guard interval is composed by zeros. However, this technique requires more complex receivers, although the complexity can be alleviated at the expense of some degradation in the power efficiency [35]. In fact, the simplicity and elegance of CP-OFDM makes it the most common implementation of OFDM in practical systems. Here, CP-OFDM systems are considered and it is always assumed that the duration of the CP is larger enough to compensate the adverse effects of the channel length and, thus, to avoid IBI and ICI. In the following, it is shown how the CP can remove the IBI and the ICI.

In the signal processing chain considered here, the CP addition is modeled as a multiplication between the time-domain samples of a given OFDM block $\mathbf{s} = [s_0\ s_1\ s_2\ \ldots\ s_{N-1}]^T \in \mathbb{C}^N$ and the $N_t \times N$ "CP addition" matrix $\mathbf{G}_a$, that is defined as

$$\mathbf{G}_a = \left[\ [\mathbf{0}\ \mathbf{I}_{N_{cp}}]^T\ \mathbf{I}_{N_t - N_{cp}}\right]^T, \tag{2.63}$$

[6]For instance, in long term evolution (LTE), the subcarrier spacing is $\Delta f = 15$ kHz and the symbol duration is $T_{\text{symb}} = 66.7\mu$s. For the normal CP, where $T_g = 5.2\mu$s, we have $T_g/T_{\text{symb}} \approx 0.08 \leq 0.25$. For the extended CP, used to deal with larger delay spreads, we have $T_g = 17\mu$s and $T_g/T_{\text{symb}} \approx 0.25$.

with $\mathbf{I}_p$ denoting the $p \times p$ identity matrix. After the CP addition, we have

$$\begin{aligned} \mathbf{x} &= \mathbf{G}_a \mathbf{s} \\ &= \mathbf{G}_a \mathbf{F}^{-1} \mathbf{S}, \end{aligned} \tag{2.64}$$

where $\mathbf{x} = [x_0\ x_1\ x_2\ \dots\ x_{N_t-1}]^T \in \mathbb{C}^{N_t}$ denotes the augmented set of time-domain samples that results from the concatenation of $\mathbf{s} = [s_0\ s_1\ s_2\ \dots\ s_{N-1}]^T \in \mathbb{C}^N$ with the CP samples. Note that the nth element of $\mathbf{x} = [x_0\ x_1\ x_2\ \dots\ x_{N_t-1}]^T \in \mathbb{C}^{N_t}$ is given by

$$x_n = \begin{cases} s_{N-N_{cp}+n}, & 0 \leq n \leq N_{cp} - 1 \\ s_{n-N_{cp}}, & N_{cp} \leq n \leq N_t - 1. \end{cases} \tag{2.65}$$

Let us consider the discrete-time version of the time-dispersive channel represented in (2.42). Having in mind (2.44), the output of the discrete-time channel is the discrete convolution between the time-domain samples of the OFDM signal $\mathbf{x} = [x_0\ x_1\ x_2\ \dots\ x_{N_t-1}]^T \in \mathbb{C}^{N_t}$ and time-domain channel samples $\mathbf{h} = [h_0\ h_1\ h_2\ \dots\ h_{N-1}]^T \in \mathbb{C}^N$, i.e.,

$$\mathbf{z} = \mathbf{h} * \mathbf{x}. \tag{2.66}$$

Therefore, the nth time-domain sample at the channel output is

$$\begin{aligned} z_n &= \sum_{j=-\infty}^{+\infty} x_j h_{n-j} \\ &= \sum_{j=n-L+1}^{n} x_j h_{n-j}. \end{aligned} \tag{2.67}$$

From (2.67), one can note that the nth sample at the channel output is not only a function of the nth data sample x_n, but also of the $L-1$ previous samples. For instance, regarding the 0th sample of the mth block, $z_0^{(m)}$, we have

$$z_0^{(m)} = \underbrace{x_{-L+1}^{(m-1)} h_{L-1} + x_{-L+2}^{(m-1)} h_{L-2} + \cdots +}_{\text{interference from } (m-1)\text{th block}} \underbrace{x_0^{(m)} h_0}_{\text{useful part}}, \tag{2.68}$$

where it is clear that samples of the $(m-1)$th block interfere with samples of the mth block, leading to the so-called IBI. However, as the receiver discards all the CP samples, the IBI can be completely eliminated provided that N_{cp} is larger than L. In the signal processing scheme considered here, the CP removal is modeled by the multiplication between $\mathbf{z} = [z_0\ z_1\ z_2\ \dots\ z_{N-1}]^T \in \mathbb{C}^N$ and the CP removal matrix $\mathbf{G}_r$, which is an $N_t \times N_t$ diagonal matrix defined as

$$\mathbf{G}_r = \operatorname{diag}\left([\underbrace{0\ 0 \dots 0}_{N_{cp}}\ \underbrace{1\ 1 \dots 1}_{N}] \right). \tag{2.69}$$

Therefore, also considering the additive white Gaussian noise (AWGN), the signal for detection purposes is given by

$$\mathbf{r} = \mathbf{G}_r \mathbf{z} + \nu, \tag{2.70}$$

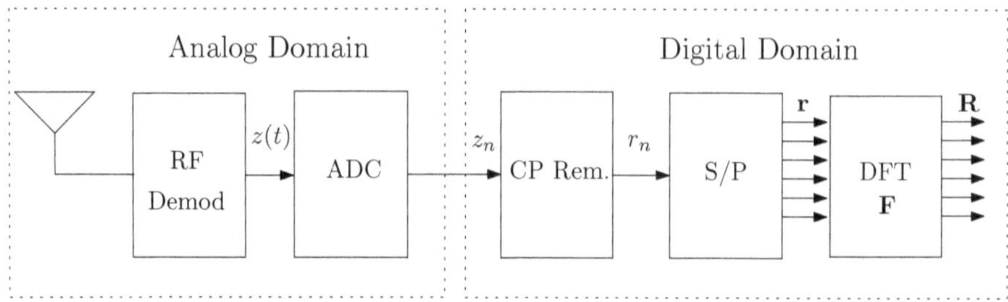

Figure 2.10 OFDM receiver structure.

where $\mathbf{r} = [r_0 \; r_1 \; r_2 \; ... \; r_{N-1}]^T \in \mathbb{C}^N$ denotes the time-domain received samples without the CP and $\nu = [\nu_0 \; \nu_1 \; \nu_2 \; ... \; \nu_{N-1}]^T \in \mathbb{C}^N$ represents the array of time-domain AWGN samples. The real and the imaginary parts of these noise samples are modeled by a Gaussian random variable with zero mean and variance σ_ν^2, i.e., $\nu_n \sim \mathcal{N}(0, \sigma_\nu^2)$. Thus, the power of the noise samples is $\mathbb{E}[|\nu_n|^2] = 2\sigma_\nu^2$.

The OFDM receiver structure is shown in Fig. 2.10. After the CP elimination, all received signal samples $\mathbf{r} = [r_0 \; r_1 \; r_2 \; ... \; r_{N-1}]^T \in \mathbb{C}^N$ of the mth block are only a function of the data samples of that block, appearing that there is no channel memory, i.e., no ISI. For this reason, in matrix notation, we can write

$$\begin{bmatrix} r_0 \\ r_1 \\ \vdots \\ r_{N-1} \end{bmatrix} = \underbrace{\begin{bmatrix} h_0 & 0 & \cdots & & 0 & h_{L-1} & \cdots & h_2 & h_1 \\ h_1 & h_0 & 0 & \cdots & & 0 & h_{L-1} & \cdots & h_2 \\ \vdots & \ddots & \ddots & & & & \ddots & \ddots & \vdots \\ 0 & \ddots & 0 & h_{L-1} & \cdots & h_0 & \cdots & & 0 \\ & & & & & \ddots & \ddots & \ddots & \\ 0 & \cdots & & 0 & h_{L-1} & \cdots & h_1 & h_0 & 0 \\ 0 & \cdots & & 0 & & h_{L-1} & \cdots & h_1 & h_0 \end{bmatrix}}_{\mathbf{h}^c} \cdot \begin{bmatrix} s_1 \\ s_2 \\ \vdots \\ s_{N-1} \end{bmatrix} + \begin{bmatrix} \nu_1 \\ \nu_2 \\ \vdots \\ \nu_{N-1} \end{bmatrix}. \tag{2.71}$$

Clearly, from (2.71) one can note that $\mathbf{h}^c$ is a circulant matrix. Recalling the fact that these types of matrices can be diagonalized with the DFT matrix represented in (2.18), we may decompose $\mathbf{h}^c$ as[7]

$$\mathbf{h}^c = \mathbf{F}^{-1}\mathbf{H}\mathbf{F}. \tag{2.72}$$

[7]Here, a slight abuse in the notation is committed since, for simplicity, we are denoting the diagonal matrix composed by the elements of **H** also as **H**.

Therefore,

$$\mathbf{r} = \mathbf{F}^{-1}\mathbf{HFs} + \nu. \tag{2.73}$$

By applying the DFT to both sides of (2.73), we obtain

$$\begin{aligned}\mathbf{R} &= \mathbf{HFs} + \mathbf{F}\nu \\ &= \mathbf{HS} + \mathbf{N},\end{aligned} \tag{2.74}$$

where $\mathbf{H}$ is a diagonal matrix whose diagonal elements are composed of the DFT of the channel coefficients $\mathbf{h} = [h_0\ h_1\ h_2\ \ldots\ h_{N-1}]^T \in \mathbb{C}^N$ (the DFT of the first column of $\mathbf{h}^c$), i.e.,

$$\mathbf{H} = \begin{bmatrix} H_0 & 0 & 0 & \cdots & 0 \\ 0 & H_1 & 0 & \cdots & 0 \\ 0 & 0 & H_2 & \cdots & 0 \\ 0 & 0 & \cdots & \ddots & \vdots \\ 0 & 0 & 0 & 0 & H_{N-1} \end{bmatrix}, \tag{2.75}$$

and $\mathbf{N} = [N_0\ N_1\ N_2\ \ldots\ N_{N-1}]^T \in \mathbb{C}^N$ represents the noise samples in the frequency domain, i.e., the DFT of the block $\nu = [\nu_0\ \nu_1\ \nu_2\ \ldots\ \nu_{N-1}]^T \in \mathbb{C}^N$. After the DFT operation, the noise samples still have a Gaussian distribution, however, their real and imaginary parts have variance

$$\sigma_N^2 = \sigma_\nu^2 N_u O^2. \tag{2.76}$$

and $\mathbb{E}[|N_k|^2] = 2\sigma_N^2$. From (2.74), the received signal for the kth subcarrier can be written as

$$R_k = S_k H_k + N_k, \tag{2.77}$$

where H_k and N_k represent the channel frequency response (see (2.58)) and the noise component of the kth subcarrier, respectively. In fact, from (2.77), one can note that the existence of a CP larger than the delay spread and the invariability of the channel during the symbol time, allows the elimination of the IBI and ICI. Put in other words, this means that the frequency-selective channel is converted into a set of N parallel, flat-fading channels. The equivalent, subcarrier-level model for an OFDM system in these conditions is shown in Fig. 2.11[8].

The next task done by the receiver is the equalization. It is worth to mention that as each subcarrier only experiences a complex "gain" H_k, there is no need to perform heavy and complex equalization procedures and simple one-tap equalizers can be employed before the detection. In conventional OFDM schemes, these blocks work individually on each subcarrier as can be seen in Fig. 2.12, which shows the equalization and detection blocks. Denoting the equalization factor for a given subcarrier as F_k, the data estimated on the kth subcarrier is

$$\hat{S}_k = F_k R_k. \tag{2.78}$$

[8]Thorough this book we will denote this model as the "linear OFDM model".

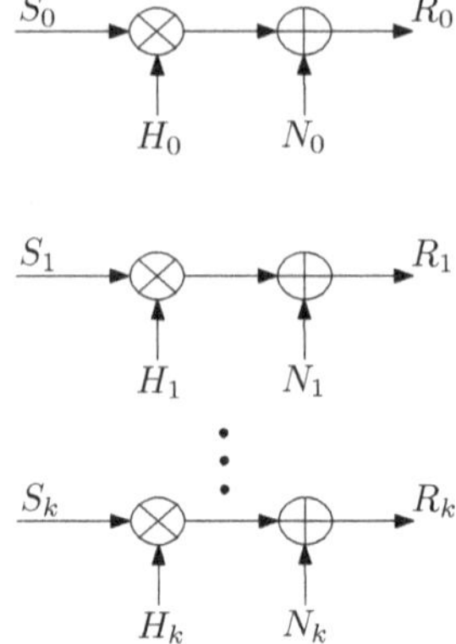

Figure 2.11 Equivalent, subcarrier-level model for an OFDM system without IBI and ICI.

The simplest equalization procedure involves a single-channel inversion and is known as zero-forcing (ZF) technique. In that technique, the equalization factor is

$$F_k = \frac{1}{H_k} = \frac{H_k^*}{|H_k|^2}, \tag{2.79}$$

and the equalized symbol of the kth subcarrier is hence

$$\begin{aligned} \hat{S}_k &= R_k F_k \\ &= S_k + \frac{N_k}{H_k}. \end{aligned} \tag{2.80}$$

Although very simple, this equalization procedure may lead to poor results due to the undesired noise enhancement associated with the subcarriers with the deepest fades. Note that in those subcarriers, the noise is severely amplified by the factor

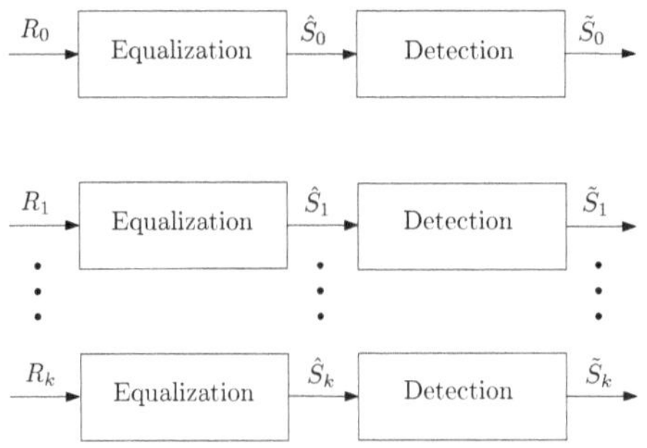

Figure 2.12 Equalization and detection in conventional OFDM.

$1/H_k$, not to mention the problems that may arise in situations where $H_k \approx 0$, since in those situations is very difficult to invert the channel and obtain the corresponding equalization factor.

Regarding ZP-OFDM schemes, more complex equalization methods such as minimum mean square error (MMSE) techniques can be employed. In these techniques, the subcarrier's SNR is taken into account for the computation of the equalization factor [36].

After the equalization process, the symbol of the kth subcarrier, $\hat{S}_k$, is sent to the decision device that outputs the estimated data symbol $\tilde{S}_k$. In conventional OFDM systems, the detection is also made on a subcarrier basis. This means that the detection of the kth data symbol is made only with information of the kth subcarrier. For instance, the typical detection procedure is based on a hard-decision on each subcarrier. Contrarily to the conventional detection (see Fig. 2.12), the optimum detection, which will be introduced in the next chapter, involves the information of the entire block (i.e., of all subcarriers), since it is a block-by-block detection [33]. In the following, a set of BER performance results regarding conventional OFDM transmissions under different scenarios is presented. In all those scenarios, perfect channel knowledge and perfect time and frequency synchronization are assumed.

As a consequence of (2.77), the BER of a given OFDM system can be obtained by averaging the BER of all subcarriers. Having in mind that only N_u out of N subcarriers are used for data transmission and denoting the BER associated with the kth subcarrier as $P_{b,k}$, we have

$$P_b = \frac{1}{N_u} \sum_{k=1}^{N_u} P_{b,k}. \tag{2.81}$$

Note that for the case of ideal AWGN channels, $P_{b,k}$ does not change with k, which means that $P_b = P_{b,k}$. Under these conditions, it can be shown that the BER of an OFDM system employing M-QAM constellations and a Gray mapping rule is given by [33]

$$P_b \approx \frac{2}{\log_2(\sqrt{M})} \left(1 - \frac{1}{\sqrt{M}}\right) Q\left(\sqrt{\frac{3\log_2(M)E_b}{(M-1)N_0}}\right), \tag{2.82}$$

where $Q(x)$ represents the well-known Q-function, that gives the tail probability of the standard normal distribution and can be computed as [37]

$$Q(x) = \frac{1}{\sqrt{2\pi}} \int_x^{\infty} \exp\left(-\frac{u^2}{2}\right) du, \tag{2.83}$$

and E_b denotes the average bit energy. Note that for the particular case of QPSK constellations, where two bits are transmitted per subcarrier, the average bit energy is

$$E_b = \frac{1}{2N_u} \sum_{k=0}^{N-1} \mathbb{E}\left[|S_k|^2\right] = \frac{2N_u\sigma_S^2}{2N_u} = \sigma_S^2. \tag{2.84}$$

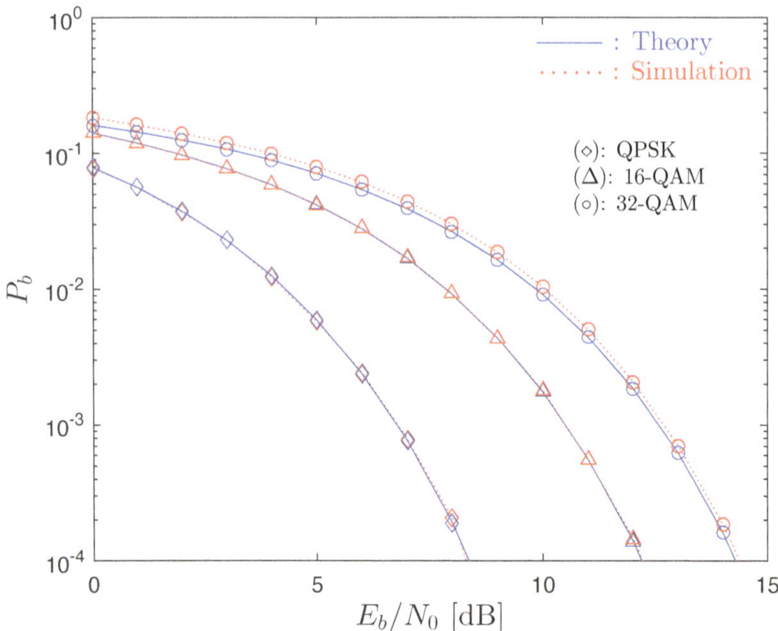

Figure 2.13 Simulated and theoretical BER of an OFDM system for ideal AWGN channels and different constellations.

This means that, when normalized QPSK constellations (i.e., $M = 4$) are employed, we have $\sigma_S = 1$ and $E_b = 1$. Under these conditions, from (2.82), the approximate BER is given by

$$P_b \approx Q\left(\sqrt{\frac{2E_b}{N_0}}\right). \tag{2.85}$$

Fig. 2.13 shows the simulated and theoretical BER (obtained with (2.82)) of an OFDM transmission in ideal AWGN channels, considering $N_u = 256$, $O = 4$ and for different constellations. From the results depicted in the figure, it can be pointed out that (2.82) is a very tight approximation for the BER, especially, for large values of E_b/N_0. As expected, the BER is higher when larger constellations are employed. For a target BER of $P_b = 10^{-4}$, an E_b/N_0 of approximately 8.3 dB is required if QPSK constellations are employed. However, the required value of E_b/N_0 to achieve that BER increases to approximately 12.2 dB and 14.3 dB when 16-QAM and 32-QAM constellations are considered, respectively. It should also be mentioned that this BER represents the "ideal" performance, i.e., the performance that could be obtained in ideal conditions, i.e., in a channel does not present impairments such as fading and/or ICI.

As can be observed from (2.77), for frequency-selective channels with uncorrelated Rayleigh fading, the received signal for the kth subcarrier is dependent on the channel frequency response of that subcarrier. Therefore, due to the random nature of H_k, the SNR experienced on the kth subcarrier is a random quantity that depends on the channel gain $|H_k|^2$, which means that the BER of the kth subcarrier is also a random quantity, i.e., $P_{b,k}(|H_k|^2)$. As seen before in this section, the channel gain for each subcarrier can be modeled by Gamma distribution with unitary average value (see Fig. 2.7). Thus, its distribution is given by

$$p\left(|H_k|^2\right) = \exp\left(-|H_k|^2\right), \quad |H_k|^2 > 0. \tag{2.86}$$

Under these conditions, the average BER for a given subcarrier can be obtained as

$$P_{b,k} = \mathbb{E}\left[P_{b,k}\left(|H_k|^2\right)\right] = \int_0^{+\infty} P_{b,k}\left(|H_k|^2\right) p\left(|H_k|^2\right) d|H_k|^2. \tag{2.87}$$

where $P_{b,k}\left(|H_k|^2\right)$ represents the BER for the kth subcarrier, associated with a given channel realization $\mathbf{H}$. As the distribution of the channel gain is independent of k, the average BER is equal to all subcarriers, i.e., $P_b = P_{b,k}$. After straightforward but lenghty manipulations, it can be shown that the BER for OFDM transmissions in frequency-selective channels is given by [33]

$$P_b \approx \frac{2}{\log_2\left(\sqrt{M}\right)}\left(1 - \frac{1}{\sqrt{M}}\right)\left(1 - \sqrt{\frac{K_M}{1+K_M}}\right), \tag{2.88}$$

where $K_M = \frac{3\log_2(M)E_b}{2(M-1)N_0}$. Since the channel has an average unitary gain, it was considered that the average bit energy averaged over several channel realizations is still E_b. Thorough this work, both (2.82) and (2.88) are used as BER references for uncoded OFDM scenarios, regarding both ideal AWGN and frequency-selective channels, respectively. Fig. 2.14 shows the average BER of an OFDM system considering frequency-selective channels with $L = 32$ multipath components and different QAM constellations. From this figure, one can note that (2.88) is very accurate, especially, in the asymptotic region (i.e., for large values of E_b/N_0) and/or when small constellations are employed. As expected, the BER in frequency-selective channels is considerably higher than in ideal AWGN channels. Even with QPSK constellations, to obtain a target BER of $P_b = 10^{-3}$, we need approximately $E_b/N_0 = 23.8$ dB, and this value increases to around $E_b/N_0 = 26.8$ dB and $E_b/N_0 = 28.9$ dB when 16-QAM and 32-QAM constellations are employed, respectively.

In practice, to avoid this very large BER, OFDM is typically used in conjugation with some process of channel coding, which results in the so-called coded OFDM (COFDM) [38, 39]. The use of channel coding allows to substantially improve the performance of OFDM systems in strongly frequency-selective channels. One of the most popular channel codes for OFDM is the 64-state convolutional code with rate $C_r = 1/2$[9]. The wide use of convolutional codes can be justified by the possibility that

[9] In the IEEE 802.11a wireless LAN standard, a 64-state convolutional code is used. The rate of this code can be $C_r = 1/2$, $C_r = 2/3$ or $C_r = 3/4$.

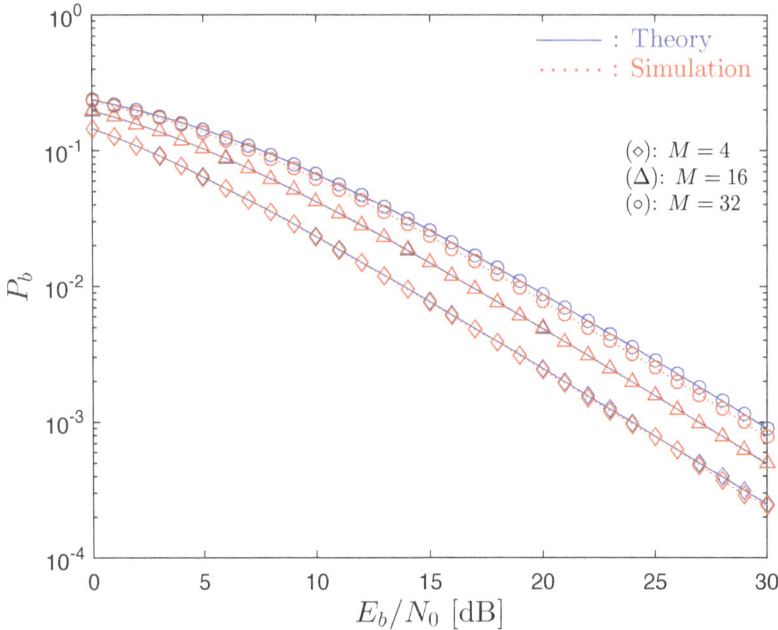

Figure 2.14 Average BER of an OFDM system considering frequency-selective channels with $L = 32$ and different constellations.

they offer to realize the decoding operation in real-time through the use of the Viterbi algorithm [40]. Fig. 2.15 presents the BER of both an OFDM and a COFDM system, considering both ideal AWGN and frequency-selective channels with $L = 32$. The channel coding is made with a 64-state convolutional code with rate $C_r = 1/2$ and the signals have $N_u = 128$ and $O = 4$. From the results shown in the figure, one can clearly note the use of a channel code can substantially improve the performance of OFDM systems. This is especially important in scenarios where the channel presents a frequency-selective behavior since, in these cases, the BER is highly conditioned by the subcarriers that are in deep fade, which can considerably reduce system's performance. In fact, for a target BER of $P_b = 10^{-3}$, the use of channel coding gives rise to a gain of about 19 dB relatively to the uncoded OFDM transmission. Regarding ideal AWGN channels and a target BER of $P_b = 10^{-4}$, there is a gain of approximately 7 dB relatively to the uncoded OFDM scenario.

2.3 PAPR PROBLEM

One of the major problems of multicarrier modulations is the very large envelope fluctuations of their waveforms, that are constituted by the summation of several modulated subcarriers (see (2.2)). Therefore, very high power peaks can be produced

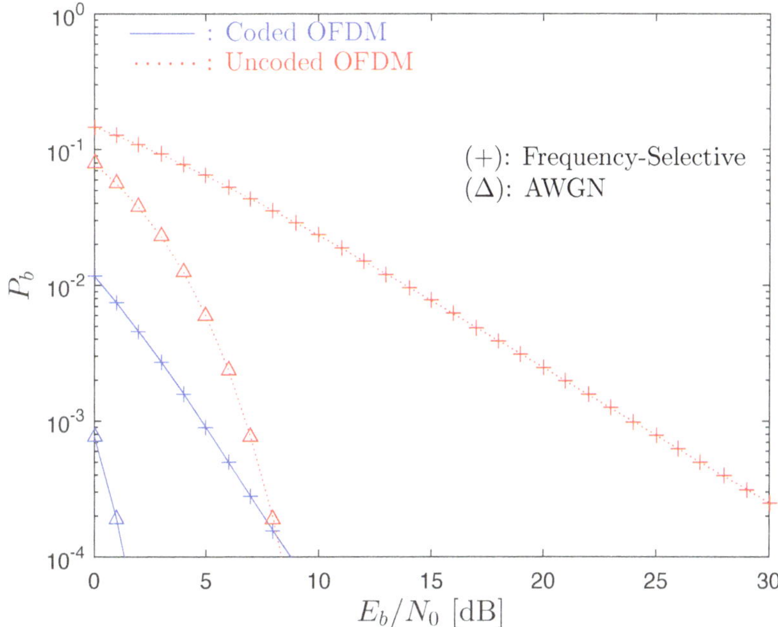

Figure 2.15 BER for both coded and uncoded OFDM systems considering both ideal AWGN and frequency-selective channels.

at the IDFT output, leading to signals with large dynamic range. Signals with such characteristics lead to difficulties in the amplification, quantization and other operations that are done in the transmission chain of MC systems, since the devices associated with such operations should present a large linear dynamic range.

There are different metrics to quantify the high power peaks and large envelope fluctuations of multicarrier signals, but the most consensual is the so-called PAPR. The PAPR is defined as the ratio between the maximum instantaneous power (peak power), $P_{\text{in,max}}$, and the average power of a given OFDM block $\mathbf{s} = [s_0\ s_1\ s_2\ \ldots\ s_{N-1}]^T \in \mathbb{C}^N$, $P_{\text{in,avg}}$, i.e.,

$$\text{PAPR} = \frac{P_{\text{in,max}}}{P_{\text{in,avg}}} = \frac{\max\left(|s_n|^2\right)}{2\sigma^2}. \tag{2.89}$$

Fig. 2.16 shows both the instantaneous power and the average power of the samples of an OFDM signal with $N_u = 256$ useful subcarriers and $O = 1$. From the figure, it is clear that very large peaks can occur within a block. However, it should be mentioned that the peaks associated with the discrete-time version of the OFDM signal may not coincide with the peaks of the continuous-time signal, since we can have peaks between the sampling times. In fact, the PAPR of the discrete time-domain samples

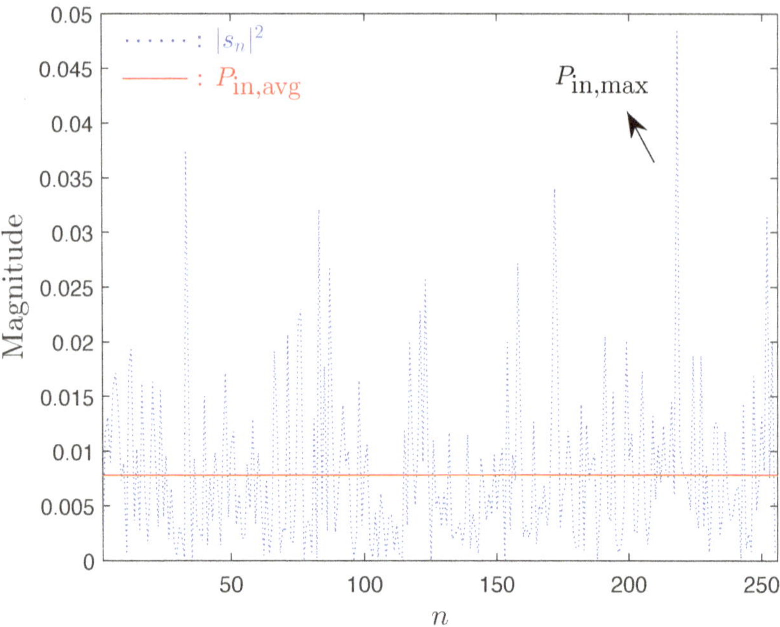

Figure 2.16 Instantaneous power and average power of the samples of a given OFDM signal.

is a lower bound of the true PAPR, that cannot be obtained with the OFDM samples taken at the Nyquist rate. However, it can be shown that the accurate PAPR values can be obtained with the discrete-time OFDM samples at the IDFT output, provided that the oversampling factor is at least $O = 4$ [41].

The peak power that yields the maximum PAPR value is related to the adopted constellation and it is proportional to the number of subcarriers, i.e., it can be substantially deviated from the average power, though it can be shown that its occurrence is an extremely rare event since it requires that all the subcarriers are phase aligned. For instance, regarding QPSK constellations ($M = 4$), there are M^2 sequences that can generate the maximum peak power. As the total number of possible sequences is M^{N_u}, the occurrence of this peak power has a very low probability. In fact, it can be shown that the probability of having the maximum peak power decays exponentially with the number of subcarriers [42].

As mentioned before, as the number of subcarriers increases, the corresponding OFDM signals become approximate Gaussian. Under these conditions, the PAPR metric can be seen as a random variable with a given statistical characterization. Therefore, to evaluate and measure the envelope fluctuations of a given OFDM system, it is better to analyze the distribution of the PAPR, rather than just compute its maximum value, since the distribution gives more information about the probability of having a given PAPR.

One of the most common ways to measure the PAPR is to analyze its complementary cumulative density function (CCDF). The CCDF of the PAPR tells us what is the probability of having a PAPR higher than a given threshold, and can be obtained through the cumulative density function (CDF) as

$$\begin{aligned}\mathrm{CCDF}(\mathrm{PAPR}) &= P(\mathrm{PAPR} > X)\\ &= 1 - P(\mathrm{PAPR} \leq X)\\ &= 1 - \mathrm{CDF}(\mathrm{PAPR}),\end{aligned} \tag{2.90}$$

with $P(\cdot)$ representing the probability of a given event and

$$\mathrm{CDF}(\mathrm{PAPR}) = P(\mathrm{PAPR} \leq X). \tag{2.91}$$

Note also that as N_u increases, the Gaussian approximation for the real and imaginary parts of s_n holds (see (2.26)). Therefore, the amplitude of the nth time-domain sample, $|s_n|$, has a Rayleigh distribution. Moreover, the instantaneous power associated with the nth sample has Chi squared distribution with two degrees of freedom. Under these conditions, the CCDF of the PAPR can be bounded as [43]

$$\mathrm{CCDF}(\mathrm{PAPR})_{\mathrm{bound}} = 1 - (1 - \exp(-\mathrm{PAPR}))^{\beta N_u}, \tag{2.92}$$

with $\beta = 1$ when the time-domain samples s_n are uncorrelated. However, it should be referred that when there is oversampling and the time-domain samples are correlated, β is chosen to be higher than 1, but lower than the oversampling factor. Fig. 2.17 shows the simulated and theoretical CCDF (obtained with (2.92)) of the PAPR of OFDM signals with QPSK constellations, $O = 4$ and different values of N_u. From the figure, it can be concluded that the bound of (2.92) is precise, presenting an error of approximately 1 dB in the worst case. Additionally, it can be noted that regardless of N_u, the PAPR is always higher than 5 dB, which demonstrates how accentuated the envelope fluctuations can be, even for a low to moderate number of subcarriers. As expected, the PAPR increases with the number of subcarriers. For instance, although the PAPR can be lower than 6.5 dB when $N_u = 64$, it is impossible to have those PAPR values when much higher values of N_u are considered. In addition, the probability of exceeding a PAPR of 10 dB is only 0.3% when $N_u = 64$, but increases to 0.7% when $N_u = 128$ and can even reach approximately 7.1% when $N_u = 1024$. In fact, although a rare event, the PAPR can exceed 12.5 dB when $N_u = 1024$. This means that even though the event of having the maximum PAPR is very unlikely, there are always considerable differences between the peak power and the average power, which means that OFDM signals present very large envelope fluctuations, which undoubtedly constitutes one of its major drawbacks.

2.3.1 AMPLIFICATION ISSUES

The main consequence of the very large PAPR of OFDM signals is the existence of severe amplification issues that lead to a high sensitivity to nonlinear distortions. In fact, the linearity requirements of a power amplifier designed for multicarrier signals

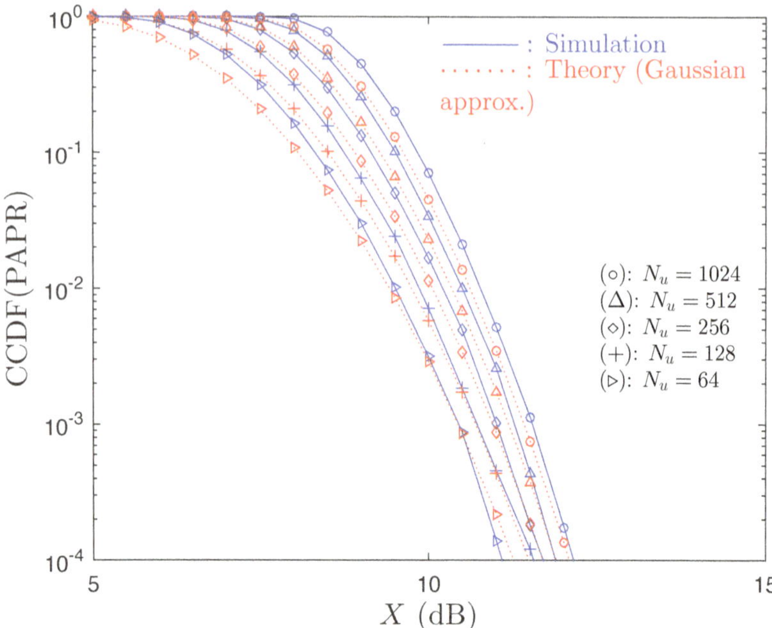

Figure 2.17 Simulated CCDF of the PAPR of an OFDM system with QPSK constellations, $O = 4$ and different values of N_u.

are very high. These high linearity requirements preclude the use of highly efficient, low-cost nonlinear power amplifiers such as the ones of class D, E and F, that have a theoretical maximum efficiency of 100% [44]. For a linear amplification of signals with large dynamic ranges, linear power amplifiers such as the ones of class A or B should be used. However, these power amplifiers have a maximum theoretical efficiency of 50% and 75%, respectively [44]. In addition, their effective efficiency may be considerably lower, since they may incur in a large power penalty to accommodate the very large PAPR of multicarrier signals and assure a linear amplification. This effect is illustrated in Fig. 2.18, which shows a nonlinear amplification curve as well as different operation points of the amplifier. As can be seen in the figure, the ideal amplification curve is linear, i.e., the signal at the amplifier output is a scaled version of its input. However, in practice, linear power amplifiers only present this behavior for a limited range of inputs, usually denoted as the linear region. On the other hand, in the nonlinear region, the input signal saturates the amplifier and the amplified signal becomes a nonlinear function of the input, which leads to the existence of nonlinear distortion effects.

There are several ways to measure the efficiency of an amplifier. One of the most common measures is the power-added efficiency, which is computed by the ratio

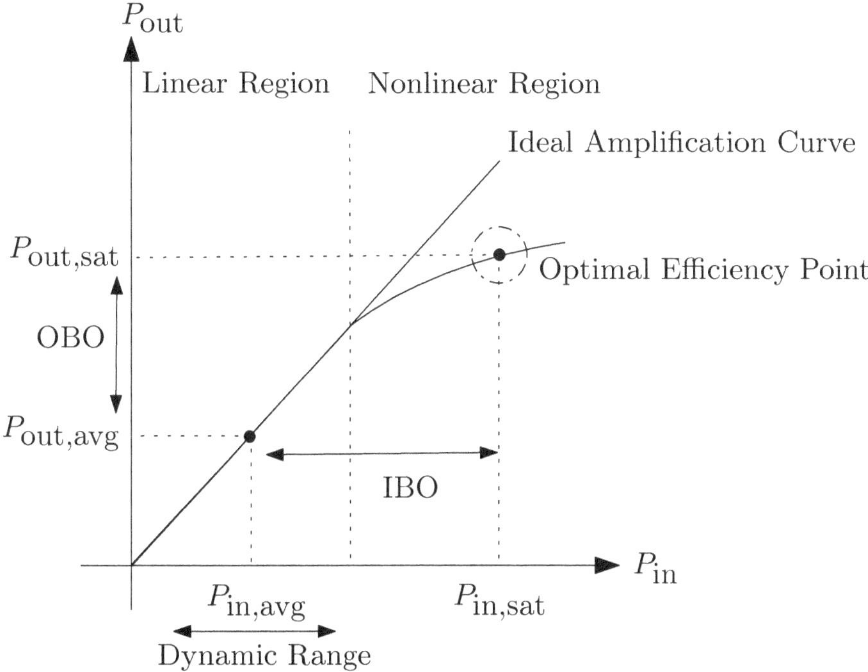

Figure 2.18 Amplification curve of a nonlinear amplifier.

between the difference between the RF output and input powers to the DC input power. However, for amplifiers with large gains, the drain efficiency yields a good approximation of the amplification efficiency. The drain efficiency is computed as the ratio between the output power and the power consumed by the amplifier. For commercial linear power amplifiers, the maximum drain efficiency typically ranges from 20% to 30% [45] and is obtained when the amplifier is driven to work near the saturation region. However, it decreases sharply as the input power decreases, i.e., as the operation point is deviated from the optimal efficiency point. For constant or quasi-constant envelope signals this is not a severe limitation, since the amplifier can work in the nonlinear region of the amplification curve without distorting the output signals. However, for non-constant envelope signals such as OFDM signals, this constitutes a severe problem since, to prevent nonlinear distortion effects, the amplifier cannot work on its maximal efficiency point, which means that the amplification efficiency may decrease substantially.

It is widely known that the nonlinear distortion effects can significantly impair the amplified signals. Due to the intermodulation, these distortion effects take place both in the in-band region and in the out-of-band region. The distortion in the in-band region leads to BER degradations and even to high irreducible error floors. On the other hand, the distortion that appears in the out-of-band region leads to large ACI levels, that are usually regulated by the maximum allowed adjacent channel power ratio (ACPR). Therefore, to avoid the problems created by the nonlinear distortion, the amplifier should be driven to work within its linear region, which means that an output back-off (OBO) should take place. The OBO is defined as the ratio between the output saturation power and the average power of the output signal, i.e.,

$$\mathrm{OBO} = \frac{P_{\mathrm{out,sat}}}{P_{\mathrm{out,avg}}}. \tag{2.93}$$

Its counterpart, regarding the input power, is the input back-off (IBO). The IBO is defined as the ratio between the input power that leads to the amplifier's saturation and the average power of the input signal, i.e.,

$$\mathrm{IBO} = \frac{P_{\mathrm{in,sat}}}{P_{\mathrm{in,avg}}}. \tag{2.94}$$

It should be mentioned that if no ACPR is allowed, then the power amplifier should be fully backed off, which means that the IBO should be at least equal to the PAPR. This means that the back-off may assume very large values, drastically reducing the effective amplification efficiency. Indeed, for an amplifier with a given theoretical maximum drain efficiency of $\eta_{\mathrm{a,max}}$, the average effective efficiency is given by

$$\eta_{\mathrm{a}} = \eta_{\mathrm{a,max}} \frac{1}{\mathrm{OBO}}, \tag{2.95}$$

i.e., the effective amplification efficiency is inversely proportional to the adopted OBO. Therefore, the larger the required IBO and, consequently, the adopted OBO, the larger the deviation relatively to the optimal efficiency point, i.e., the lower the energy efficiency of the power amplifier. For instance, considering an OFDM signal with a PAPR around 10 dB and a class A amplifier, whose $\eta_{\mathrm{a,max}} = 50\%$, the effective efficiency may be lower than 5%, which is a prohibitively low value. Additionally, it is important to note that as the power amplifier is one of the major energy consumers of wireless networks[10], an inefficient amplification process can substantially compromise the energy efficiency of the entire system. This problem is specially serious in mobile devices such as cellular phones, since in these battery-powered devices the power efficiency is crucial to increase the battery lifetime.

For a signal with a given PAPR, better amplification efficiency can only be achieved if stronger nonlinear distortion effects are tolerated [45]. Essentially, there is a trade-off between robustness to nonlinear distortion effects and power efficiency, which makes the amplification of multicarrier signals a very challenging task. Commonly, to maximize the amplification efficiency, the PAPR of OFDM signals is reduced prior to the amplification process. In these conditions, the required IBOs are lower which also means that lower OBOs can be considered. It should be stressed

[10]In cellular networks, for instance, a large fraction of the energy is consumed at the BS, from which 50% to 80% can be spent in the amplification process.

that a PAPR reduction also allows us to work with quantizers with lower resolutions as well as to reduce the sensibility to phase noise in multicarrier optical signals, which means that, in general, a lower PAPR leads to a lower sensitivity to nonlinear distortion effects. In the next subsection, some techniques to mitigate the large PAPR of multicarrier signals are presented.

2.3.2 PAPR REDUCING TECHNIQUES

With the wide deployment of OFDM modulations in many wireless communications systems, the need to solve their amplification problems become a very important subject. In fact, this challenge aroused the interest of many researchers and led to the development of several techniques that aim to mitigate the very large PAPR of signals associated with both single-input, single-output (SISO) and MIMO-OFDM systems [46]. Without a considerable PAPR reduction in the baseband signal, the amplification efficiency of an OFDM system becomes very low. However, when the PAPR is lower, the linearity requirements are lower too, which means that the amplification efficiency can be higher. Therefore, due to the important role of the amplification efficiency in the system, the overall system's energy efficiency may increase.

Although the PAPR-reduction techniques share a common goal, they act very differently. According to [6], which presents a taxonomy of the PAPR-reduction techniques, they can be divided into three main categories: signal distortion techniques, multiple signaling and probabilistic techniques and coding techniques. Their capability to reduce the envelope fluctuations can be measured in several ways but it is usually evaluated by the CCDF of the PAPR of the resultant signal [6]. Depending on their nature, the PAPR-reducing techniques may bring disadvantages such as increased complexity of the transceivers, introduction of redundant information that leads to lower useful bit-rates, reduction of transmitted signal power, BER degradation and spectral spreading of the transmitted signals. Therefore, trade-offs between capability of PAPR reduction and one (or more) of the drawbacks aforementioned may arise, challenging the system designer.

Signal Distortion Techniques

Signal distortion techniques are employed before the amplification process. They allow an effective PAPR reduction and, consequently, the use of highly efficient, nonlinear power amplifiers, such as the ones of classe D or E. However, these techniques introduce both in-band and out-of-band nonlinear distortion in the transmitted signals. As a consequence, both the BER and the spectral efficiency can be degraded, respectively. The signal distortion techniques can be divided into four types: clipping and filtering techniques [47], peak windowing techniques [48], peak cancellation techniques [49] and companding techniques [50].

In the clipping techniques, that are considered the most simple and effective techniques to reduce the PAPR of multicarrier signals, the amplitude peaks higher than a certain threshold are clipped to that threshold. To allow for higher PAPR reduction, the OFDM signal may be oversampled before the clipping operation [51]. When the

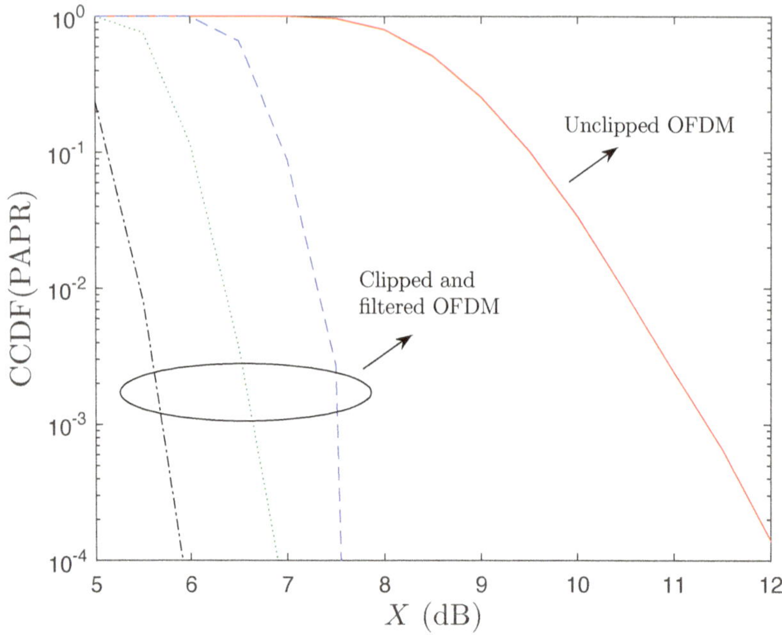

Figure 2.19 CCDF of the PAPR of a clipped and filtered OFDM signal considering different clipping levels.

clipping operation is applied to the envelope of an OFDM signal, these techniques are known as envelope clipping techniques [51,52]. On the other hand, when the clipping operation is applied separately regarding the real and imaginary parts of an OFDM signal, we have Cartesian or I-Q clipping techniques [7]. To maintain the original bandwidth of the signal, the out-of-band distortion can be eliminated with a post-filtering operation leading to the well-known clipping and filtering techniques [53]. Fig. 2.19 shows the CCDF of the PAPR of a clipped and filtered OFDM signal for $N_u = 512$, $O = 4$ and different clipping levels. From the results depicted in this figure it can be noted that the clipping and filtering technique can substantially reduce the PAPR of OFDM signals. As expected, the capability to reduce the PAPR increases when the clipping level decreases. However, it should be noted that the filtering operation may cause some regrowth of envelope fluctuations, which can be higher when the OFDM signal is not oversampled. An alternative is to perform the clipping and filtering several times in an iterative way, leading to the so-called iterative clipping and filtering technique [54]. In iterative clipping and filtering techniques, the signal is clipped and filtered several times, till the corresponding PAPR decreases to the desired value. As shown in the previous section, the occurrence of a very high power peak (that leads to the maximum PAPR value) is a rare event and, for this reason, these techniques can be designed in such a way that the resulting nonlinear distortion does not compromise significantly the performance [55]. However, the impact of the nonlinear distortion effects on the performance is not negligible and must be evaluated carefully.

Instead of being clipped, when the peak windowing techniques are employed, the high amplitude peaks are attenuated by a window function, which leads to signals with lower nonlinear distortions when compared to the ones that are clipped. The window is built in such a way that its minimum values are aligned with the high amplitude peaks of the OFDM signal. The multiplication of the OFDM signal by the generated window results in a signal with a lower PAPR [48]. Commonly, Hamming or Kaiser-Bessel windows are used due to their good spectral properties.

In peak cancellation techniques, a peak cancellation window is generated, scaled and subtracted from the original OFDM signal. To generate such a window, a peak detector must analyze the OFDM signal to find the positions of the high-amplitude peaks. This is made by comparing the OFDM signal with predetermined thresholds, and must be made in a symbol-by-symbol basis [49]. It should be noted that the peak cancellation technique can be seen as a particular case of the clipping and filtering techniques as demonstrated in [56, 57].

Regarding the companding transform PAPR-reducing methods, the original OFDM signal is companded using specific functions that can be classified regarding their linearity and symmetry properties. A common transform function (that is nonlinear and asymmetric) is the μ-law. In the μ-law transform function, the high peaks of the signal are maintained but the lower ones are enhanced. Thus, although the average power increases, the PAPR reduces too [50].

Multiple Signaling Techniques

The main idea behind multiple signaling techniques is to create several representations of the same OFDM signal. Between these replicas of the same signal, the one that presents the lowest PAPR is chosen for amplification and transmission. The main drawbacks of multiple signaling techniques are the increased complexity to generate the different symbols that represent the same information as well as the need to transmit side information to the receiver, which reduces the useful data rate.

In selective mapping [58], different OFDM symbols are obtained through the multiplication of the original signal by different phase sequences. Between the generated symbols, the one with lower PAPR is then transmitted. It should be mentioned that the receiver must know what is the phase sequence that leads to the OFDM symbol with lower PAPR.

In interleaved OFDM [59], multiple OFDM symbols are generated by permuting and reordering the elements of the original OFDM symbol. Once again, side information must be transmitted, since the receiver needs to know what was the interleaver considered in the transmitter.

In partial transmit sequences (PTS), the OFDM block is divided up into several sub-blocks, transformed to the time-domain through an IDFT and weighted by a phase factor. The goal is to find the set of phase-factors that minimize the PAPR of the combined block [60]. In MIMO-OFDM systems, the PTS technique is applied for each stream and the sets of phase factors are equal to all antennas, which allows to reduce the maximum PAPR among the T antennas and to reduce the amount of side information that must be transmitted per stream.

The tone injection technique allows the mitigation of the high PAPR by mapping the constellation points onto a set of different points in an expanded constellation and hence the "best" point between them can be chosen before the IDFT, to reduce the PAPR of the transmitted signal [61]. In [61], it is also proposed a technique named tone reservation, where a set of tones is reserved to reduce the PAPR. The reserved subcarriers are determined with a convex optimization. The capability to reduce the PAPR depends on the number of reserved tones as well as on their location in the frequency-domain. The tone reservation technique can be easily extended to MIMO-OFDM systems [46].

In [62], the constellation shapping technique is proposed. The main idea of this technique is to reduce the PAPR by dynamically change the constellation without significantly degrade the BER.

Coding Techniques

In addition to the capability of correcting and detecting errors, coding schemes can reduce the high PAPR of OFDM signals without significantly increase their complexity.

Different from the conventional linear block coding schemes, where the redundant bits are dedicated to error correction to achieve BER improvements, these bits can be allocated to reduce the PAPR of OFDM signals [63]. For instance, large PAPR values can be avoided if the sequences that lead to those PAPR values are avoided too. This can be done by block coding techniques where the original data words are transformed into a set of code words with lower PAPR values. Due to the existence of redundant bits, however, this approach involves lower useful data rates and exhaustive searches to find the best codes, which also means more complexity.

There are schemes that combine block coding techniques with Golay complementary sequences and have good capabilities of error correction and PAPR reduction [64]. However, due to their very high complexity, their use is limited to OFDM systems that use a small number of subcarriers. In [65], a turbo coding scheme that can also reduce the PAPR of OFDM signals is proposed.

3 Nonlinear Distortion in Multicarrier Systems

Nonlinear distortion is one of the main impairments that can be encountered in multicarrier communications. This distortion arises from hardware limitations of electronic devices that compose the transceivers that might not be able to linearly handle all signals at their input without a substantial increase in the complexity and/or considerable energy inefficiencies. An example of such devices is the power amplifier.

In fact, depending on the specific system, nonlinear distortion can be specially severe and very difficult (or impossible) to avoid, contributing substantially to performance degradation. As seen in the previous chapter, multicarrier systems are very sensitive to nonlinearities due to their large envelope fluctuations. For this reason, it is very likely that they face nonlinear distortion effects. This distortion typically results from a nonlinear amplification or from the use of the most simple and promising PAPR reducing techniques (see Subsection (2.3.2) of previous chapter) – the clipping techniques - that are employed prior to the amplification process. For this reason, study the impact of nonlinear distortion effects on the performance of multicarrier systems is of extreme importance, not only to quantify the corresponding performance penalty, but also to aid the development of ways to mitigate it.

In the following section, the main analytical models to characterize memoryless nonlinearities are presented. These results are then employed in Section 3.2, where the analytical characterization of nonlinearly distorted Gaussian signals is made regarding both the time-domain and the frequency-domain. In Section 3.3 an efficient, low-complexity method for characterizing nonlinearly distorted multicarrier signals is presented.

3.1 MEMORYLESS NONLINEARITIES

In order to quantify and predict the unwanted effects caused by the existence of nonlinearities in a given system, it is important to have tools that allow to capture their behavior. These tools help to solve the trade-offs faced by the system designer, as for instance the trade-off between linearity and energy efficiency that arises in the amplification of multicarrier signals explained in the previous chapter. Moreover, if the system designers can accurately characterize nonlinear systems, they can develop mechanisms that might mitigate the impact of nonlinear distortion on the system's performance [66]. Examples of such mechanisms are the so-called linearization techniques. The main idea of these techniques is to submit the signal to a nonlinear operation before the amplification process. This may allow to cancel the nonlinear distortion effects by the compound effect of the two nonlinearities [67,68].

DOI: 10.1201/9781032708744-3

The study of nonlinear devices, in general, can be carried out through circuit-level models or physical models that provide large accuracy. Regarding high power amplifiers, for instance, these models can be too complex since they involve a full characterization of the amplifier's internal composition, which may lead to the use of complicated analytical tools such as nonlinear differential equations, which may be intractable from a practical point of view. Moreover, the simulation time of these physical models may be very large for a system-level simulation. For this reason, empirical models, also known as "black-box" models have been largely considered to capture the effect of nonlinearities associated with high power amplifiers, although these models are built with no knowledge of the amplifiers structure [69]. Due to their simplicity, these models can be put into the simulation environment without significantly augment its complexity, i.e., they can be used to predict and simulate the nonlinear distortion effects, giving rise to a good trade-off between complexity and accuracy [66].

In this section, we present the main mathematical tools and models that are usually employed to characterize the common nonlinear devices that can be encountered in multicarrier communication systems. This comprises "black-box" models for both baseband and bandpass memoryless nonlinearities, i.e., nonlinearities whose output signals do not depend on the bandwidth of the input signal. Put differently, their output at a given time instant depends only on their input at that time instant. These frequency-independent nonlinearities are also known as zero-memory nonlinearities.

3.1.1 BASEBAND NONLINEARITIES

The memoryless baseband nonlinearities are memoryless nonlinearities that operate in baseband signals. In this chapter, we consider that these baseband signals can be both real-valued or complex-valued multicarrier signals. In the latter case, the nonlinearities operate separately on the real and imaginary parts of the signal and are denoted throughout this document as Cartesian nonlinearities.

Real-valued Nonlinearities

Let us consider a baseband, real-valued multicarrier signal $s(t)$. When a memoryless nonlinearity described by the function $f(\cdot)$ has at its input $s(t)$, it yields

$$y(t) = f(s(t)). \tag{3.1}$$

Under some conditions, the nonlinearity output $y(t)$ can be characterized by a Taylor series [70], resulting

$$y(t) = a_0 + a_1 s(t) + a_2 s^2(t) + a_3 s^3(t) + \cdots + a_n s^n(t) = \sum_{n=0}^{\infty} a_n s^n(t), \tag{3.2}$$

where the coefficients a_n can be obtained by fitting a polynomial function to a given nonlinear characteristic. Commonly, as long as the input signal does not present a large dynamic range and the nonlinearity is relatively "smooth", the power series in

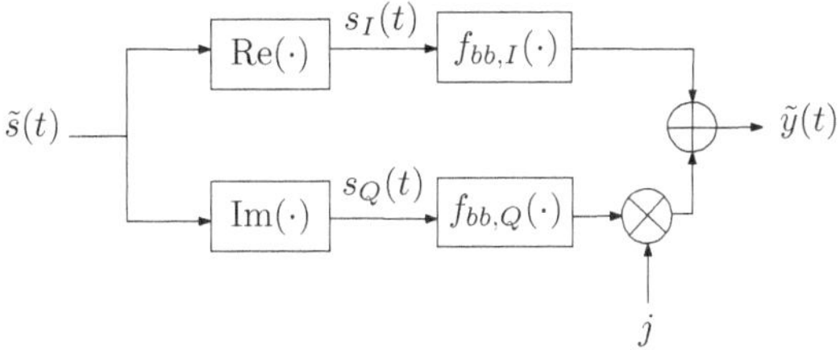

Figure 3.1 Model of a Cartesian nonlinearity.

(3.2) can be truncated considering a relatively low number of terms without leading to a significant loss of accuracy.

Cartesian Nonlinearities

In Cartesian nonlinearities, also known as "I-Q" nonlinearities, the real and imaginary parts of a complex signal are submitted to memoryless nonlinearities. Therefore, for a given complex-valued, multicarrier signal defined as

$$\tilde{s}(t) = s_I(t) + js_Q(t), \tag{3.3}$$

the nonlinearity yields

$$\begin{aligned}\tilde{y}(t) &= f_{bb}(\tilde{s}(t)) \\ &= f_{bb,I}(s_I(t)) + jf_{bb,Q}(s_Q(t)),\end{aligned} \tag{3.4}$$

where $f_{bb,I}(\cdot)$ and $f_{bb,Q}(\cdot)$ denote the nonlinearities operating on the real and imaginary parts of $\tilde{s}(t)$, respectively. The model of this nonlinearity is depicted in Fig. 3.1. An example of a Cartesian nonlinearity is the one associated with baseband clipping techniques for PAPR reduction of multicarrier signals. In that case, the nonlinearities operating in the real and imaginary parts are equal, i.e., $f_{bb,I}(\cdot) = f_{bb,Q}(\cdot) = f(\cdot)$. Fig. 3.2 shows the nonlinear function that represents the aforementioned clipping operation considering a normalized clipping level of $s_M = 1.0$, as well as its polynomial approximations obtained by the least-squares method and considering different degrees. As expected, the accuracy increases with the order of the polynomial approximation. In addition, due to the quasi-linear behavior of $f(s)$ for low values of s, it can be seen that the nonlinearity can be approximated by a polynomial function with low degree in that zone.

3.1.2 BANDPASS NONLINEARITIES

Although accurate to model baseband, memoryless nonlinearities, provided that the number of polynomial terms is adequate, the power series represented in (3.2) cannot, however, capture the memory effects of a baseband nonlinearity. In addition,

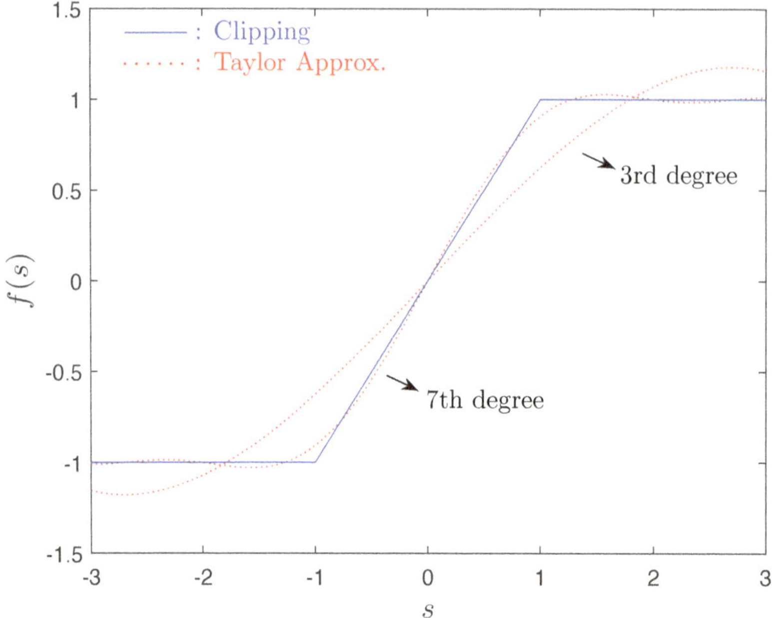

Figure 3.2 Nonlinear function of a clipping operation and its polynomial approximations considering polynomials with different degrees.

besides the importance of characterizing baseband nonlinearities, it is also of extreme importance to characterize the output of nonlinear systems excited by bandpass signals, such as RF and microwave power amplifiers. Due to electrical and electrothermal effects, these power amplifiers might present memory effects, since the output at a given time instant depends not only on current but also on past inputs.

In order to capture memory effects, the nonlinearities with memory are usually characterized by behavioral models. These behavioral models are commonly Volterra models since they are based on the Volterra series [5], which is a powerful mathematical formulation proposed by Wiener in the 1940s for the characterization of nonlinear systems [71]. The Volterra series is based on the previous work of Volterra on functional series expansion [72], and constitutes the most embracing tool to characterize the input-output relation of nonlinear systems, being adequate to model nonlinearities with and without memory, regardless the nature of their input signals, that can be either baseband or bandpass [66, 73, 74]. In the following, it is shown how these behavioral, high-level models can be obtained through the Volterra series.

The Volterra series is defined as an infinite sum of multidimensional integrals and can be expressed as

$$y(t) = \sum_{m=0}^{\infty} \int_{-\infty}^{\infty} \cdots \int_{-\infty}^{\infty} u_m(\tau_1, \cdots, \tau_m) \times \prod_{j=1}^{m} s(t - \tau_j) d_{\tau_1} d_{\tau_2} \cdots d_{\tau_m}, \tag{3.5}$$

where $u_m(\tau_1,..,\tau_m)$ is the nth dimensional impulse response of the system (also known as Volterra kernel of order m).

It should be noted that when (3.5) is truncated to $m=1$, it yields (omitting the subscript m and assuming a null 0th order Volterra kernel, i.e., a null DC component)

$$y(t)=\int_{-\infty}^{\infty}u(\tau)s(t-\tau)d\tau, \tag{3.6}$$

i.e., the input-output relation of linear systems with memory, that is given by the convolution of the input signal and the impulse response of the system (see for instance (2.44), which describes the output of a frequency-selective channel modeled as a linear, time-invariant system). However, to obtain the nonlinear behavior of a given system, higher-order kernels (with $m>1$) should be taken into account. Nevertheless, it is clear that when m increases, the complexity of Volterra series becomes very high, not to mention the difficulty of obtaining those high-order kernels. For this reason, it is not common to obtain the output of a nonlinear system through the Volterra series at least, in its original form. Usually, resorting to some assumptions, the Volterra series can degenerate in the aforementioned high-level, behavioral models, that can be used to characterize bandpass nonlinearities. In the following, it is shown how these high-level models can be derived from the Volterra series.

Let us start by defining a general bandpass, multicarrier signal (this signal can be, for instance, the bandpass OFDM signal of (2.41)) as

$$s(t)=\mathrm{Re}(\tilde{s}(t)\exp(j2\pi f_c t)), \tag{3.7}$$

where f_c is the carrier frequency and $\tilde{s}(t)$ is the corresponding complex envelope of the bandpass signal. This complex envelope has bandwidth $B_s \ll f_c$, and is defined as

$$\tilde{s}(t)=r(t)\exp(j\theta(t)), \tag{3.8}$$

where $r(t)=|\tilde{s}(t)|$ represents the absolute value of $\tilde{s}(t)$ and $\theta(t)=\arg(\tilde{s}(t))$ denotes its argument. A possible spectrum of (3.8) and (3.9) is depicted in Fig. 3.3. Note that, by replacing (3.8) in (3.7), we may also express the bandpass signal as

$$\begin{aligned} s(t)&=\mathrm{Re}(r(t)\exp(j\theta(t))\exp(j2\pi f_c t))\\ &=r(t)\cos(\underbrace{2\pi f_c t+\theta(t)}_{\psi(t)})\\ &=r(t)\cos(\psi(t)), \end{aligned} \tag{3.9}$$

where $\psi(t)$ is the argument associated to the bandpass multicarrier signal. Let us consider that the bandpass signal represented in (3.9) is submitted to a bandpass nonlinearity. At the nonlinearity output, the nonlinearly distorted signal is formed by several harmonics that are centered along multiples of the carrier frequency. However, by making use of the fact that the carrier frequency f_c is much higher than the bandwidth of the baseband signal (i.e., $B_s \ll f_c$), the spectral components created

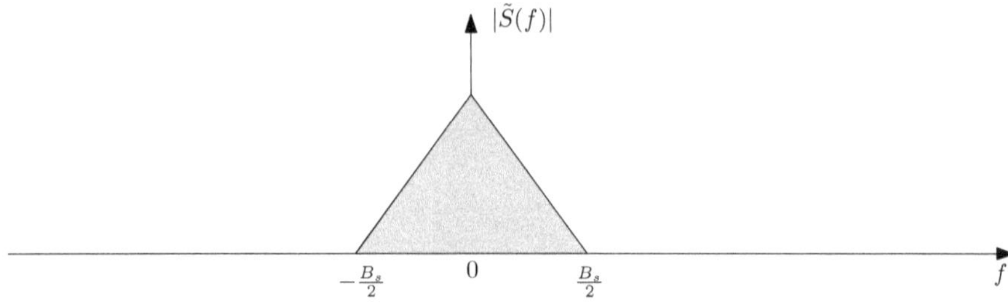

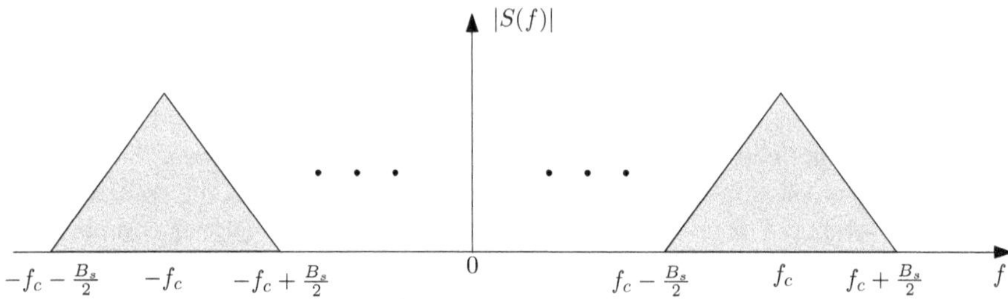

Figure 3.3 Amplitude spectrum of a given baseband signal $\tilde{s}(t)$ and the spectrum of its bandpass version $s(t)$.

along multiples of f_c can be neglected, since they almost do not contribute for the spectrum of interest that is located around the carrier frequency. This can be thought as if we could have a zonal filter that removes those harmonics centered at DC, $\pm 2f_c$, $\pm 3f_c$ and so forth. Under these conditions, the bandpass nonlinearity output can be written as

$$y(t) = \mathrm{Re}(\tilde{y}(t)\exp(j2\pi f_c t)). \tag{3.10}$$

In fact, this means that the Volterra series can be formulated to express the relation between the complex envelopes of the signals at the input and at the output of the bandpass nonlinearity [73–75], since this relation is enough to characterize the bandpass output signal of (3.10). This equivalence is clearly depicted in Fig. 3.4, which shows how the bandpass signals $s(t)$ and $y(t)$ can be characterized through a baseband nonlinearity.

The equivalent baseband Volterra series is given by [73, 75]

$$\begin{aligned} \tilde{y}(t) = \sum_{m=0}^{\infty} \int_{-\infty}^{\infty} \cdots \int_{-\infty}^{\infty} \tilde{u}_{2m+1}(\tau_1, \cdots, \tau_{2m+1}) \\ \times \prod_{j=1}^{m+1} \tilde{s}(t-\tau_j) \prod_{j=m+2}^{2m+1} \tilde{s}^*(t-\tau_j) d_{\tau_1} \cdots d_{\tau_{2m+1}}, \end{aligned} \tag{3.11}$$

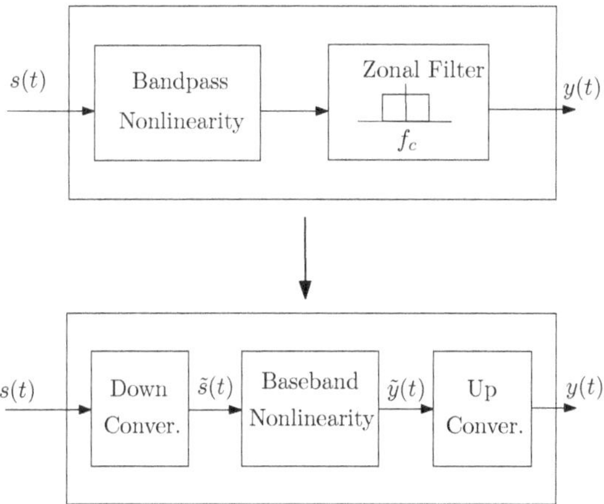

Figure 3.4 Characterization of a bandpass nonlinearity considering only its input and output complex envelopes.

where $\tilde{u}_{2m+1}(\tau_1,\cdots,\tau_{2m+1})$ represents the baseband equivalent of the Volterra kernel of order $2m+1$, defined as

$$\tilde{u}_{2m+1}(\tau_1,\cdots,\tau_{2m+1}) = \frac{1}{2^{2m}}\binom{2m+1}{m} u_{2m+1}(\tau_1,\cdots,\tau_{2m+1})\times \exp\left(-j2\pi f_c\left(\sum_{i=1}^{m+1}\tau_i - \sum_{i=m+2}^{2m+1}\tau_i\right)\right), \tag{3.12}$$

with the binomial coefficient given by

$$\binom{a}{b} = \frac{a!}{b!(a-b)!}. \tag{3.13}$$

Note that due to the bandpass nature of the nonlinearity, only the odd power terms appear in (3.11), since the even-order terms fall outside the band of interest centered around f_c. This equation is usually named complex baseband representation of a bandpass nonlinearity, since the nonlinearity is completely described in the baseband.

Let us define the multidimensional Fourier transform of the Volterra kernel of order m, $u_m(\tau_1,\cdots,\tau_m)$, as

$$U_m(f_1,\cdots,f_m) = \int_{-\infty}^{\infty}\cdots\int_{-\infty}^{\infty} u_m(\tau_1,\cdots,\tau_m) \times \exp(-j2\pi(f_1\tau_1+f_2\tau_2+\cdots f_m\tau_m))\,d_{\tau_1}d_{\tau_2}\cdots d_{\tau_m}. \tag{3.14}$$

Additionally, let us define a tuple $T_{2m+1} = (t_0, \cdots, t_{2m+1})$, where the ith element t_i is given by the ith element of the sum

$$\left(\sum_{i=1}^{m+1} 1 - \sum_{i=m+2}^{2m+1} 1 \right) = t_1 + t_2 + \cdots + t_{2m+1}, \tag{3.15}$$

which means that

$$T_{2m+1} = \begin{cases} (-1), & m = 0 \\ (1,1,-1) & m = 1 \\ (1,1,1,-1,-1) & m = 2 \\ \vdots & \vdots \\ (\underbrace{1,1,....,1}_{k+1}, \underbrace{-1,-1,...,-1}_{k}) & m = k, \end{cases} \tag{3.16}$$

i.e., in a tuple of dimension $2m+1$, the first $m+1$ points have the value 1 and the remaining m points have the value -1. Under these conditions, by making use of the delay property of the multidimensional Fourier transform, one can write the frequency-domain version of $\tilde{u}_{2m+1}(\tau_1, \cdots, \tau_{2m+1})$ as

$$\begin{aligned} \tilde{U}_{2m+1}(f_1, \cdots, f_{2m+1}) = & \frac{1}{2^{2m}} \binom{2m+1}{m} \\ & \times U_{2m+1}\left((f_1, \cdots, f_m) + f_c T_{2m+1}\right). \end{aligned} \tag{3.17}$$

As mentioned before, real electronic devices such as high-power amplifiers have a frequency-dependent behavior, i.e., present memory effects. In fact, although the frequency-dependent behavior associated with the existence of memory cannot be physically removed, if the circuit's time constants are much smaller than the inverse of the maximum frequency of the input signal, then they can be considered almost negligible. Therefore, if the input signal of a given high-power amplifier is narrow comparatively to the frequency response of the Volterra kernels, then the memory effects might be considered as "short-term" memory effects. Under these conditions, the following approximation can be taken into account

$$\tilde{s}(t - \tau_i) \approx \tilde{s}(t). \tag{3.18}$$

By considering this approximation, the baseband Volterra series represented in (3.11) can be rewritten as

$$\begin{aligned} \tilde{y}(t) \approx & \sum_{m=0}^{\infty} \int_{-\infty}^{\infty} \cdots \int_{-\infty}^{\infty} \tilde{u}_{2m+1}(\tau_1, \cdots, \tau_{2m+1}) d_{\tau_1} d_{\tau_2} \cdots d_{\tau_{2m+1}} |\tilde{s}(t)|^{2m} \tilde{s}(t) \\ = & \sum_{m=0}^{\infty} \int_{-\infty}^{\infty} \cdots \int_{-\infty}^{\infty} \tilde{u}_{2m+1}(\tau_1, \cdots, \tau_{2m+1}) \exp\left(-j2\pi \left(0\tau_1 + 0\tau_2 + \cdots 0\tau_{2m+1}\right)\right) \\ & d_{\tau_1} d_{\tau_2} \cdots d_{\tau_{2m+1}} |\tilde{s}(t)|^{2m} \tilde{s}(t) \\ = & \sum_{m=0}^{\infty} \tilde{U}_{2m+1} \underbrace{(0,0,\cdots,0)}_{2m+1} |\tilde{s}(t)|^{2m} \tilde{s}(t) \\ = & \sum_{m=0}^{\infty} \frac{1}{2^{2m}} \binom{2m+1}{m} U_{2m+1}\left(f_c T_{2m+1}\right) |\tilde{s}(t)|^{2m} \tilde{s}(t). \end{aligned} \tag{3.19}$$

From (3.19), it is clear that there is a polynomial relation between the output complex envelope $\tilde{y}(t)$ and the input complex envelope $\tilde{s}(t)$. Furthermore, since $|\tilde{s}(t)| = r(t)$, we may write

$$\tilde{y}(t) = \left(\sum_{m=0}^{\infty} K_m r^{2m}(t) \right) \tilde{s}(t) \tag{3.20}$$

$$= \left(\sum_{m=0}^{\infty} K_m r^{2m}(t) \right) r(t) \exp(j\theta(t)), \tag{3.21}$$

where

$$K_m = \frac{1}{2^{2m}} \binom{2m+1}{m} U_{2m+1}\left(f_c T_{2m+1}\right). \tag{3.22}$$

Therefore,

$$\tilde{y}(t) = \underbrace{\left(\sum_{m=0}^{\infty} K_m r^{2m+1}(t) \right)}_{f_{bp}(r(t))} \exp(j\theta(t))$$

$$= f_{bp}(r(t)) \exp(j\theta(t)), \tag{3.23}$$

where $f_{bp}(r(t))$ is denoted as the bandpass nonlinear function, defined as (omitting the dependence with t)

$$f_{bp}(r) = A(r) \exp(j\Theta(r)), \tag{3.24}$$

with $A(r)$ and $\Theta(r)$ denoting the so-called amplitude modulation/amplitude modulation (AM/AM) and amplitude modulation/phase modulation (AM/PM) conversion functions, respectively. The AM/AM function $A(r)$ describes the variation of the amplitude of the output signal with the amplitude of the input signal, while the AM/PM function describes the impact of the input amplitude on the phase of the output complex envelope. The nonlinear complex function represented in (3.24) usually defines the behavioral, high-level model of a bandpass nonlinearity.

It is very important to note that, although the physical device presents short-term memory effects, its complex baseband representation is memoryless. For this reason, these types of nonlinearities are commonly known as bandpass memoryless nonlinearities throughout the literature [73, 74, 76].

Depending on the nature of the AM/PM conversion function $\Theta(r)$, two types of bandpass memoryless nonlinearities can be identified. On one hand, we have strictly memoryless bandpass nonlinearities when the device is physically memoryless and the Volterra kernels are multidimensional Dirac delta functions, i.e.,

$$u_{2m+1}(\tau_1, \cdots, \tau_{2m+1}) \propto \delta(\tau_1, \cdots, \tau_{2m+1}). \tag{3.25}$$

As a consequence, for this type of nonlinearity, the AM/PM conversion function is constant, since the coefficients K_m are real-valued and the phase of the output complex envelope does not depend on the amplitude of the input complex envelope.

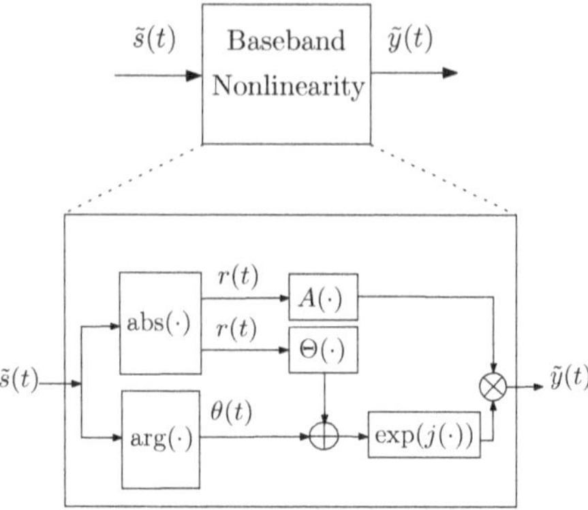

Figure 3.5 Block model for a bandpass memoryless nonlinearity.

On the other hand, if the coefficients K_m are complex-valued, the AM/PM conversion function is, in fact, a function of r. Under these conditions, we have quasi-memoryless bandpass nonlinearities. These nonlinearities can model amplifiers that present short-term memory effects, i.e., memory effects small enough so that it is possible that the behavioral model of these devices can be memoryless. Note also that systems with "long-term" memory effects (i.e., systems with wideband inputs, where the memory effects are high) cannot be solely characterized by AM/AM and AM/PM conversion functions, and the Volterra series should be considered in its original form as represented in (3.11) [76].

Fig. 3.5 shows the "internal composition" of a baseband nonlinearity that models bandpass devices (see also Fig. 3.4). The bandpass signal at the output of the nonlinearity can hence be written by substituting (3.23) into (3.10), resulting

$$\begin{aligned} y(t) &= \mathrm{Re}(f_{bp}(r(t))\exp(j\theta(t))\exp(j2\pi f_c t)) \\ &= \mathrm{Re}(f_{bp}(r(t))\exp(j\psi(t))). \end{aligned} \tag{3.26}$$

As demonstrated in Appendix A, the baseband model for representing bandpass nonlinearities given in (3.26) can also be obtained by expanding the bandpass signal in a Fourier series, provided that it is sufficiently narrow to be considered approximately periodic in the short term (i.e., $f_c \gg B_s$) [8, 30]. In the following, we show some models of the common high-power amplifiers employed in multicarrier systems.

Two types of high-power amplifiers are considered: the solid-state power amplifiers (SSPAs) and the traveling wave tube amplifiers (TWTAs). Thorough literature, one can find different models to characterize these amplifiers. Among them, there are polynomial-based models that try to fit measured characteristics. However, more general models are typically considered. For instance, the TWTAs are usually characterized by the Saleh's model [77], that defines the AM/AM conversion function of the bandpass nonlinearity as

$$A(r) = 2\frac{\left(\frac{r}{s_M}\right)}{1+\left(\frac{r}{s_M}\right)^2}, \tag{3.27}$$

and the AM/PM conversion function as

$$\Theta(r) = 2\frac{\theta_M\left(\frac{r}{s_M}\right)^2}{1+\left(\frac{r}{s_M}\right)^2}, \tag{3.28}$$

where s_M is maximum value of the input which agrees with the output envelope at saturation and θ_M is the phase rotation when the input envelope is s_M. For solid state power amplifiers (SSPAs) modeling, the Ghorbani's model [78] can be employed. In that model, the AM/AM and AM/PM conversion functions are given by

$$A(r) = \frac{a_0 r^{a_1}}{1+a_2 r^{a_1}} + a_3 r \tag{3.29}$$

and

$$\Theta(r) = \frac{b_0 r^{b_1}}{1+b_2 r^{b_1}} + b_3 r, \tag{3.30}$$

where a_i and b_i $(0 \leq i \leq 3)$ are parameters obtained from physical measurements. Nevertheless, the Rapp's model [79] is, however, more common to model SSPAs. This model considers that the phase distortion is negligible so that the AM/PM conversion function is approximately zero ($\Theta(r) \approx 0$). On the other hand, it considers that the AM/AM conversion function is given by

$$A(r) = \frac{r}{\sqrt[2p]{\left(1+\left(\frac{r}{s_M}\right)^{2p}\right)}}, \tag{3.31}$$

where s_M is the value of the output in the saturation region and the parameter p controls the smoothness between the linear and the nonlinear regions. Fig. 3.6 shows the AM/AM conversion function of a TWTA obtained with the Saleh's model, as well the AM/AM conversion function of an SSPA obtained with the Rapp's model considering different values of p. In both cases, the saturation amplitude is $s_M = 1.0$. From the results depicted in the figure, it can be clearly noted that, when p increases, the AM/AM conversion function of the SSPA becomes more closer to an ideal envelope clipping function.

3.2 CHARACTERIZATION OF NONLINEARLY DISTORTED GAUSSIAN SIGNALS

It is widely known that the impact of nonlinearities on the performance of multicarrier communication systems can be carried out by simulation, more concretely,

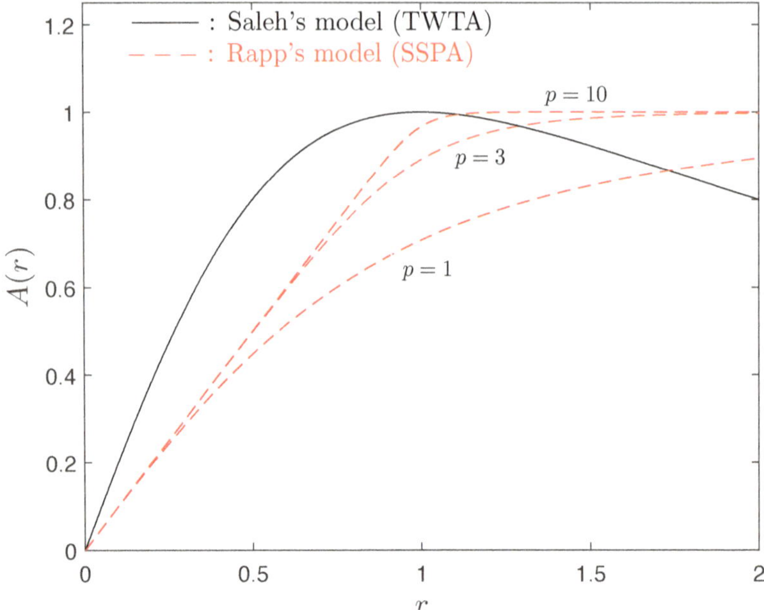

Figure 3.6 AM/AM conversion function of a TWTA and an SSPA considering $s_M = 1.0$.

by considering Monte Carlo simulations. However, even when employing the low-complexity, high-level models for nonlinearities characterization that were presented in the previous section, these Monte Carlo simulations may require large computation times to achieve accurate results.

As mentioned in Chapter 2, multicarrier signals such as OFDM signals present a Gaussian nature due to the CLT, provided that the number of subcarriers is large. Taking advantage of that Gaussian nature, all well-known analytical tools to characterize Gaussian signals submitted to nonlinear systems can be employed to statistically characterize nonlinear distortion effects in multicarrier systems, regarding both the time-domain and the frequency-domain. Therefore, the large computation times associated with Monte Carlo simulations can be avoided with the use of such analytical tools.

In the following subsections, we use the aforementioned analytical tools for obtaining closed-form expressions for both the autocorrelation and the PSD of nonlinearly distorted multicarrier signals. The analytical expressions for the PSD are obtained based on what we will call the "truncated IMP" approach. The analysis of nonlinearly distorted, baseband multicarrier signals is made in Subsection 3.2.1. The autocorrelation of bandpass, nonlinearly distorted multicarrier signals is presented in Subsection 3.2.2. Although these results are not entirely new, they are useful to introduce and motivate the concept of "equivalent nonlinearities" presented in Section 3.3.

3.2.1 BASEBAND MULTICARRIER SIGNALS

Real-valued Nonlinearities

Here, we focus on the statistical characterization of baseband, real-valued multicarrier signals that are submitted to memoryless nonlinearities. The multicarrier input signal $s(t)$ is modeled by a stationary, Gaussian random process with variance σ^2. Due to that stationarity, the autocorrelation of $s(t)$ only depends on the time lag τ between two observation moments and is defined as

$$R_{s,bb}(\tau) = \mathbb{E}[s(t)s(t-\tau)]. \tag{3.32}$$

Under these conditions, the average power of the signal can be obtained by fixing $\tau = 0$, i.e.,

$$R_{s,bb}(0) = \mathbb{E}[s^2(t)] = \sigma^2. \tag{3.33}$$

Naturally, the random variable that results from a given observation moment is Gaussian distributed (see (2.26)). If $s(t)$ is submitted to a nonlinearity such as those characterized in Subsection 3.1.1, the corresponding output is (see (3.1))

$$y(t) = f(s(t)). \tag{3.34}$$

By taking advantage of the Gaussian nature of $s(t)$, the Bussgang's theorem [10, 80] can be used for the characterization of the nonlinearity output. As is widely known, the Bussgang's theorem, which can also be obtained as a special case of the Price's theorem [81], can be employed to obtain the statistical characterization of a nonlinearly distorted Gaussian signal. The theorem states that the output of a memoryless nonlinearity driven by a Gaussian signal can be divided in two uncorrelated components: one that is proportional to the input signal and another that concentrates the nonlinear distortion effects. Under these conditions, in addition to (3.34), we can express the nonlinearly distorted version of $s(t)$ as

$$y(t) = \alpha_{bb}s(t) + d(t), \tag{3.35}$$

where $d(t)$ is the nonlinear distortion component uncorrelated with the input signal, i.e.,

$$\mathbb{E}[s(t)d(t)] = 0, \tag{3.36}$$

and α_{bb} is a scale factor that relates the cross-correlation between the input and output signals and the autocorrelation of the input signal, being defined as

$$\alpha_{bb} = \frac{\mathbb{E}[s(t)f(s(t))]}{\mathbb{E}[s(t)^2]}. \tag{3.37}$$

Due to the stationarity of the input signal, the dependence with t can be omitted, resulting

$$\alpha_{bb} = \frac{\mathbb{E}[sf(s)]}{\mathbb{E}[s^2]} = \frac{\mathbb{E}[sf(s)]}{\sigma^2} = \frac{1}{\sigma^2} \int_{-\infty}^{+\infty} sf(s)p(s)ds. \tag{3.38}$$

The average power at the nonlinearity output is

$$P_{\text{nl}} = \mathbb{E}[f^2(s)] = \int_{-\infty}^{+\infty} f^2(s)p(s)ds. \tag{3.39}$$

Moreover, as the average power of the useful component of the nonlinearly distorted signal is

$$P_{\text{nl,u}} = \alpha_{bb}^2 \sigma^2, \tag{3.40}$$

it is easy to note that the average power of the nonlinear distortion term is hence

$$P_{\text{nl,d}} = P_{\text{nl}} - P_{\text{nl,u}}. \tag{3.41}$$

In the following, an expression for the autocorrelation of nonlinearly distorted multicarrier signals is presented. Let us start by defining two Gaussian random variables, obtained from the random process $s(t)$ at two different time instants separated by τ. These random variables are defined as $s_1 = s(t_1)$ and $s_2 = s(t_1 - \tau)$. Under these conditions, (3.32) can be written as

$$R_{s,bb}(\tau) = \mathbb{E}[s_1 s_2] = \int_{-\infty}^{+\infty}\int_{-\infty}^{+\infty} s_1 s_2 p(s_1, s_2) ds_1 ds_2, \tag{3.42}$$

with $p(s_1, s_2)$ denoting the joint PDF between s_1 and s_2, that can be expressed as

$$p(s_1, s_2) = \frac{1}{2\pi\sigma^2\sqrt{1-\rho_{bb}^2(\tau)}} \exp\left(-\frac{s_1^2 + s_2^2 - 2\rho_{bb}(\tau)s_1 s_2}{2\sigma^2(1-\rho_{bb}^2(\tau))}\right), \tag{3.43}$$

where

$$\rho_{bb} = \rho_{bb}(\tau) = \frac{R_{s,bb}(\tau)}{R_{s,bb}(0)} = \frac{R_{s,bb}(\tau)}{\sigma^2}, \tag{3.44}$$

is the correlation factor between s_1 and s_2. On the other hand, the autocorrelation of the nonlinearly distorted signal is given by

$$R_{y,bb}(\tau) = \mathbb{E}[f(s_1)f(s_2)] = \int_{-\infty}^{+\infty}\int_{-\infty}^{+\infty} f(s_1)f(s_2)p(s_1, s_2) ds_1 ds_2. \tag{3.45}$$

However, by applying the so-called Mehler's formula [82], the joint PDF of s_1 and s_2 represented in (3.43) can be expressed as a function of their marginal densities

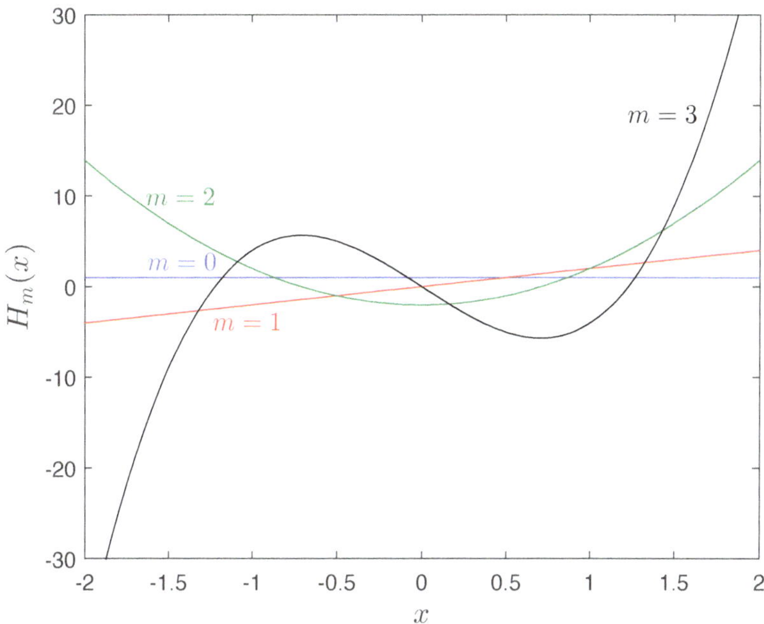

Figure 3.7 Hermite polynomials $H_m(x)$ for different values of m.

($p(s_1)$ and $p(s_2)$, respectively) and of the Hermite polynomials [30]. Therefore, we have

$$p(s_1,s_2) = p(s_1)p(s_2)\sum_{m=0}^{+\infty}\frac{\rho_{bb}^m(\tau)}{2^m m!}H_m\left(\frac{s_1}{\sqrt{2}\sigma}\right)H_m\left(\frac{s_2}{\sqrt{2}\sigma}\right), \tag{3.46}$$

where

$$H_m(x) = (-1)^m \exp\left(x^2\right)\frac{d^m}{dx^m}\left(\exp\left(-x^2\right)\right), \tag{3.47}$$

represents the Hermite polynomial of order m [83]. Note that, by fixing $m = 0$, we have $H_0(x) = 1$, for $m = 1$, we have $H_1(x) = 2x$ and so on. In fact, these polynomials are orthogonal with respect to the Gaussian PDF represented in (2.26). Fig. 3.7 depicts the first four Hermite polynomials, i.e., from $H_0(x)$ to $H_3(x)$. By replacing (3.46) in (3.45), we obtain

$$\begin{aligned} R_{y,bb}(\tau) &= \int_{-\infty}^{+\infty}\int_{-\infty}^{+\infty} f(s_1)f(s_2)p(s_1)p(s_2)H_m\left(\frac{s_1}{\sqrt{2}\sigma}\right)H_m\left(\frac{s_2}{\sqrt{2}\sigma}\right)ds_1ds_2 \\ &= \sum_{m=0}^{+\infty}\frac{\rho_{bb}^m(\tau)}{2^m m!}\int_{-\infty}^{+\infty} f(s_1)p(s_1)H_m\left(\frac{s_1}{\sqrt{2}\sigma}\right)ds_1 \\ &\int_{-\infty}^{+\infty} f(s_2)p(s_2)H_m\left(\frac{s_2}{\sqrt{2}\sigma}\right)ds_2. \end{aligned} \tag{3.48}$$

This means that we can rewrite the output autocorrelation as

$$R_{y,bb}(\tau) = \sum_{m=0}^{+\infty} \frac{\rho_{bb}^m(\tau)}{2^m m!} \left(\int_{-\infty}^{+\infty} f(s)p(s)H_m\left(\frac{s}{\sqrt{2}\sigma}\right) ds \right)^2. \tag{3.49}$$

By defining the power of the IMP of order m as

$$P_m^{bb} = \frac{1}{2^m m!} \left(\int_{-\infty}^{+\infty} f(s)p(s)H_m\left(\frac{s}{\sqrt{2}\sigma}\right) ds \right)^2, \tag{3.50}$$

we can rewrite (3.49) as

$$\begin{aligned} R_{y,bb}(\tau) &= \sum_{m=0}^{+\infty} \rho_{bb}^m(\tau) P_m^{bb} \\ &= \sum_{m=0}^{+\infty} \left(\frac{R_{s,bb}(\tau)}{\sigma^2} \right)^m P_m^{bb}. \end{aligned} \tag{3.51}$$

Having in mind that the nonlinearity output $y(t)$ is separable into two uncorrelated components (see (3.35)), we can also write

$$R_{y,bb}(\tau) = |\alpha_{bb}|^2 R_{s,bb}(\tau) + R_{d,bb}(\tau), \tag{3.52}$$

where $R_{d,bb}(\tau)$ is the autocorrelation of the nonlinear distortion component that is defined as

$$R_{d,bb}(\tau) = \mathbb{E}[d(t)d(t-\tau)] = P_0^{bb} + \sum_{m=2}^{\infty} \left(\frac{R_{s,bb}(\tau)}{\sigma^2} \right)^m P_m^{bb}. \tag{3.53}$$

By taking into account the Wiener-Khinchin's theorem [84], the PSD of the nonlinearly distorted signal can be obtained as the Fourier transform $R_{y,bb}(\tau)$, i.e.,

$$\begin{aligned} G_{y,bb}(f) &= \mathscr{F}(R_{y,bb}(\tau)) \\ &= \sum_{m=0}^{+\infty} \frac{P_m^{bb}}{\sigma^{2m}} \mathscr{F}\left(R_{s,bb}^m(\tau)\right) \\ &= \sum_{m=0}^{+\infty} \frac{P_m^{bb}}{\sigma^{2m}} \underbrace{\left(G_{s,bb}(f) * G_{s,bb}(f) * \cdots * G_{s,bb}(f)\right)}_{m \text{ times}}, \end{aligned} \tag{3.54}$$

where $G_{s,bb}(f) = \mathscr{F}(R_{s,bb}(\tau))$ is the PSD of the input signal $s(t)$. Note that (3.54) can also be written as

$$G_{y,bb}(f) = |\alpha_{bb}|^2 G_{s,bb}(f) + G_{d,bb}(f), \tag{3.55}$$

where $G_{d,bb}(f)$ is the PSD of the distortion component $d(t)$, that is given by

$$\begin{aligned} G_{d,bb}(f) &= \mathscr{F}(R_{d,bb}(\tau)) \\ &= P_0^{bb}\delta(f) + \sum_{m=2}^{+\infty} \frac{P_m^{bb}}{\sigma^{2m}} \underbrace{(G_{s,bb}(f) * G_{s,bb}(f) * \cdots * G_{s,bb}(f))}_{m \text{ times}}. \end{aligned} \tag{3.56}$$

From (3.54) and (3.56), one can also note that $P_{\text{nl}} = R_{y,bb}(0)$ and $P_{\text{nl,d}} = R_{d,bb}(0)$. It should also be mentioned that the Hermite polynomials are even for m even and odd for m odd (see Fig. 3.7). Therefore, for odd nonlinearities such as the clipping represented in Fig. 3.2, only the odd-order IMPs are used to characterize the corresponding nonlinearly distorted Gaussian signal. Under those conditions, the autocorrelation of the nonlinearly distorted signal is

$$R_{y,bb}(\tau) = \sum_{\gamma=0}^{+\infty} \left(\frac{R_{s,bb}(\tau)}{\sigma^2}\right)^{2\gamma+1} P_{2\gamma+1}^{bb}, \tag{3.57}$$

where, according to (3.50), we have

$$P_{2\gamma+1}^{bb} = \frac{1}{2^{2\gamma+1}(2\gamma+1)!} \left(\int_{-\infty}^{+\infty} f(s)p(s)H_{2\gamma+1}\left(\frac{s}{\sqrt{2}\sigma}\right) ds \right)^2. \tag{3.58}$$

Therefore, the PSD of the nonlinearly distorted signal can be written as

$$G_{y,bb}(f) = \sum_{\gamma=0}^{+\infty} \frac{P_{2\gamma+1}^{bb}}{\sigma^{2(2\gamma+1)}} \underbrace{(G_{s,bb}(f) * G_{s,bb}(f) * \cdots * G_{s,bb}(f))}_{2\gamma+1 \text{ times}}, \tag{3.59}$$

and the PSD of the nonlinear distortion component is

$$G_{d,bb}(f) = \sum_{\gamma=1}^{+\infty} \frac{P_{2\gamma+1}^{bb}}{\sigma^{2(2\gamma+1)}} \underbrace{(G_{s,bb}(f) * G_{s,bb}(f) * \cdots * G_{s,bb}(f))}_{2\gamma+1 \text{ times}}. \tag{3.60}$$

Complex-valued Nonlinearities

Here, we focus on the statistical characterization of complex signals submitted to Cartesian nonlinearities. The input signal at the Cartesian nonlinearity is expressed as

$$\tilde{s}(t) = s_I(t) + js_Q(t), \tag{3.61}$$

where $s_I(t) = \text{Re}(\tilde{s}(t))$ and $s_Q(t) = \text{Im}(\tilde{s}(t))$ are the real and imaginary parts of $\tilde{s}(t)$, respectively. We assume that $\tilde{s}(t)$ is modeled by a complex Gaussian random process, where $s_I(t)$ and $s_Q(t)$ are stationary Gaussian random processes with the same statistical properties, i.e., both Gaussian distributed with zero mean and variance σ^2. The autocorrelation of the input signal is

$$\begin{aligned} R_{\tilde{s},bb}(\tau) &= \mathbb{E}[\tilde{s}(t)\tilde{s}^*(t-\tau)] \\ &= \mathbb{E}[(s_I(t)+js_Q(t))(s_I(t-\tau)-js_Q(t-\tau))] \\ &= \mathbb{E}[s_I(t)s_I(t-\tau)] - j\mathbb{E}[s_I(t)s_Q(t-\tau)] + j\mathbb{E}[s_Q(t)s_I(t-\tau)] \\ &+ \mathbb{E}[s_Q(t)s_Q(t-\tau)]. \end{aligned} \tag{3.62}$$

By noting that $\mathbb{E}[s_I(t)s_Q(t-\tau)] = -\mathbb{E}[s_Q(t)s_I(t-\tau)] \stackrel{\Delta}{=} R_{\tilde{s},Q}(\tau)$ and $\mathbb{E}[s_I(t)s_I(t-\tau)] = \mathbb{E}[s_Q(t)s_Q(t-\tau)] \stackrel{\Delta}{=} R_{\tilde{s},I}(\tau)$, we can rewrite (3.62) as

$$R_{\tilde{s},bb}(\tau) = 2R_{\tilde{s},I}(\tau) + j2R_{\tilde{s},Q}(\tau). \tag{3.63}$$

The average power of the signal is $R_{\tilde{s},bb}(0) = 2R_{\tilde{s},I}(0) = 2\sigma^2$, since $R_{\tilde{s},Q}(0) = 0$. Let us consider that the complex signal $\tilde{s}(t)$ is submitted to a Cartesian nonlinearity. Under these conditions, both the real and the imaginary parts are submitted to a memoryless nonlinearity as explained in Subsection 3.1.1. As demonstrated in (3.4), at the Cartesian nonlinearity output, we have

$$\begin{aligned} \tilde{y}(t) &= f_{bb}(\tilde{s}(t)) \\ &= f_{bb,I}(s_I(t)) + jf_{bb,Q}(s_Q(t)). \end{aligned} \tag{3.64}$$

For the sake of simplicity, we focus on Cartesian nonlinearities whose nonlinearities operating on the real and on the imaginary parts of the input signal are equal, i.e. $f_{bb,I}(\cdot) = f_{bb,Q}(\cdot) = f(\cdot)$, with $f(\cdot)$ denoting an odd memoryless nonlinearity (the generalization for other cases is straightforward). Having in mind the Bussgang's theorem [10], one can also decompose the nonlinearity output as a sum of uncorrelated useful and distortion components, i.e.,

$$\tilde{y}(t) = \alpha_{bb}\tilde{s}(t) + \tilde{d}(t), \tag{3.65}$$

where α_{bb} is given by (3.38) and $\tilde{d}(t)$ represents the nonlinear distortion component. The autocorrelation of the nonlinearly distorted signal is

$$\begin{aligned} R_{\tilde{y},bb}(\tau) &= \mathbb{E}[f(s_I)(t)f(s_I(t-\tau))] + \mathbb{E}[f(s_Q(t))f(s_Q(t-\tau))] \\ &- j\mathbb{E}[f(s_I(t))f(s_Q(t-\tau))] + j\mathbb{E}[f(s_Q(t))f(s_I(t-\tau))]. \end{aligned} \tag{3.66}$$

This means that $R_{\tilde{y},bb}(\tau)$ can be written as the sum of four autocorrelations. These autocorrelations involve equally distributed Gaussian random variables submitted to the same memoryless nonlinearity $f(\cdot)$. As previously seen, each one of these autocorrelations can be written as in (3.57), where $P_{2\gamma+1}^{bb}$ is equal for all autocorrelations. After some lengthy but straightforward manipulations, it can be shown that (3.66) can be written as [8, 30]

$$R_{\tilde{y},bb}(\tau) = 2\sum_{\gamma=0}^{+\infty} P_{2\gamma+1}^{bb} \frac{(\mathrm{Re}(R_{\tilde{s},bb}(\tau)))^{2\gamma+1} + j(\mathrm{Im}(R_{\tilde{s},bb}(\tau)))^{2\gamma+1}}{(R_{\tilde{s},bb}(0))^{2\gamma+1}}$$
$$= \sum_{\gamma=0}^{+\infty} \frac{P_{2\gamma+1}^{bb}}{2^{2\gamma}\sigma^{2(2\gamma+1)}} (\mathrm{Re}(R_{\tilde{s},bb}(\tau)))^{2\gamma+1} + j(\mathrm{Im}(R_{\tilde{s},bb}(\tau)))^{2\gamma+1}. \quad (3.67)$$

Additionally, by taking use of the following relations

$$\mathrm{Re}(R_{\tilde{s},bb}(\tau)) = \frac{R_{\tilde{s},bb}(\tau) + R^*_{\tilde{s},bb}(\tau)}{2}, \quad (3.68)$$

and

$$\mathrm{Im}(R_{\tilde{s},bb}(\tau)) = \frac{R_{\tilde{s},bb}(\tau) - R^*_{\tilde{s},bb}(\tau)}{2j}, \quad (3.69)$$

we can rewrite (3.67) as a function of the input signal autocorrelation, $R_{\tilde{s},bb}(\tau)$, i.e.,

$$R_{\tilde{y},bb}(\tau) = \sum_{\gamma=0}^{+\infty} \frac{P_{2\gamma+1}^{bb}}{2^{4\gamma+1}\sigma^{2(2\gamma+1)}} \left((R_{\tilde{s},bb}(\tau) + R^*_{\tilde{s},bb}(\tau))^{2\gamma+1} + \frac{j}{j^{2\gamma+1}} (R_{\tilde{s},bb}(\tau) - R^*_{\tilde{s},bb}(\tau))^{2\gamma+1} \right)$$
$$= \sum_{\gamma=0}^{+\infty} \frac{P_{2\gamma+1}^{bb}}{2^{4\gamma+1}\sigma^{2(2\gamma+1)}} \left((R_{\tilde{s},bb}(\tau) + R^*_{\tilde{s},bb}(\tau))^{2\gamma+1} + (-1)^{\gamma} (R_{\tilde{s},bb}(\tau) - R^*_{\tilde{s},bb}(\tau))^{2\gamma+1} \right). \quad (3.70)$$

Once again, by considering the Wiener-Khinchin's theorem [84], the PSD of the nonlinearly distorted signal at the Cartesian nonlinearity output can be obtained by the Fourier transform of (3.70), i.e.,

$$G_{\tilde{y},bb}(f) = \mathscr{F}(R_{\tilde{y},bb}(\tau))$$
$$= \sum_{\gamma=0}^{+\infty} \frac{P_{2\gamma+1}^{bb}}{2^{4\gamma+1}\sigma^{2(2\gamma+1)}} \mathscr{F}\left((R_{\tilde{s},bb}(\tau) + R^*_{\tilde{s},bb}(\tau))^{2\gamma+1} + (-1)^{\gamma} (R_{\tilde{s},bb}(\tau) - R^*_{\tilde{s},bb}(\tau))^{2\gamma+1} \right)$$
$$= \sum_{\gamma=0}^{+\infty} \frac{P_{2\gamma+1}^{bb}}{2^{4\gamma+1}\sigma^{2(2\gamma+1)}} \Big(\underbrace{(G_{\tilde{s},bb}(f) + G_{\tilde{s},bb}(-f)) * \cdots * (G_{\tilde{s},bb}(f) + G_{\tilde{s},bb}(-f))}_{2\gamma+1 \text{ times}}$$
$$+ (-1)^{\gamma} \underbrace{(G_{\tilde{s},bb}(f) - G_{\tilde{s},bb}(-f)) * \cdots * (G_{\tilde{s},bb}(f) - G_{\tilde{s},bb}(-f))}_{2\gamma+1 \text{ times}} \Big), \quad (3.71)$$

where $G_{\tilde{s},bb}(f)$ denotes PSD of the input signal. Taking into account (3.65), we can also write

$$G_{\tilde{y},bb}(f) = |\alpha_{bb}|^2 G_{\tilde{s},bb}(f) + G_{\tilde{d},bb}(f), \quad (3.72)$$

where $G_{\tilde{d},bb}(f)$ represents the PSD of the nonlinear distortion component, given by

$$G_{\tilde{d},bb}(f) = \sum_{\gamma=1}^{+\infty} \frac{P_{2\gamma+1}^{bb}}{2^{4\gamma+1}\sigma^{2(2\gamma+1)}} \Big(\underbrace{(G_{\tilde{s},bb}(f)+G_{\tilde{s},bb}(-f)) * \cdots * (G_{\tilde{s},bb}(f)+G_{\tilde{s},bb}(-f))}_{2\gamma+1 \text{ times}}$$
$$+ (-1)^{\gamma} \underbrace{(G_{\tilde{s},bb}(f)-G_{\tilde{s},bb}(-f)) * \cdots * (G_{\tilde{s},bb}(f)-G_{\tilde{s},bb}(-f))}_{2\gamma+1 \text{ times}} \Big). \tag{3.73}$$

3.2.2 BANDPASS MULTICARRIER SIGNALS

In this subsection, we focus on the statistical characterization of nonlinearly distorted bandpass signals. Let us start by consider a bandpass signal represented in (3.7), whose the complex envelope is given by

$$\begin{aligned} \tilde{s}(t) &= r(t)\exp(j\theta(t)) \\ &= s_I(t) + js_Q(t). \end{aligned} \tag{3.74}$$

We also consider that this complex envelope is modeled by a complex-valued, stationary Gaussian random process. Therefore, its real and imaginary parts, $s_I(t)$ and $s_Q(t)$, respectively, have the same statistical properties, i.e., they are Gaussian-distributed with zero mean and variance σ^2 and their PDF is given by (2.26). The absolute value of the complex envelope, $r(t)$, is modeled by random variable with Rayleigh distribution, i.e., $r \sim \text{Rayleigh}(\sigma)$, that has the following PDF

$$p(r) = \frac{r}{\sigma^2}\exp\left(\frac{-r^2}{2\sigma^2}\right). \tag{3.75}$$

It should be mentioned that the autocorrelation of the complex envelope of the input signal $\tilde{s}(t)$ can be computed as in (3.63). Additionally, as demonstrated in Subsection 3.1.2, the effect of a given bandpass nonlinearity can be completely described by the AM/AM and AM/PM conversion functions, that operate in the complex envelope of the bandpass signal. Therefore, recalling (3.23), the complex envelope at the output of a bandpass nonlinearity can be written as

$$\begin{aligned} \tilde{y}(t) &= f_{bp}(r(t))\exp(j\theta(t)) \\ &= A(r(t))\exp(j(\Theta(r(t))+\theta(t))) \\ &= A(r(t))\exp(j\varphi(t)), \end{aligned} \tag{3.76}$$

where $\varphi(t) = \Theta(r(t)) + \theta(t)$ represents the argument of the complex envelope of the nonlinearly distorted, bandpass signal. By considering the Bussgang's theorem [10, 80], we also can write the complex envelope represented in (3.76) as the sum of uncorrelated useful and distortion components, resulting

$$\tilde{y}(t) = \alpha_{bp}\tilde{s}(t) + \tilde{d}(t), \tag{3.77}$$

where the scale factor α_{bp} is given by (due to the stationarity, the dependence with t can be omitted)

$$\alpha_{bp} = \frac{\mathbb{E}[\tilde{s}\tilde{y}^*]}{\mathbb{E}[|\tilde{s}|^2]} = \frac{\mathbb{E}[r\exp(j\theta)f_{bp}^*(r)\exp(-j\theta)]}{\mathbb{E}[r^2]}$$
$$= \frac{\mathbb{E}[rf_{bp}^*(r)]}{\mathbb{E}[r^2]} = \frac{1}{2\sigma^2}\int_0^{+\infty} rf_{bp}^*(r)p(r)dr. \tag{3.78}$$

In the following, we define $\tilde{s}_1 = \tilde{s}(t) = s_{1,I} + js_{1,Q}$ and $\tilde{s}_2 = \tilde{s}(t-\tau) = s_{2,I} + js_{2,Q}$. Moreover, the dependence with t is omitted, i.e., the input and output complex envelopes are expressed as $\tilde{s} = r\exp(j\theta) = s_I + js_Q$ and $\tilde{y} = A(r)\exp(j\varphi)$, respectively. The autocorrelation of the nonlinearly distorted signal is

$$R_{\tilde{y},bp}(\tau) = \mathbb{E}[\tilde{y}(t)\tilde{y}^*(t-\tau)] \tag{3.79}$$
$$= \mathbb{E}[f_{bp}(\tilde{s}_1)f_{bp}^*(\tilde{s}_2)]$$
$$= \mathbb{E}[f_{bp}(s_{1,I} + js_{1,Q})f_{bp}(s_{2,I} - js_{2,Q})]$$
$$= \int_{-\infty}^{+\infty}\int_{-\infty}^{+\infty}\int_{-\infty}^{+\infty}\int_{-\infty}^{+\infty} f_{bp}(s_{1,I} + js_{1,Q})f_{bp}(s_{2,I} - js_{2,Q})$$
$$p(s_{1,I}, s_{1,Q}, s_{2,I}, s_{2,Q})ds_{1,I}ds_{1Q}ds_{2,I}ds_{2,Q}. \tag{3.80}$$

Let us define an array composed of the real and imaginary parts of $\tilde{s}_1$ and $\tilde{s}_2$ as $\mathbf{q} = [s_{1,I}\, s_{1,Q}\, s_{2,I}\, s_{2,Q}]^T$. Since we are assuming that $\mathbb{E}[s_{1,I}] = \mathbb{E}[s_{1,Q}] = \mathbb{E}[s_{2,I}] = \mathbb{E}[s_{2,Q}] = 0$, the covariance of $\mathbf{q}$ can be defined as

$$\mathbf{Q} = \mathbb{E}[\mathbf{q}\mathbf{q}^T] \tag{3.81}$$
$$= \begin{bmatrix} \mathbb{E}[s_{1,I}s_{1,I}] & \mathbb{E}[s_{1,I}s_{1,Q}] & \mathbb{E}[s_{1I}s_{2,I}] & \mathbb{E}[s_{1,I}s_{2,Q}] \\ \mathbb{E}[s_{1,Q}s_{1,I}] & \mathbb{E}[s_{1,Q}s_{1,Q}] & \mathbb{E}[s_{1Q}s_{2,I}] & \mathbb{E}[s_{1,Q}s_{2,Q}] \\ \mathbb{E}[s_{2,I}s_{1,I}] & \mathbb{E}[s_{2,I}s_{1,Q}] & \mathbb{E}[s_{2I}s_{2,I}] & \mathbb{E}[s_{2,I}s_{2,Q}] \\ \mathbb{E}[s_{2,Q}s_{1,I}] & \mathbb{E}[s_{2,Q}s_{1,Q}] & \mathbb{E}[s_{2Q}s_{2,I}] & \mathbb{E}[s_{2,Q}s_{2,Q}] \end{bmatrix}.$$

After some lengthy but straightforward manipulations, it can be shown that the determinant of the covariance matrix $\mathbf{Q}$ is [8]

$$\det(\mathbf{Q}) = \sigma^8\left(1 - \frac{|R_{\tilde{s},bp}(\tau)|^2}{4\sigma^4}\right)^2$$
$$= \sigma^8\left(1 - \rho_{bp}^2\right)^2, \tag{3.82}$$

where ρ_{bp} denotes the correlation coefficient (i.e., the normalized autocorrelation of the complex Gaussian process $\tilde{s}$), defined as

$$\rho_{bp} = \frac{|R_{\tilde{s},bp}(\tau)|}{R_{\tilde{s},bp}(0)} = \frac{|R_{\tilde{s},bp}(\tau)|}{2\sigma^2}. \tag{3.83}$$

The joint PDF of array $\mathbf{q}$ is given by [85]

$$\begin{aligned} p(q) &= \frac{1}{(2\pi)^2\sqrt{\det(\mathbf{Q})}}\exp(-\frac{1}{2}\mathbf{q}^T\mathbf{Q}^{-1}\mathbf{q}) \\ &= \frac{1}{4\pi^2\sigma^4\left(1-\rho_{bp}^2\right)^2}\exp(A). \end{aligned} \tag{3.84}$$

Let us now consider polar coordinates and redefine $\tilde{s}_1$ and $\tilde{s}_2$ as $\tilde{s}_1 = r_1\exp(j\theta_1)$ and $\tilde{s}_2 = r_2\exp(j\theta_2)$, respectively. In this case, the parameter A in (3.84) can be defined as [86]

$$\begin{aligned} A &= -\frac{\sigma^2\left(r_1^2+r_2^2\right) - |R_{\tilde{s},bp}(\tau)|r_1r_2\cos\left(\theta_1-\theta_2+\arg\left(R_{\tilde{s},bp}(\tau)\right)\right)}{2\sigma^4(1-\rho_{bp}^2)} \\ &= -\frac{r_1^2+r_2^2-2\rho_{bp}r_1r_2\cos\left(\theta_1-\theta_2+\phi\right)}{\rho_{bp,0}}, \end{aligned} \tag{3.85}$$

where $\phi = \arg\left(R_{\tilde{s},bp}(\tau)\right)$ and $\rho_{bp,0} = 2\sigma^2(1-\rho_{bp}^2)$. Using (3.84) and (3.85) in (3.79) and applying the considered polar coordinate transform, we have

$$\begin{aligned} R_{\tilde{y},bp}(\tau) = \frac{1}{2\pi^2\sigma^2\rho_{bp,0}} \int_0^{+\infty}\int_0^{+\infty}\int_0^{2\pi}\int_0^{2\pi} f_{bp}(r_1)f_{bp}^*(r_2)\exp(j(\theta_1-\theta_2)) \\ \exp\left(-\frac{r_1^2+r_2^2-2\rho_{bp}r_1r_2\cos\left(\theta_1-\theta_2+\phi\right)}{\rho_{bp,0}}\right) r_1r_2dr_1dr_2d\theta_1d\theta_2. \end{aligned} \tag{3.86}$$

By defining $\theta' = \theta_1-\theta_2+\phi$, we may write

$$\begin{aligned} R_{\tilde{y},bp}(\tau) &= \frac{1}{2\pi^2\sigma^2\rho_{bp,0}} \int_0^{+\infty}\int_0^{+\infty} f_{bp}(r_1)f_{bp}^*(r_2)r_1r_2\exp\left(-\frac{r_1^2+r_2^2}{\rho_{bp,0}}\right) \\ &\left(\int_0^{2\pi}\int_0^{2\pi}\exp\left(j(\theta'-\phi)\right)\exp\left(\frac{2\rho_{bp}r_1r_2\cos(\theta')}{\rho_{bp,0}}\right)d\theta_1d\theta_2\right)dr_1dr_2 \\ &= \frac{\exp(-j\phi)}{2\pi^2\sigma^2\rho_{bp,0}} \int_0^{+\infty}\int_0^{+\infty} f_{bp}(r_1)f_{bp}^*(r_2)r_1r_2\exp\left(-\frac{r_1^2+r_2^2}{\rho_{bp,0}}\right) \\ &\left(\int_0^{2\pi}\int_0^{2\pi}\exp(j\theta')\exp\left(\frac{2\rho_{bp}r_1r_2\cos(\theta')}{\rho_{bp,0}}\right)d\theta_1d\theta_2\right)dr_1dr_2. \end{aligned} \tag{3.87}$$

Let us now consider the modified Bessel function of first kind, $I_n(z)$, that is defined as (see [83], equation (9.6.19))

$$I_n(z) = \frac{1}{\pi}\int_0^{\pi}\cos(n\theta')\exp\left(z\cos(\theta')\right)d\theta'. \tag{3.88}$$

By replacing $n = 1$ and $z = \frac{2\rho_{bp}r_1r_2}{\rho_{bp,0}}$ in (3.88), we have

$$I_1\left(\frac{2\rho_{bp}r_1r_2}{\rho_{bp,0}}\right) = \frac{1}{\pi}\int_0^{\pi}\cos(\theta')\exp\left(\frac{2\rho_{bp}r_1r_2}{\rho_{bp,0}}\cos(\theta')\right)d\theta'. \tag{3.89}$$

Moreover, since the integrand is a periodic function, we also may write

$$\frac{1}{\pi}\int_0^{2\pi}\exp(j\theta')\exp(z\cos(\theta'))d\theta' = \frac{1}{\pi}\int_0^{2\pi}(\cos(\theta') + j\sin(\theta'))\exp\left(z\cos(\theta')\right)d\theta'. \tag{3.90}$$

Under these conditions, the double integral in (3.87) can be written as

$$\begin{aligned}&\int_0^{2\pi}\int_0^{2\pi}\cos(\theta_1-\theta_2+\phi)\exp\left(\frac{2\rho_{bp}r_1r_2\cos(\theta_1-\theta_2+\phi)}{\rho_{bp,0}}\right)d\theta_1d\theta_2\\ &= 4\pi^2I_1\left(\frac{2\rho_{bp}r_1r_2}{\rho_{bp,0}}\right).\end{aligned} \tag{3.91}$$

Therefore, (3.87) becomes

$$\begin{aligned}R_{\tilde{y},bp}(\tau) = \frac{2\exp(-j\phi)}{\sigma^2\rho_{bp,0}}&\int_0^{+\infty}\int_0^{+\infty}f(r_1)f^*(r_2)r_1r_2\exp\left(-\frac{r_1^2+r_2^2}{\rho_{bp,0}}\right)\\ &I_1\left(\frac{2\rho_{bp}r_1r_2}{\rho_{bp,0}}\right)dr_1dr_2.\end{aligned} \tag{3.92}$$

Using the Laguerre polynomial series expansion (see [87], equation (5.11.3.7)), we can write

$$\begin{aligned}&\sum_{\gamma=0}^{+\infty}\frac{\gamma}{(\alpha+1)_\gamma}t^\gamma L_\gamma^{(\alpha)}(x)L_\gamma^{(\alpha+n)}(y) = \Gamma(\alpha+1)(1-t)^{-n-1}(txy)^{-\frac{\alpha}{2}}\\ &\exp\left(\frac{x+y}{t-1}t\right)\sum_{\gamma=0}^{n}(-1)^\gamma\binom{n}{\gamma}\left(\frac{tx}{y}\right)^{\frac{\gamma}{2}}I_{\gamma+\alpha}\left(\frac{2\sqrt{txy}}{1-t}\right),\end{aligned} \tag{3.93}$$

where $L_\gamma^{(\alpha)}(x)$ denotes the generalized Laguerre polynomial of order γ that is defined through the Rodrigues formula as [82]

$$L_\gamma^{(\alpha)}(x) = \frac{1}{\gamma!}x^{-\alpha}\exp(x)\frac{d^\gamma}{dx^\gamma}\left(\exp(-x)x^{\gamma+\alpha}\right). \tag{3.94}$$

Considering $n = 0$, $\alpha = 1$, and

$$(A)_k = \frac{\Gamma(A+k)}{\Gamma(A)}, \tag{3.95}$$

where, $\Gamma(\cdot)$ represents the Gamma function defined as

$$\Gamma(A) = (A-1)!, \tag{3.96}$$

we note that

$$(\alpha+1)_\gamma = 2_\gamma = \frac{\Gamma(2+\gamma)}{\Gamma(2)} = \frac{(\gamma+1)!}{1!} = (\gamma+1)!. \tag{3.97}$$

Under these conditions, (3.93) becomes

$$\sum_{\gamma=0}^{+\infty} \frac{1}{\gamma+1} t^\gamma L_\gamma^{(1)}(x) L_\gamma^{(1)}(y) = \frac{1}{(1-t)\sqrt{txy}} \exp\left(\frac{x+y}{t-1} t\right) I_1\left(\frac{2\sqrt{txy}}{1-t}\right). \tag{3.98}$$

By replacing $t = \rho_{bp}^2$, $x = \frac{r_1^2}{2\sigma^2}$ and $y = \frac{r_2^2}{2\sigma^2}$, we can rewrite (3.98) as

$$\begin{aligned}\sum_{\gamma=0}^{+\infty} \tfrac{1}{\gamma+1} \rho_{bp}^{2\gamma} L_\gamma^{(1)}\left(\tfrac{r_1^2}{2\sigma^2}\right) L_\gamma^{(1)}\left(\tfrac{r_2^2}{2\sigma^2}\right) &= \frac{1}{(1-\rho_{bp}^2)\sqrt{\rho_{bp}^2 \frac{r_1^2}{2\sigma^2}\frac{r_2^2}{2\sigma^2}}} \exp\left(\frac{\frac{r_1^2}{2\sigma^2}+\frac{r_2^2}{2\sigma^2}}{\rho^2-1}\rho_{bp}^2\right) \\ &\quad \times I_1\left(\frac{2\sqrt{\rho_{bp}^2 \frac{r_1^2}{2\sigma^2}\frac{r_2^2}{2\sigma^2}}}{1-\rho_{bp}^2}\right) \\ &= \frac{2\sigma^2}{(1-\rho_{bp}^2)\rho_{bp} r_1 r_2} \exp\left(-\frac{r_1^2+r_2^2}{\rho_{bp,0}}\rho_{bp}^2\right) I_1\left(\frac{2\rho_{bp} r_1 r_2}{\rho_{bp,0}}\right).\end{aligned} \tag{3.99}$$

Fig. 3.8 shows the plot of the polynomials $L_\gamma^{(1)}(x)$ for different values of γ. By replacing (3.99) in (3.92), we obtain

$$\begin{aligned}R_{\tilde{y},bp}(\tau) &= \frac{\rho_{bp}}{2\sigma^6} \exp(-j\phi) \int_0^{+\infty}\int_0^{+\infty} f_{bp}(r_1) f_{bp}^*(r_2) r_1^2 r_2^2 \exp\left(-\frac{(1-\rho_{bp}^2)(r_1^2+r_2^2)}{2\sigma^2(1-\rho_{bp}^2)}\right) \\ &\quad \times \frac{2\sigma^2}{(1-\rho_{bp}^2)\rho_{bp} r_1 r_2} \exp\left(-\frac{r_1^2+r_2^2}{\rho_{bp,0}}\rho_{bp}^2\right) I_1\left(\frac{2\rho_{bp} r_1 r_2}{\rho_{bp,0}}\right) dr_1 dr_2 \\ &= \frac{1}{2\sigma^6} \exp(-j\phi) \int_0^{+\infty}\int_0^{+\infty} f_{bp}(r_1) f_{bp}^*(r_2) r_1^2 r_2^2 \exp\left(\frac{r_1^2+r_2^2}{2\sigma^2}\right) \\ &\quad \times \left(\sum_{\gamma=0}^{+\infty} \frac{1}{\gamma+1} \rho_{bp}^{2\gamma+1} L_\gamma^{(1)}\left(\frac{r_1^2}{2\sigma^2}\right) L_\gamma^{(1)}\left(\frac{r_2^2}{2\sigma^2}\right)\right) dr_1 dr_2.\end{aligned} \tag{3.100}$$

Since the double integral in (3.100) is separable in two equal integrals with respect to r_1 and r_2, we can write

$$\begin{aligned}R_{\tilde{y},bp}(\tau) &= \frac{1}{2\sigma^6} \exp(-j\phi) \sum_{\gamma=0}^{+\infty} \frac{1}{\gamma+1} \rho_{bp}^{2\gamma+1} \left| \int_0^{+\infty} r^2 f_{bp}(r) \exp\left(-\frac{r^2}{2\sigma^2}\right) L_\gamma^{(1)}\left(\frac{r^2}{2\sigma^2}\right) dr \right|^2 \\ &= 2\sum_{\gamma=0}^{+\infty} P_{2\gamma+1}^{bp} \rho_{bp}^{2\gamma+1} \exp(-j\phi),\end{aligned} \tag{3.101}$$

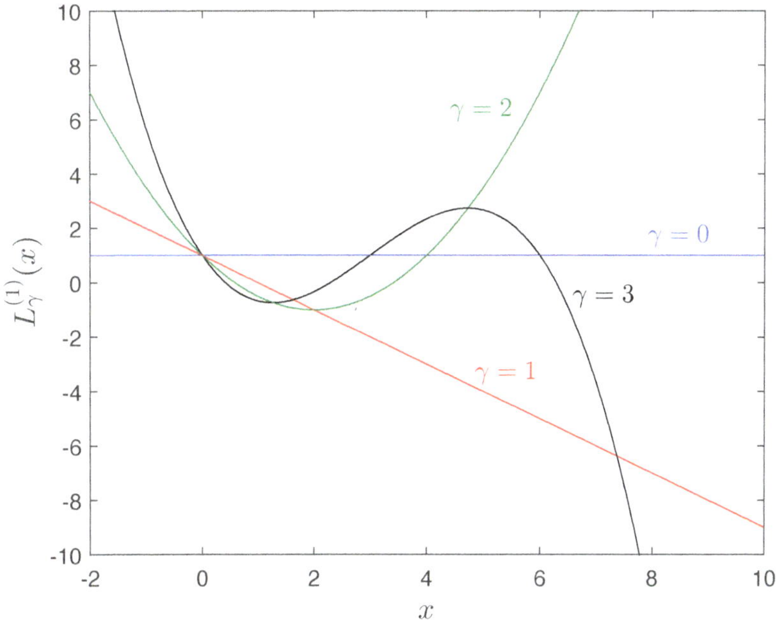

Figure 3.8 Plot of Laguerre polynomials $L_\gamma^{(1)}(x)$ for different values of γ.

where $P_{2\gamma+1}^{bp}$ is the power associated with the IMP of order $2\gamma+1$ that is defined as [8, 11, 12, 30, 88]

$$P_{2\gamma+1}^{bp} = \frac{1}{4\sigma^6(\gamma+1)} \left| \int_0^{+\infty} r^2 f_{bp}(r) \exp\left(-\frac{r^2}{2\sigma^2}\right) L_\gamma^{(1)}\left(\frac{r^2}{2\sigma^2}\right) dr \right|^2 . \tag{3.102}$$

By noting that

$$\rho_{bp}^{2\gamma+1} \exp(-j\phi) = \frac{R_{\tilde{s},bp}(\tau)^{\gamma+1} R_{\tilde{s},bp}^*(\tau)^\gamma}{R_{\tilde{s},bp}(0)^{2\gamma+1}}, \tag{3.103}$$

we can finally write the autocorrelation of the nonlinearly distorted signal as

$$R_{\tilde{y},bp}(\tau) = 2 \sum_{\gamma=0}^{+\infty} P_{2\gamma+1}^{bp} \frac{R_{\tilde{s},bp}(\tau)^{\gamma+1} R_{\tilde{s},bp}^*(\tau)^\gamma}{R_{\tilde{s},bp}(0)^{2\gamma+1}}. \tag{3.104}$$

Under these conditions, the average power of the nonlinearly distorted bandpass signal can be computed as

$$P_{\text{nl}}^{bp} = \frac{1}{2} R_{\tilde{y},bp}(0) = \sum_{\gamma=0}^{+\infty} P_{2\gamma+1}^{bp}$$
$$= \frac{1}{2} \int_{0}^{+\infty} A^2(r) p(r) dr, \tag{3.105}$$

where $A(r)$ is the AM/AM conversion function of (3.76). Having in mind the Bussgang's theorem, the total output power can also be defined as

$$P_{\text{nl}}^{bp} = P_{\text{nl,u}}^{bp} + P_{\text{nl,d}}^{bp}, \tag{3.106}$$

with

$$P_{\text{nl,d}}^{bp} = \frac{1}{2} R_{\tilde{d},bp}(0) = \sum_{\gamma=1}^{+\infty} P_{2\gamma+1}^{bp}, \tag{3.107}$$

denoting the average power of nonlinear distortion component and

$$P_{\text{nl,u}}^{bp} = |\alpha_{bp}|^2 \sigma^2 = P_1^{bp}, \tag{3.108}$$

denoting the average power of the useful component. The PSD associated with the complex envelope of the nonlinearly distorted signal at the bandpass nonlinearity output is given by the Fourier transform of (3.104), i.e.,

$$G_{\tilde{y},bp}(\tau) = 2 \sum_{\gamma=0}^{+\infty} \frac{P_{2\gamma+1}^{bp}}{R_{\tilde{s},bp}(0)^{2\gamma+1}} \mathscr{F}\left(R_{\tilde{s},bp}(\tau)^{\gamma+1} R_{\tilde{s},bp}^{*}(\tau)^{\gamma}\right)$$
$$= 2 \sum_{\gamma=0}^{+\infty} \frac{P_{2\gamma+1}^{bp}}{R_{\tilde{s},bp}(0)^{2\gamma+1}} \left(\underbrace{G_{\tilde{s},bp}(f) * \cdots * G_{\tilde{s},bp}(f)}_{\gamma+1 \text{ times}} \right)$$
$$* \left(\underbrace{G_{\tilde{s},bp}(-f) * \cdots * G_{\tilde{s},bp}(-f)}_{\gamma \text{ times}} \right), \tag{3.109}$$

where $G_{\tilde{s},bp}(f)$ is the PSD for the complex envelope of the input signal. As the output is composed of two uncorrelated components (see (3.77)), we also can write

$$G_{\tilde{y},bp}(f) = |\alpha_{bp}|^2 G_{\tilde{s},bp}(f) + G_{\tilde{d},bp}(f), \tag{3.110}$$

where $G_{\tilde{d},bp}(f)$ is the PSD associated with the complex envelope of the nonlinear distortion component, given by

$$G_{\tilde{d},bp} = 2\sum_{\gamma=1}^{+\infty} \frac{P_{2\gamma+1}^{bp}}{R_{\tilde{s},bp}(0)^{2\gamma+1}} \left(\underbrace{G_{\tilde{s},bp}(f) * \cdots * G_{\tilde{s},bp}(f)}_{\gamma+1 \text{ times}} \right)$$

$$* \left(\underbrace{G_{\tilde{s},bp}(-f) * \cdots * G_{\tilde{s},bp}(-f)}_{\gamma \text{ times}} \right). \tag{3.111}$$

From (3.109), it is clear that the bandwidth of the nonlinearly signal $\tilde{y}(t)$ increases in relation to the bandwidth of the input signal $\tilde{s}(t)$, since the former is obtained by convolving the PSD of the input signal with itself many times. In fact, this leads not only to out-of-band radiation, but also in-band radiation that can severely degrade the performance of multicarrier systems.

3.3 EQUIVALENT NONLINEARITIES

Equations (3.57), (3.70) and (3.104) allow us to analytically obtain the autocorrelation (and the corresponding PSD) of nonlinearly distorted Gaussian signals considering both real-valued and complex-valued signals submitted to baseband or bandpass nonlinearities, respectively. However, all of these equations involve an infinite summation that is not practically realizable. Therefore, in order to obtain the autocorrelation of a given nonlinearly distorted signal, one must truncate those summations considering a given value of n_γ IMPs. Under these conditions, only the contribution of a finite number of IMPs is taken into account. As aforementioned, we denote this approach as the "truncated IMP" approach.

The truncated IMP approach is relatively straightforward and can be employed for most baseband and bandpass memoryless nonlinearities that can be encountered in multicarrier transceivers. However, it should be mentioned that the accuracy of this approach is intimately related with the number of IMPs employed for the computation of the autocorrelation of the nonlinearly distorted signal. In fact, the truncated IMP approach implicitly involves a polynomial approximation for the nonlinearity, since the nonlinear distortion effects at the nonlinearity output are divided into the individual contribution of the different IMP and each IMP is associated with a given polynomial term. This means that the number of required IMPs varies with the "smoothness" associated of the nonlinearity. Therefore, the higher the desired accuracy, the higher the number of IMPs that should be considered, i.e., the higher the value of n_γ. Put in other words, the value of n_γ is strongly dependent on the degree of the polynomial approximation associated with the nonlinear characteristic. Clearly, this means that for non-smooth nonlinear characteristics (such as the ones inherent to clipping operations or low-resolution quantization processes with many discontinuities), the value of n_γ required to obtain an acceptable accuracy can be very large, leading to higher complexity. In addition, the truncated IMP approach may also present convergence issues since we may find numerical problems when obtaining the contribution of IMPs of very high order. Therefore, in some situations, increasing n_γ does not necessarily means better accuracy, which conditions the applicability of this approach in some situations.

In this section, we present a simplified method for obtaining the statistical characterization of nonlinearly distorted Gaussian signals that is based on the truncated IMP approach presented in the previous section. This approach is specially adequate to be employed in situations where the nonlinearities are severe and operate in sampled multicarrier signals. The method involves the creation of an equivalent nonlinearity that can substitute the original, non-smooth nonlinearity for the evaluation of the corresponding nonlinear distortion effects, but that leads to signals with the same spectral characterization of the ones distorted by the original nonlinearity. This equivalent nonlinearity is polynomial and has low degree. For this reason, it allows to easily obtain the statistical characterization of nonlinearly distorted Gaussian signals. In the following, the principle behind these equivalent nonlinearities is described. The baseband equivalent nonlinearities that can be employed both in real-valued and complex-valued baseband signals are introduced in Subsection 3.3.1. The bandpass equivalent nonlinearities are introduced in Subsection 3.3.2.

3.3.1 BASEBAND EQUIVALENT NONLINEARITIES

As aforementioned, the "strong" nonlinearities of some operations performed in the transmission chain of multicarrier systems require the computation of a large number of IMPs, provided that an accurate statistical characterization of the corresponding nonlinearly distorted signals is desired. In fact, the complexity of such process is related to the nonlinearity, i.e, it can be very high for severe nonlinearities with discontinuities. For this reason, the use of the truncated IMP approach in those situations usually leads to inaccurate results since, to avoid high complexity, only a small number of IMPs is taken into account for the computation of the autocorrelation of the nonlinearly distorted signal. In the following, the concept of baseband, real-valued equivalent nonlinearities is explained. In addition, it is shown how these nonlinearities can be obtained.

Regarding the truncated IMP approach, the approximate autocorrelation of a real-valued multicarrier signal submitted to a baseband nonlinearity represented by $f(\cdot)$ is obtained by truncating (3.57) to the first $n_\gamma+1$ IMPs, i.e.,

$$R_{y,bb}(\tau) \approx \sum_{\gamma=0}^{n_\gamma} \left(\frac{R_{s,bb}(\tau)}{\sigma^2} \right)^{2\gamma+1} P_{2\gamma+1}^{bb}. \tag{3.112}$$

As mentioned before, the choice of n_γ is dependent on a trade-off between accuracy and complexity, but in some situations its increase does not guarantee a better accuracy. Here, we propose to replace the conventional baseband nonlinearity, $f(\cdot)$, by a smoother, polynomial equivalent nonlinearity, $g(\cdot)$, that gives rise to signals with the same spectral characterization of the ones submitted to $f(\cdot)$. This equivalent nonlinearity can then be employed to obtain the statistical characterization of Gaussian signals submitted to baseband nonlinearities [89, 90]. In the following, the steps associated with the obtainment of this equivalent nonlinearity are described.

Let us consider a stationary, Gaussian signal $s(t)$ with bandwidth B_s and average power $R_{s,bb}(0) = \sigma^2$. In addition, let us consider that this signal is sampled at a rate $f_s = 2OB_s$ (i.e., considering an oversampling factor of O) and submitted to a real-valued, baseband nonlinearity that yieldsa sampled version of $y(t)$ (see (3.1)). As demonstrated before, by making use of the Wiener-Khinchin's theorem, the PSD of $y(t)$ can be obtained through the Fourier transform of its autocorrelation (see (3.59)). Note that, by defining the frequency-domain distribution associated with the IMP of order $2\gamma+1$ as

$$G_{y,bb}^{2\gamma+1}(f) = \frac{P_{2\gamma+1}^{bb}}{\sigma^{2(2\gamma+1)}} \underbrace{\left(G_{s,bb}(f) * G_{s,bb}(f) * \cdots * G_{s,bb}(f)\right)}_{2\gamma+1 \text{ times}}, \tag{3.113}$$

we can rewrite (3.59) as a summation of the PSDs of the different IMPs as

$$G_{y,bb}(f) = \sum_{\gamma=0}^{+\infty} G_{y,bb}^{2\gamma+1}(f). \tag{3.114}$$

Fig. 3.9 shows the PSD of a baseband, real-valued nonlinearly distorted signal considering $O = 4$, as well as the PSDs associated with IMPs of order 1, 7 and 9. It is assumed that the PSD of the signal at the nonlinearity input, $G_{s,bb}(f)$, is rectangular. From the results shown in the figure, it can be noted that the PSD associated with $\gamma = 0$, $G_{y,bb}^{1}(f)$, is proportional to the input signal's PSD, which was expected from (3.113) since, in fact, $G_{y,bb}^{1}(f) = |\alpha_{bb}|^2 G_{s,bb}(f)$ (see (3.55)), which justifies it rectangular shape. Additionally, it can also be noted that when γ increases, the corresponding PSDs of the IMPs of that order become more flat. For instance, when $\gamma = 4$, one can note that $G_{y,bb}^{9}(f)$ is almost constant. Therefore, this means that its corresponding autocorrelation, $R_{y,bb}^{9}(\tau) = \mathscr{F}^{-1}(G_{y,bb}^{9}(f))$ ($\mathscr{F}^{-1}$ denotes the inverse Fourier transform), is approximately given by a scaled Dirac delta function. This can be explained by taking into account that as

$$\max(R_{s,bb}(\tau)) = \sigma^2, \tag{3.115}$$

we have

$$\left(R_{s,bb}(\tau)/\sigma^2\right)^{2\gamma+1} \approx \delta(\tau). \tag{3.116}$$

Indeed, that approximation can be observed from the IMPs of a given order $\gamma \geq \gamma_{\max}$, where $2\gamma_{\max}+1$ is defined as the order of the IMP from which the corresponding PSD becomes approximately flat.

Fig. 3.10 shows the PSD of a baseband, real-valued nonlinearly distorted signal considering $O = 2$, as well as the PSDs for the IMPs of order 1, 3 and 5. From the figure it can be noted that when the oversampling is lower, the value of $\gamma_{\max}$ that leads to a constant PSD at the nonlinearity output is also lower. In fact, the value of

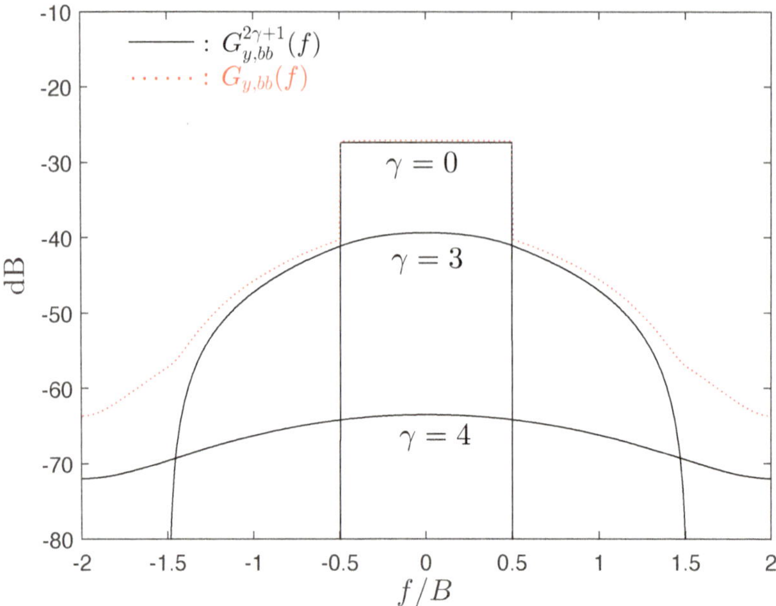

Figure 3.9 PSD of a given nonlinearly distorted signal and the PSD associated with the IMPs of order $2\gamma+1$ considering an oversampling factor of $O = 4$ and different values of γ.

$\gamma_{\max}$ is related not only to the nature of the nonlinearity (i.e., with its severeness), but also to the oversampling factor. Therefore, we can note that from a given value of $\gamma = \gamma_{\max}$, one can concentrate the effect of all IMPs of order $2\gamma+1 \geq 2\gamma_{\max}+1$ in that IMP of order $2\gamma_{\max}+1$. Under these conditions, (3.57) can be approximated as

$$\begin{aligned} R_{y,bb}(\tau) &= \sum_{\gamma=0}^{+\infty} \left(\frac{R_{s,bb}(\tau)}{\sigma^2}\right)^{2\gamma+1} P^{bb}_{2\gamma+1} \\ &\approx \sum_{\gamma=0}^{\gamma_{\max}-1} \left(\frac{R_{s,bb}(\tau)}{\sigma^2}\right)^{2\gamma+1} P^{bb}_{2\gamma+1} + \sum_{\gamma=\gamma_{\max}}^{+\infty} P^{bb}_{2\gamma+1}\delta(\tau) \\ &\approx \sum_{\gamma=0}^{\gamma_{\max}-1} \left(\frac{R_{s,bb}(\tau)}{\sigma^2}\right)^{2\gamma+1} P^{bb}_{2\gamma+1} + P^{bb,\infty}_{2\gamma+1}\delta(\tau), \end{aligned} \tag{3.117}$$

where, according to the definition of the average power of the signal at the output of the real-valued, baseband nonlinearity in (3.39), we have

$$P^{bb,\infty}_{2\gamma+1} = P_{\text{nl}} - \sum_{\gamma=0}^{\gamma_{\max}-1} P^{bb}_{2\gamma+1}. \tag{3.118}$$

This means that the autocorrelation of the nonlinearly distorted signal can be obtained by computing the total output power and the power of the first $\gamma_{\max}$ IMPs.

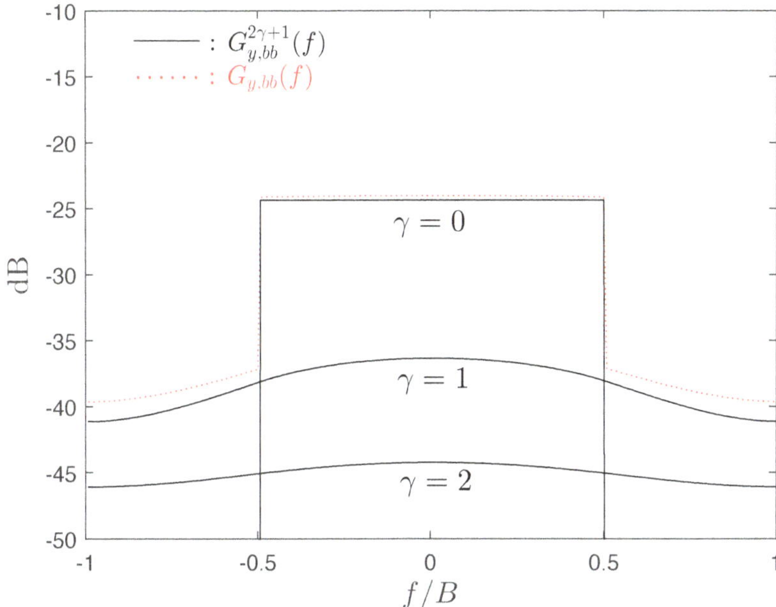

Figure 3.10 PSD of the nonlinearly distorted signal and the PSD for the IMP of order $2\gamma+1$ considering an oversampling factor of $O=2$ and different values of γ.

Therefore, this means that the infinite summation of (3.57) can be avoided, which is traduced in an effective complexity reduction for obtaining an accurate PSD. By making use of this fact, we can define an equivalent, polynomial nonlinearity, $g(\cdot)$, with degree $2\gamma_{\max}+1$, where its last polynomial term concentrates the effect of several terms of the "original nonlinearity", $f(\cdot)$. This equivalent nonlinearity is defined as

$$g(s) = \sum_{\gamma=0}^{\gamma_{\max}} c^{bb}_{2\gamma+1} s^{2\gamma+1} = c^{bb}_1 s + c^{bb}_3 s^3 + \cdots + c^{bb}_{2\gamma_{\max}+1} s^{2\gamma_{\max}+1}, \tag{3.119}$$

where $c^{bb}_{2\gamma+1}$ is the polynomial coefficient of the $(2\gamma+1)$th polynomial term. We also define the set constituted by all polynomial coefficients of $g(\cdot)$ as $c^{bb} = [c^{bb}_1 \; c^{bb}_3 \; \cdots \; c^{bb}_{2\gamma_{\max}+1}] \in \mathbb{R}^{\gamma_{\max}+1}$. Note that having in mind (3.118), we can define the IMPs of the equivalent nonlinearity as follows[1]

$$P^{bb,g}_{2\gamma+1} = \begin{cases} P^{bb,f}_{2\gamma+1}, & 0 \leq \gamma < \gamma_{\max} \\ \sum_{\gamma=\gamma_{\max}}^{+\infty} P^{bb,f}_{2\gamma+1} = P^{f}_{\text{out}} - \sum_{\gamma=0}^{\gamma_{\max}-1} P^{bb,f}_{2\gamma+1}, & \gamma = \gamma_{\max} \\ 0, & \gamma > \gamma_{\max}. \end{cases} \tag{3.120}$$

[1] We adopt the superscript f or g to distinguish between the original and the equivalent nonlinearity, respectively.

In fact, as we have the power associated with the IMPs of the equivalent nonlinearity and we know its expression (see (3.119)), we are able to obtain its polynomial coefficients $c^{bb} = [c_1^{bb}\ c_3^{bb}\ \cdots\ c_{2\gamma_{\max}+1}^{bb}] \in \mathbb{R}^{\gamma_{\max}+1}$, which allows us to completely define equivalent nonlinearity $g(\cdot)$. In order to do that, let us start by redefining (3.58) as

$$P_{2\gamma+1}^{bb,g} = \frac{\left(p_{2\gamma+1}^{bb,g}\right)^2}{2^{2\gamma+1}(2\gamma+1)!}. \tag{3.121}$$

Moreover, let us define the block $p^{bb,g} = [p_1^{bb,g}\ p_3^{bb,g}\ \ldots\ p_{2\gamma_{\max}+1}^{bb,g}]^T \in \mathbb{R}^{\gamma_{\max}+1}$ formed by the coefficients $p_{2\gamma+1}^{bb,g}$, where

$$\begin{aligned} p_{2\gamma+1}^{bb,g} &= 2^{2\gamma+1}(2\gamma+1)!\sqrt{P_{2\gamma+1}^{bb,g}} \\ &= \int_{-\infty}^{+\infty} g(s)p(s)H_{2\gamma+1}\left(\frac{s}{\sqrt{2}\sigma}\right)ds. \end{aligned} \tag{3.122}$$

By replacing (3.119) into (3.122), we have

$$\begin{aligned} p_{2\gamma+1}^{bb,g} &= \int_{-\infty}^{+\infty}\left(\sum_{\gamma'=0}^{\gamma_{\max}} c_{2\gamma'+1}^{bb} s^{2\gamma'+1}\right)p(s)H_{2\gamma+1}\left(\frac{s}{\sqrt{2}\sigma}\right)ds \\ &= \sum_{\gamma'=0}^{\gamma_{\max}} c_{2\gamma'+1}^{bb} \underbrace{\int_{-\infty}^{+\infty} s^{2\gamma'+1}p(s)H_{2\gamma+1}\left(\frac{s}{\sqrt{2}\sigma}\right)ds}_{\beta_{\gamma\gamma'}^{bb}} \\ &= \sum_{\gamma'=0}^{\gamma_{\max}} c_{2\gamma'+1}^{bb}\beta_{\gamma\gamma'}^{bb}. \end{aligned} \tag{3.123}$$

Additionally, by defining the $\gamma_{\max} \times \gamma_{\max}$ square matrix β^{bb} as

$$\beta^{bb} = \begin{bmatrix} \beta_{0,0}^{bb} & \beta_{0,1}^{bb} & \cdots & \beta_{0,\gamma_{\max}}^{bb} \\ \vdots & \ddots & \cdots & \vdots \\ \beta_{\gamma_{\max},0}^{bb} & \cdots & \cdots & \beta_{\gamma_{\max},\gamma_{\max}}^{bb} \end{bmatrix}, \tag{3.124}$$

we can write (3.123) in matrix notation, resulting

$$p^{bb,g} = c^{bb}\beta^{bb}. \tag{3.125}$$

Therefore, as the set $p^{bb,g} = [p_1^{bb,g}\ p_3^{bb,g}\ \ldots\ p_{2\gamma_{\max}+1}^{bb,g}]^T \in \mathbb{R}^{\gamma_{\max}+1}$ and the matrix β^{bb} are both known, the polynomial coefficients of the equivalent nonlinearity $c^{bb} = [c_1^{bb}\ c_3^{bb}\ \cdots\ c_{2\gamma_{\max}+1}^{bb}] \in \mathbb{R}^{\gamma_{\max}+1}$ can be computed as

$$c^{bb} = \beta^{-1}p^{bb,g}. \tag{3.126}$$

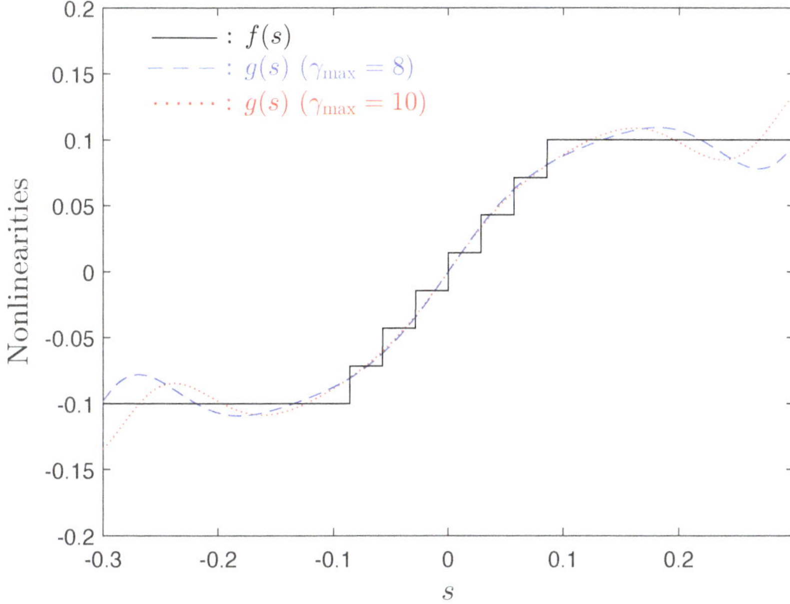

Figure 3.11 Nonlinearity associated with a quantization operation as well as its equivalent nonlinearities obtained for different values of $\gamma_{\max}$.

Fig. 3.11 shows the nonlinear characteristic of a quantizer with $n_b = 3$ bits of resolution as well as its corresponding equivalent nonlinearities considering different values of $\gamma_{\max}$. It should be mentioned that this equivalent nonlinearity is polynomial, but it does not constitute a polynomial approximation for the original nonlinearity.

For the case of Cartesian nonlinearities, this approach of replacing the original nonlinearity, $f_{bb}(\cdot)$, with the equivalent nonlinearity can also be employed to obtain a simple and accurate statistical characterization of the corresponding nonlinearly distorted signals. However, for Cartesian nonlinearities, both the nonlinearities operating on the real and imaginary parts of the signal $f_{bb,I}(\cdot)$ and $f_{bb,Q}(\cdot)$, respectively, should be replaced by the corresponding equivalent nonlinearities. Nevertheless, if $f_{bb,I}(\cdot) = f_{bb,Q}(\cdot) = f(\cdot)$, then only one equivalent nonlinearity should be computed.

3.3.2 BANDPASS EQUIVALENT NONLINEARITIES

The concept of equivalent nonlinearities can also be employed for bandpass nonlinearities [91, 92]. In this section, we consider bandpass nonlinearities characterized by the AM/AM and AM/PM baseband model represented in (3.24), and it is shown how we can obtain their corresponding equivalent nonlinearities.

Let us start by considering the complex envelope $\tilde{s}(t)$ of a given bandpass signal (see (3.74)). Besides of having a Gaussian nature, we assume that $\tilde{s}(t)$ has a rectangular PSD, bandwidth B_s and average power $2\sigma^2$. This complex envelope is sampled at rate $f_s = 2OB_s$ (i.e., considering an oversampling factor of O) and submitted to a bandpass, memoryless nonlinearity that yields $\tilde{y}(t) = A(r(t))\exp(j\varphi(t))$.

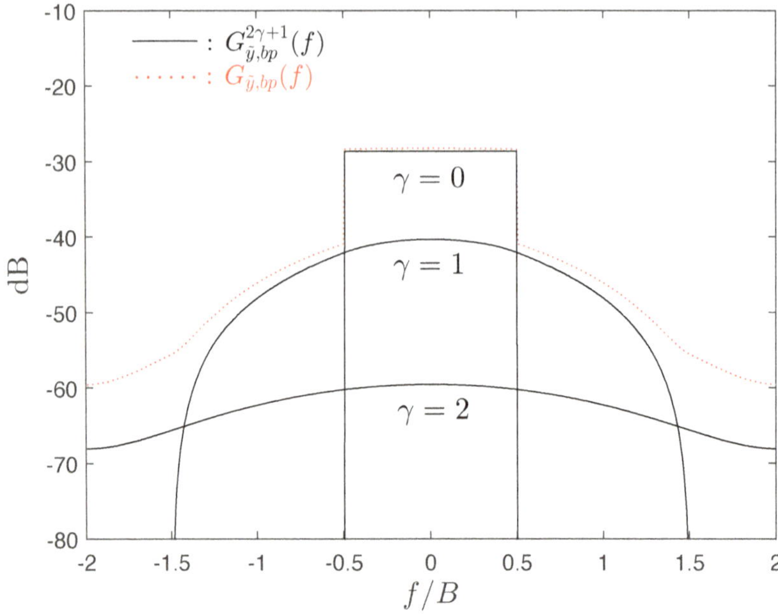

Figure 3.12 PSD of the nonlinearly distorted signal and the PSD associated with the IMP of order $2\gamma+1$ considering an oversampling factor of $O = 4$.

As in the case of baseband nonlinearities, when bandpass nonlinearities are considered, we also can decompose the PSD of the nonlinearly distorted signals (see (3.109)) as the sum of the individual PSDs of the different IMPs, resulting

$$G_{\tilde{y},bp} = 2\sum_{\gamma=0}^{+\infty} G^{2\gamma+1}_{\tilde{y},bp}(f), \tag{3.127}$$

where

$$G^{2\gamma+1}_{\tilde{y},bp}(f) = \frac{P^{bp}_{2\gamma+1}}{R_{\tilde{s},bp}(0)^{2\gamma+1}} \left(\underbrace{G_{\tilde{s},bp}(f) * \cdots * G_{\tilde{s},bp}(f)}_{\gamma+1 \text{ times}} \right) * \left(\underbrace{G_{\tilde{s},bp}(-f) * \cdots * G_{\tilde{s},bp}(-f)}_{\gamma \text{ times}} \right). \tag{3.128}$$

Fig. 3.12 shows the PSD of the nonlinearly distorted signal considering $O = 2$, as well as the PSDs for the IMPs of order 1, 3 and 5. From the figure, one can note that when $\gamma = 2$, the corresponding PSD $G^{5}_{\tilde{y},bp}(f)$ is approximately flat. This means that from a given order $\gamma = \gamma_{\max}$, all the PSDs associated with those IMPs (i.e., the ones with order $\gamma \geq \gamma_{\max}$), are approximately constant and can be concentrated in only one IMP. Therefore, the autocorrelation of the nonlinearly distorted signal can be approximated by

$$R_{\tilde{y},bp}(\tau) \approx 2 \sum_{\gamma=0}^{\gamma_{\max}-1} P_{2\gamma+1}^{bp} \frac{R_{\tilde{s},bp}(\tau)^{\gamma+1} R_{\tilde{s},bp}^{*}(\tau)^{\gamma}}{R_{\tilde{s},bp}(0)^{2\gamma+1}} + 2P_{2\gamma+1}^{bp,\infty}\delta(\tau), \tag{3.129}$$

where

$$P_{2\gamma+1}^{bp,\infty} = P_{\text{nl}}^{bp} - \sum_{\gamma=0}^{\gamma_{\max}-1} P_{2\gamma+1}^{bp}, \tag{3.130}$$

with P_{nl}^{bp} defined as in (3.106). Note that this is formally equivalent to submit the complex envelope of the bandpass signal to an equivalent, polynomial nonlinearity $g_{bp}(\cdot)$, with degree $\gamma_{\max}$, defined as

$$g_{bp}(r) = \sum_{\gamma=0}^{\gamma_{\max}} c_{2\gamma+1}^{bp} r^{2\gamma+1} = c_1^{bp} r + c_3^{bp} r^3 + \cdots + c_{2\gamma_{\max}+1}^{bp} r^{2\gamma_{\max}+1}, \tag{3.131}$$

where $c^{bp} = [c_1^{bp}\ c_3^{bp}\ \cdots\ c_{2\gamma_{\max}+1}^{bp}] \in \mathbb{R}^{\gamma_{\max}+1}$ denote the set of polynomial coefficients of the equivalent, bandpass nonlinearity $g_{bp}(\cdot)$. The IMPs of this nonlinearity are defined as

$$P_{2\gamma+1}^{bp,g} = \begin{cases} P_{2\gamma+1}^{bp,f}, & 0 \le \gamma < \gamma_{\max} \\ \sum_{\gamma=\gamma_{\max}}^{+\infty} P_{2\gamma+1}^{bp,f} = P_{\text{out},bp}^{f} - \sum_{\gamma=0}^{\gamma_{\max}-1} P_{2\gamma+1}^{bp,f}, & \gamma = \gamma_{\max} \\ 0, & \gamma > \gamma_{\max}. \end{cases} \tag{3.132}$$

Let us define the IMP of order $2\gamma+1$ as

$$P_{2\gamma+1}^{bp,g} = \frac{\left(p_{2\gamma+1}^{bp,g}\right)^2}{4\sigma^6(\gamma+1)}. \tag{3.133}$$

In addition, let us define the block $p^{bp,g} = [p_1^{bp,g}\ p_3^{bp,g}\ \ldots\ p_{2\gamma_{\max}+1}^{bp,g}]^T \in \mathbb{R}^{\gamma_{\max}+1}$, formed by the coefficients $p_{2\gamma+1}^{bp,g}$, where

$$p_{2\gamma+1}^{bp,g} = \int_0^{+\infty} g_{bp}(r) r^2 p(r) L_{\gamma}^{(1)}\left(\frac{r^2}{2\sigma^2}\right) dr. \tag{3.134}$$

By replacing (3.131) into (3.134), we obtain

$$\begin{aligned} p_{2\gamma+1}^{bp,g} &= \int_{-\infty}^{+\infty} \left(\sum_{\gamma'=0}^{\gamma_{\max}} c_{2\gamma'+1}^{bp} r^{2\gamma'+1}\right) r^2 p(r) L_{\gamma}^{(1)}\left(\frac{r^2}{2\sigma^2}\right) dr \\ &= \sum_{\gamma'=0}^{\gamma_{\max}} c_{2\gamma'+1}^{bp} \underbrace{\int_{-\infty}^{+\infty} r^{2\gamma'+1} r^2 p(r) L_{\gamma}^{(1)}\left(\frac{r^2}{2\sigma^2}\right) dr}_{\beta_{\gamma\gamma'}^{bp}} \\ &= \sum_{\gamma'=0}^{\gamma_{\max}} c_{2\gamma'+1}^{bp} \beta_{\gamma\gamma'}^{bp}. \end{aligned} \tag{3.135}$$

Additionally, by defining the $\gamma_{\max} \times \gamma_{\max}$ square matrix β^{bp} as

$$\beta^{bp} = \begin{bmatrix} \beta^{bp}_{0,0} & \beta^{bp}_{0,1} & \cdots & \beta^{bp}_{0,\gamma_{\max}} \\ \vdots & \ddots & \cdots & \vdots \\ \beta^{bp}_{\gamma_{\max},0} & \cdots & \cdots & \beta^{bp}_{\gamma_{\max},\gamma_{\max}} \end{bmatrix}, \tag{3.136}$$

we can rewrite (3.135) in matrix notation as

$$p^{bp,g} = c^{bp}\beta^{bp}. \tag{3.137}$$

Therefore, as the both set $p^{bp,g} = [p_1^{bp,g}\ p_3^{bp,g}\ \ldots\ p_{2\gamma_{\max}+1}^{bp,g}]^T \in \mathbb{R}^{\gamma_{\max}+1}$ and the matrix β^{bp} are known, the polynomial coefficients for the equivalent nonlinearity $c^{bp} = [c_1^{bp}\ c_3^{bp}\ \cdots\ c_{2\gamma_{\max}+1}^{bp}] \in \mathbb{R}^{\gamma_{\max}+1}$ can be computed as

$$c^{bp} = \beta^{-1} p^{bp,g}. \tag{3.138}$$

Fig. 3.13 shows the nonlinear function of an envelope clipping operation, as well as its corresponding equivalent nonlinearities obtained for $\gamma_{\max} = 3$ and $\gamma_{\max} = 9$.

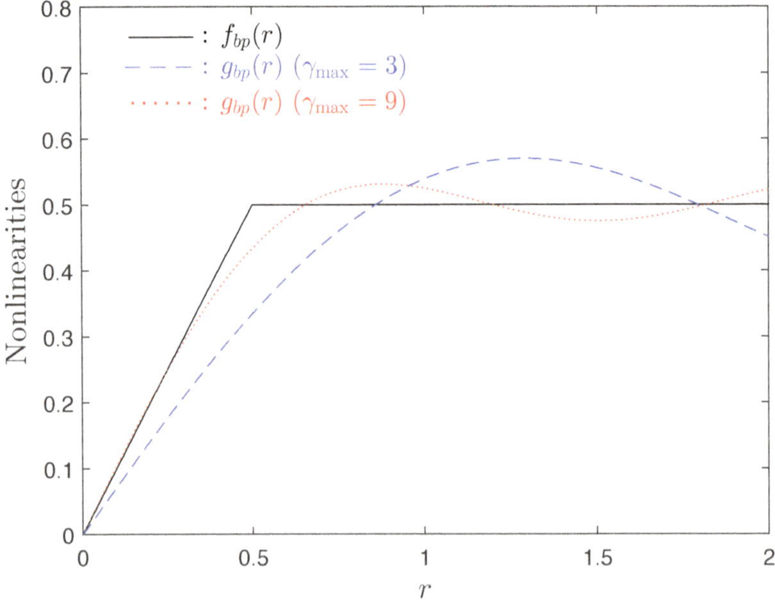

Figure 3.13 Nonlinearity of an envelope clipping operation as well as its equivalent nonlinearity computed with different values of $\gamma_{\max}$.

4 Optimum Detection of Nonlinear Multicarrier Schemes

In this chapter, we analyze the optimum detection of nonlinear multicarrier schemes. To motivate the use of the optimum detection, the conventional approaches commonly employed for detection of nonlinear multicarrier schemes are briefly described in the Subsection 4.1. In fact, the optimum detection arises as an alternative to those conventional approaches that does not consider the nonlinear distortion as an additional noise term that leads to performance degradations, but instead as useful information that can be used to improve the performance of multicarrier schemes.

This was noticed in [14], where the gains associated with the optimum detection of OFDM schemes with nonlinear distortion effects were studied, and it was verified that the nonlinear distortion gives an additional diversity effect that can be used to improve the performance in frequency-selective channels. More recently, it was demonstrated that the optimum detection of nonlinearly distorted multicarrier signals provide potential asymptotic gains relatively to linear OFDM transmissions, even in ideal AWGN channels, considering different nonlinearities, and considering different multicarrier schemes such as DMT schemes [15, 93–95]. These potential asymptotic gains were studied analytically and it was also verified that even sub-optimum receivers present performance improvements relatively to the conventional detection of linear, multicarrier schemes.

In order to understand the motivation behind the existence of these potential gains, we firstly present in Section 4.1 the motivation behind the need of optimum detection schemes. The theoretical principle associated with the optimum detection considering a linear multicarrier scheme is presented in Section 4.2. Besides the motivation for the optimum detection of nonlinearly distorted signals, which includes an introduction for this type of detection, it presents an analytical method for obtaining the asymptotic optimum performance for different nonlinearities and different transmission scenarios. It is shown that the optimum detection present potential asymptotic gains relatively to the conventional detection not only when nonlinear multicarrier schemes are considered, but also in comparison with linear, multicarrier schemes. After that, Section 4.3 deals with the asymptotic performance associated with the optimum detection considering both linear and nonlinear multicarrier schemes. Then, theoretical expressions for the potential asymptotic gains considering different multicarrier schemes, different nonlinearities and both for ideal AWGN and frequency-selective channels are presented in Subsections 4.4 and 4.5, respectively.

DOI: 10.1201/9781032708744-4

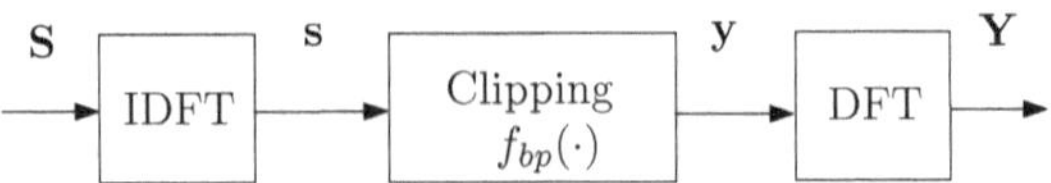

Figure 4.1 Equivalent, subcarrier-level model for a nonlinear OFDM system with an envelope clipping operation.

4.1 MOTIVATION AND CONVENTIONAL APPROACHES

As seen in the last section, nonlinearities introduce both in-band and out-of-band radiation on the transmitted signals. Although the out-of-band nonlinear distortion can be mitigated through the use of proper filtering techniques, the in-band distortion cannot be easily removed. Conventionally, the in-band nonlinear distortion is considered as an additional noise term that is added to the AWGN. This is a consequence of the Bussgang's theorem [10], which allows us to separate the nonlinearly distorted signals into uncorrelated useful and distortion components. In fact, although the nonlinear distortion is not Gaussian in the time-domain, it can be shown that it is approximately Gaussian at the subcarrier level [8], which reinforces the idea of considering it as noisy term that can substantially degrade the performance of multicarrier systems. In the following, the impact of nonlinear distortion effects on the performance of conventional receivers based on a simple hard-decision, which works in a subcarrier-by-subcarrier basis (see Subsection 2.2.2), is shown. Without loss of generality, the specific case of a bandpass OFDM signal submitted to a bandpass nonlinearity is analyzed. More concretely, the nonlinear distortion effects associated with an envelope clipping are considered. The equivalent, subcarrier-level scenario for this nonlinear OFDM scheme is depicted in Fig. 4.1.

Each OFDM signal has N_u QPSK symbols ($M = 4$), oversampling factor O (see (2.23)) and is represented by the frequency-domain block $\mathbf{S} = [S_0\ S_1\ S_2\ \ldots\ S_{N-1}]^T \in \mathbb{C}^N$. Its corresponding time-domain version is $\mathbf{s} = [s_0\ s_1\ s_2\ \ldots\ s_{N-1}]^T \in \mathbb{C}^N$. As mentioned before, the time-domain samples s_n have a complex Gaussian distribution. The variance of its real and imaginary parts is σ^2 (see (2.26)). These time-domain samples are submitted to an envelope clipping operation (see Fig. 3.5 and Fig. 3.13), whose the corresponding AM/PM conversion function is null and the AM/AM conversion function is given by[1]

$$f_{bp}(r_n) = A(r_n) = \begin{cases} r_n, & r_n \leq s_M \\ s_M, & r_n > s_M, \end{cases} \tag{4.1}$$

where $r_n = |s_n|$. According to the Bussgang's theorem [10], the nonlinearly distorted signal is given by the sum of two uncorrelated terms, i.e.,

$$\mathbf{y} = \alpha_{bp}\mathbf{s} + \mathbf{d}, \tag{4.2}$$

[1] Although the clipping level (also known as "saturation" level) is s_M, we usually refer to the normalized clipping level s_M/σ since the magnitude of the nonlinear distortion effects and their corresponding impact on the performance is conditioned by that value.

where $\mathbf{d} = [d_0\ d_1\ d_2\ ...\ d_{N-1}]^T \in \mathbb{C}^N$ is a block that gathers the nonlinear distortion components and α_{bp} is a scale factor given by (3.78). For the nth time-domain sample, we have

$$y_n = \alpha_{bp} s_n + d_n, \tag{4.3}$$

where d_n represents the nonlinear distortion term associated with the nth time-domain sample. On the other hand, as the detection in OFDM is made in the frequency-domain, one must obtain the frequency-domain version of (4.2), that is

$$\begin{aligned} \mathbf{Y} &= \mathbf{F}\mathbf{y} \\ &= \alpha_{bp}\mathbf{S} + \mathbf{D}, \end{aligned} \tag{4.4}$$

where $\mathbf{D} = [D_0\ D_1\ D_2\ ...\ D_{N-1}]^T \in \mathbb{C}^N$ is a block formed by the frequency-domain version of the nonlinear distortion terms. Therefore,

$$Y_k = \alpha_{bp} S_k + D_k, \tag{4.5}$$

where D_k is the nonlinear distortion term associated with the kth subcarrier and S_k is the QPSK data symbol transmitted on the kth subcarrier. From (4.5), one can note that the received symbol associated with the kth subcarrier is a complex-scaled replica of the transmitted data symbol plus an additive nonlinear distortion term. This means that the constellation of the received symbols can be considerably different from the original QPSK constellation, leading to a large number of errors at the detection and an increased BER, that can even be irreducible. This effect is illustrated in Fig. 4.2, which depicts the constellation of the received nonlinearly distorted OFDM symbols Y_k, considering different normalized clipping levels s_M/σ (note that, for $s_M/\sigma = +\infty$, we have an ideal, linear transmission). From the results shown in the figure, it can be noted that the lower the normalized clipping level, the higher the magnitude of the nonlinear distortion effects. It can also be seen that the constellation is shrunken according to the scale factor α_{bp} and that the higher the magnitude of the nonlinear distortion effects, the higher is dimension of the "cloud" that is formed around its central point. It should be mentioned that if the bandpass nonlinearity has a non-null AM/PM conversion function, then the scale factor α_{bp} is complex, which means that, besides being shrunken, the constellation is also rotated (naturally, this does not apply to clipping functions, since they do not have AM/PM conversion). Due these unwanted effects on the received symbol's constellation, it is easy to conclude that the nonlinear distortion can severely degrade the detection, increasing substantially the BER. This performance degradation can be confirmed in Fig. 4.3, which shows the simulated BER associated with a conventional receiver and a nonlinear OFDM transmission with $N_u = 256$, $O = 4$ and different normalized clipping levels s_M/σ. From the figure, one can note that the BER associated with the conventional receivers when nonlinear distortion effects are high can be very poor. For instance, for a target BER of $P_b = 10^{-3}$, the degradation is around 2 dB when $s_M/\sigma = 1.5$, but can reach approximately 6 dB when $s_M/\sigma = 1.0$. In fact, for lower clipping levels, we can even have an accentuated error floor where the BER becomes irreducible, regardless of the value of E_b/N_0.

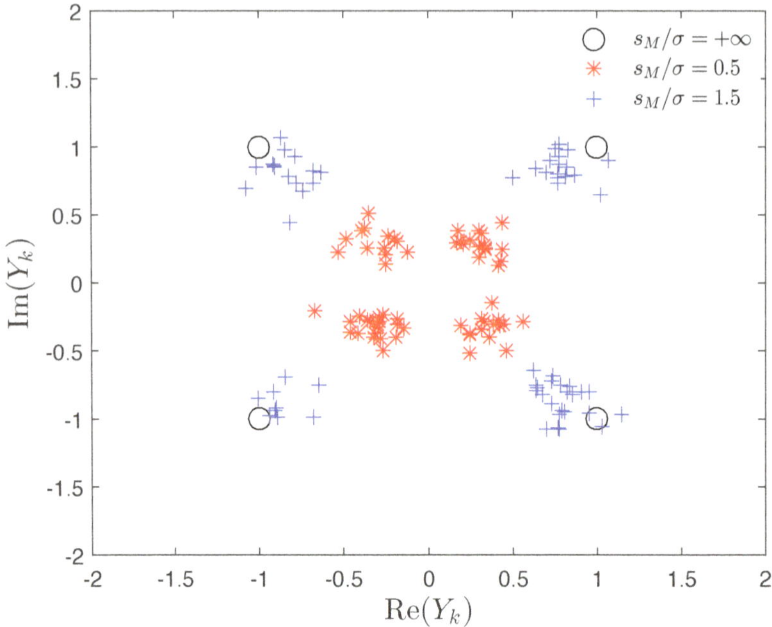

Figure 4.2 Constellation of the received symbols Y_k considering an envelope clipping operation with different normalized clipping levels s_M/σ.

To avoid this large degradation on the performance of OFDM, the Bussgang noise cancellation (BNC) receivers (also denoted as Bussgang receivers) were proposed [13, 96, 97]. The main goal of these receivers is to estimate and cancel this distortion at the receiver side. If these receivers could perfectly remove the nonlinear distortion from the received signal, the performance penalty would be only restricted to the degradation that comes from the fraction of the transmission power wasted in the nonlinear distortion component. However, this is an ideal situation since, in practice, it is very difficult to estimate the nonlinear distortion component. In fact, particularly at low SNR, the performance of BNC receivers is relatively poor due to error propagation issues, and can be even worse than the one associated with the conventional OFDM receivers based on a simple hard-decision process [98].

4.2 PRINCIPLE

In general, the task of a receiver is to output a signal from the transmission signals set, given the observed channel output. When referring to the optimum receiver, the optimization criterion is the probability of error. Thus, the goal of the optimum receiver is to maximize the probability of a correct decision or, equivalently, minimize

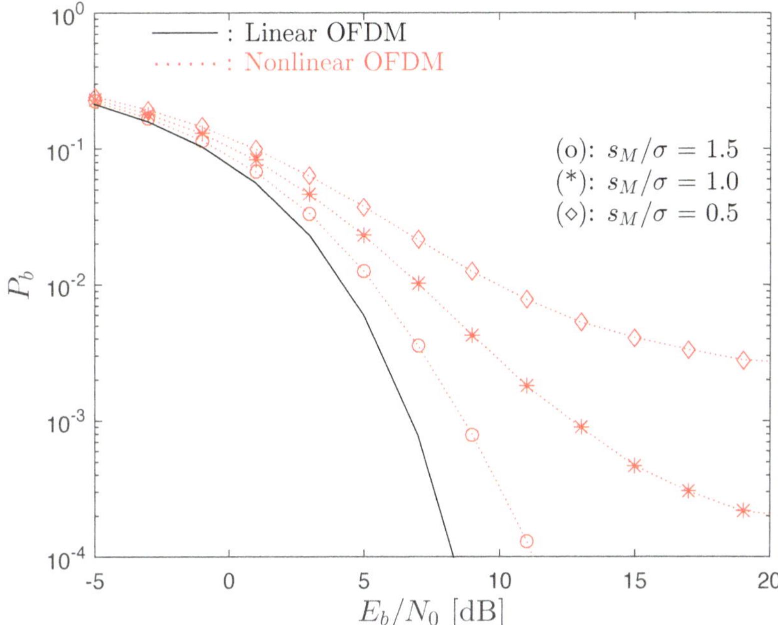

Figure 4.3 Simulated BER associated with a nonlinear OFDM transmission for different normalized clipping levels.

the error probability. In this subsection, the principle behind the optimum detection (also known as maximum likelihood (ML) detection) is presented.

After the demodulation process performed by the DFT, the received OFDM signal in the frequency-domain is represented by the block $\mathbf{R} = [R_0\ R_1\ R_2\ \ldots\ R_{N-1}]^T \in \mathbb{C}^N$ where, taking into account (2.77), we can write

$$\mathbf{R} = \mathbf{H}\mathbf{S} + \mathbf{N}. \tag{4.6}$$

For the sake of simplicity, we will start with an OFDM transmission over an ideal AWGN channel (the extension to other cases will be done later). Under these conditions, the received signal can be expressed as

$$\mathbf{R} = \mathbf{S} + \mathbf{N}. \tag{4.7}$$

Generally speaking, one can say that the probability of error is minimized if the a posteriori probability is maximized. In fact, given that $\mathbf{R}$ is observed, the detected signal $\tilde{\mathbf{S}} = [\tilde{S}_0\ \tilde{S}_1\ \tilde{S}_2\ \ldots\ \tilde{S}_{N-1}]^T \in \mathbb{C}^N$ is the one that maximizes the probability that the sequence $\mathbf{S}^{(m)} = [S_0^{(m)}\ S_1^{(m)}\ S_2^{(m)}\ \ldots\ S_{N-1}]^T \in \mathbb{C}^N$ was transmitted, with $\mathbf{S}^{(m)}$ denoting the mth possible transmitted message. Note that the superscript m ranges from $m = 1$ to the maximum number of sequences that can be generated in an OFDM system with N_u subcarriers employing M-QAM constellations, that is M^{N_u}. Put differently, the optimum receiver selects the signal $\tilde{\mathbf{S}}$ as the possible transmitted sequence $\mathbf{S}^{(m)}$ that results from the following maximization

$$\tilde{\mathbf{S}} = \arg\max_{m} \left(P\left(\mathbf{S}^{(m)} | \mathbf{R} \right) \right), \tag{4.8}$$

where $P\left(\mathbf{S}^{(m)}|\mathbf{R}\right)$ denotes the conditional probability that $\mathbf{S}^{(m)}$ was transmitted given that $\mathbf{R}$ is observed. This receiver is known as the maximum a posteriori (MAP) receiver. By taking advantage of the Bayes's rule, one can also write the conditional probability of observed $\mathbf{S}^{(m)}$ given $\mathbf{R}$ as [33]

$$P\left(\mathbf{S}^{(m)} | \mathbf{R} \right) = \frac{p\left(\mathbf{R} | \mathbf{S}^{(m)} \right) P\left(\mathbf{S}^{(m)} \right)}{p(\mathbf{R})}. \tag{4.9}$$

However, as $P(\mathbf{R})$ is independent of $\mathbf{S}$, the MAP decision rule can be rewritten as

$$\tilde{\mathbf{S}} = \arg\max_{m} \left(p\left(\mathbf{R} | \mathbf{S}^{(m)} \right) P\left(\mathbf{S}^{(m)} \right) \right). \tag{4.10}$$

Additionally, if the transmitted messages are equally probable, then $P\left(\mathbf{S}^{(m)}\right)$ is constant and does not influence (4.10). Under these conditions, we end up with the ML detection rule. The ML receiver makes its decision by solving the following maximization

$$\tilde{\mathbf{S}} = \arg\max_{m} \left(p\left(\mathbf{R} | \mathbf{S}^{(m)} \right) \right). \tag{4.11}$$

As mentioned before, each element of $\mathbf{N}$ has a complex Gaussian distribution, i.e., $N_k \sim \mathscr{CN}(0, 2\sigma_N^2)$, with σ_N^2 denoting the variance of the real and imaginary parts of each noise sample (see (2.76)). In fact, $\mathbf{N}$ can be seen as a complex, circularly symmetric Gaussian vector, whose corresponding PDF is given by

$$p(\mathbf{N}) = \frac{1}{\pi^N \det(\mathbf{C}_N)} \exp(-\mathbf{N}^H \mathbf{C}_N^{-1} \mathbf{N}), \tag{4.12}$$

where $\mathbf{C}_N$ is the covariance matrix associated with the vector $\mathbf{N}$ and $\det(\mathbf{C}_N)$ is its determinant. As $\mathbf{C}_N = 2\sigma_N^2 \mathbf{I}_N$, we have

$$p(\mathbf{N}) = \frac{1}{(2\pi\sigma_N^2)^N} \exp\left(-\frac{||\mathbf{N}||^2}{2\sigma_N^2} \right), \tag{4.13}$$

where $||\mathbf{N}||^2$ is the squared Euclidean norm of the vector $\mathbf{N}$. As the received signal $\mathbf{R}$ can be written as the sum of a deterministic and a random vector (see (4.7)), the conditional PDF of observed $\mathbf{R}$, given that $\mathbf{S}^{(m)}$ was transmitted, is

$$p\left(\mathbf{R} | \mathbf{S}^{(m)} \right) = \frac{1}{(2\pi\sigma_N^2)^N} \exp\left(-\frac{\left|\left| \mathbf{R} - \mathbf{S}^{(m)} \right|\right|^2}{2\sigma_N^2} \right). \tag{4.14}$$

Under these conditions, the ML rule presented in (4.11) can be written as

$$\tilde{\mathbf{S}} = \arg\max_{m} \left(\frac{1}{(2\pi\sigma_N^2)^N} \exp\left(-\frac{\left\| \mathbf{R} - \mathbf{S}^{(m)} \right\|^2}{2\sigma_N^2} \right) \right). \tag{4.15}$$

Clearly, the constant term $\frac{1}{(2\pi\sigma_N^2)^N}$ does not affect the above maximization and can be ignored. In addition, taking advantage of the monotonic characteristic of the logarithmic function, we can rewrite the ML decision rule as

$$\begin{aligned} \tilde{\mathbf{S}} &= \arg\max_{m} \left(\ln\left(\exp\left(-\frac{\left\| \mathbf{R} - \mathbf{S}^{(m)} \right\|^2}{2\sigma_N^2} \right) \right) \right) \\ &= \arg\max_{\mathbf{S}} \left(-\frac{\left\| \mathbf{R} - \mathbf{S}^{(m)} \right\|^2}{2\sigma_N^2} \right) \\ &= \arg\min_{\mathbf{S}} \left(\left\| \mathbf{R} - \mathbf{S}^{(m)} \right\|^2 \right). \end{aligned} \tag{4.16}$$

Note that the quantity $\left\| \mathbf{R} - \mathbf{S}^{(m)} \right\|^2$ represents the squared Euclidean norm between $\mathbf{R}$ and $\mathbf{S}^{(m)}$, i.e., the squared Euclidean distance between them. Therefore, this means that the ML receiver selects, among all the possible transmitted signals $\mathbf{S}^{(m)}$, the one that is closest to the received signal $\mathbf{R}$, i.e., the one that presents the minimum squared Euclidean distance relatively to the received signal $\mathbf{R}$. For this reason, the complexity associated with the optimum receiver can be very high, even with a low-to-moderate number of subcarriers and/or small constellations since, before taking a decision for a size-M constellation, it should perform M^{N_u} calculations of the squared Euclidean distance, in order to verify what is the sequence $\mathbf{S}^{(m)}$ closest to the received signal.

4.3 PERFORMANCE ANALYSIS

4.3.1 LINEAR MULTICARRIER SCHEMES

In this section, the performance associated with the optimum detection of linear multicarrier schemes in ideal AWGN channels is analyzed. In fact, although the specific scenario of a linear OFDM transmission is considered, the conclusions are valid for other multicarrier schemes such as DMT schemes.

Let us define $\mathscr{D}_m$ as the decision region associated with the data sequence $\mathbf{S}^{(m)}$, i.e., the region in which all received signals $\mathbf{R}$ are detected as $\mathbf{S}^{(m)}$. In addition, let us define $\mathscr{D}_m^c$ as the corresponding complementary region to $\mathscr{D}_m$. Under these conditions, the exact BER associated with the multicarrier scheme can be obtained as

$$P_b = \sum_m P\left(\mathbf{S}^{(m)}\right) P\left(\mathbf{R} \notin D_m | \mathbf{S}^{(m)}\right)$$
$$= \sum_m P\left(\mathbf{S}^{(m)}\right) P(\mathrm{e}|m), \tag{4.17}$$

where $P(\mathrm{e}|m)$ is the probability of having an error, given that the sequence $\mathbf{S}^{(m)}$ was transmitted. Indeed, given that $\mathbf{S}^{(m)}$ is transmitted, there is an error if the received signal $\mathbf{R}$ does not belong to the decision region of $\mathbf{S}^{(m)}$, i.e., does not lie in the decision region $\mathscr{D}_m$ or, equivalently, if the received signal belongs to the complementary decision region $\mathscr{D}_m^c$. Therefore, the probability of error, given that $\mathbf{S}^{(m)}$ is transmitted, can be obtained by the following multidimensional integral

$$P(\mathrm{e}|m) = \int_{\mathscr{D}_m^c} p\left(\mathbf{R}|\mathbf{S}^{(m)}\right) d\mathbf{R}. \tag{4.18}$$

Note that the event $\mathrm{e}|m$ is composed by the union of all error events, i.e.,

$$\mathrm{e}|m = \bigcup_{\substack{m' \\ m' \neq m}} \mathrm{e}^{(m')}|m, \tag{4.19}$$

where $\mathrm{e}^{(m')}|m$ represents the event of detecting $\mathbf{S}^{(m')}$ given that $\mathbf{S}^{(m)}$ was transmitted. As these error events are mutually exclusive, we can rewrite (4.19) as

$$\mathrm{e}|m = \sum_{\substack{m' \\ m' \neq m}} \mathrm{e}^{(m')}|m. \tag{4.20}$$

Under these conditions, we can redefine $P(\mathrm{e}|m)$ as

$$P(\mathrm{e}|m) = \sum_{\substack{m' \\ m' \neq m}} \int_{\mathscr{D}_{m'}} p\left(\mathbf{R}|\mathbf{S}^{(m)}\right) d\mathbf{R}. \tag{4.21}$$

By replacing the above equation in (4.17), the optimum receiver's BER can be rewritten as

$$P_b = \sum_m P\left(\mathbf{S}^{(m)}\right) \sum_{\substack{m' \\ m' \neq m}} \int_{\mathscr{D}_{m'}} p\left(\mathbf{R}|\mathbf{S}^{(m)}\right) d\mathbf{R}. \tag{4.22}$$

Although this expression gives the optimum receiver's exact BER, it is very complex due to the multidimensional integrals and the large number of possible transmitted sequences that should be analyzed. Instead of computing (4.22), the optimum performance is commonly obtained by considering some approximations. In fact, one can define both upper and lower performance bounds that are associated with "pessimistic" and "optimistic" BERs, respectively. Under some conditions, these bounds might provide an accurate insight on the truth BER of the optimum receiver, while, at the same time, avoid a very large complexity.

In the following, we present an approximation of the exact optimum receiver's BER that involves a combination of a lower and an upper bound. The lower bound is related to the assumption that an error event is only associated with single-bit errors, i.e., all errors occur when there is one and only one-bit error. Clearly, this is an optimistic assumption since we can have, although with lower probability, more than one bit error in a given block. It should be mentioned that although this is an optimistic approximation, it is quite tight, specially, in the asymptotic region, i.e., for large E_b/N_0 values. The upper bound, on the other hand, constitutes a pessimistic approximation. It is based on the so-called pairwise union bound. The idea behind this approximation is to upper bound an error event by the union of all the pairwise error probability (PEP) of two signals that belong to the transmission set. In this bound, instead of integrating the distribution $p\left(\mathbf{R}|\mathbf{S}^{(m)}\right)$ over the decision region $\mathscr{D}_{m'}$ (see (4.22)), a larger decision region $\mathscr{D}_{mm'} = \left\{p\left(\mathbf{R}|\mathbf{S}^{(m')}\right) > p\left(\mathbf{R}|\mathbf{S}^{(m)}\right)\right\}$ is considered. Note that, in this larger region, it is more likely that $\mathbf{S}^{(m')}$ was transmitted. Under these conditions, we have

$$P_b = \sum_m P\left(\mathbf{S}^{(m)}\right) \sum_{\substack{m' \\ m' \neq m}} \int_{\mathscr{D}_{m'}} p\left(\mathbf{R}|\mathbf{S}^{(m)}\right) d\mathbf{R} \leq \sum_m P\left(\mathbf{S}^{(m)}\right) \cdot \sum_{\substack{m' \\ m' \neq m}} \int_{\mathscr{D}_{mm'}} p\left(\mathbf{R}|\mathbf{S}^{(m)}\right) d\mathbf{R}, \tag{4.23}$$

where

$$P\left(S_{m \to m'}\right) = \int_{\mathscr{D}_{mm'}} p\left(\mathbf{R}|\mathbf{S}^{(m)}\right) d\mathbf{R}, \tag{4.24}$$

is the PEP between m and m', which represents the probability of estimating $\mathbf{S}^{(m')}$ given that $\mathbf{S}^{(m)}$ was transmitted. The union upper bound based on the PEPs can hence be written as

$$P_b \leq \sum_m P\left(\mathbf{S}^{(m)}\right) \sum_{\substack{m' \\ m' \neq m}} P\left(S_{m \to m'}\right). \tag{4.25}$$

Note that, by assuming that the transmitted sequences are equally probable, i.e., $P\left(\mathbf{S}^{(m)}\right) = M^{-N_u}$, we have

$$P_b \leq \sum_m M^{-N_u} \sum_{\substack{m' \\ m' \neq m}} P\left(S_{m \to m'}\right). \tag{4.26}$$

Note also that the event $S_{m \to m'}$ happens when $\mathbf{S}^{(m')}$ differs at least one bit from $\mathbf{S}^{(m)}$. Let us denote $\Phi(\mu, m)$ as the set of all sequences $\mathbf{S}^{(m')}$ that differ from $\mathbf{S}^{(m)}$ in μ bits and $P\left(S_{m \xrightarrow{\mu} m'}\right)$ as the probability that the detected sequence $\mathbf{S}^{(m')}$ differs from $\mathbf{S}^{(m)}$ in μ bits. Under these conditions, we can write

$$P_b \leq \sum_{m=1}^{M^{N_u}} M^{-N_u} \sum_{\mu=1}^{\log_2(M)N_u} \frac{\mu}{\log_2(M)N_u} \sum_{\Phi(\mu,m)} P\left(S_{m \xrightarrow{\mu} m'}\right). \tag{4.27}$$

Moreover, it can be shown that, in ideal AWGN channels, the PEP associated with sequences that differ in μ bits is related to the squared Euclidean distance between them [33]. In the following, we denote this squared Euclidean distance as $D^{2\,(l)}_{m,m'}(\mu)$ (the superscript (l) is related to the linear characteristic of the transmission). Therefore, the PEP associated with sequences that differ in μ bits can be written as

$$P\left(S_{m \xrightarrow{\mu} m'}\right) = Q\left(\sqrt{\frac{D^{2\,(l)}_{m,m'}(\mu)}{2N_0}}\right), \tag{4.28}$$

where $Q(\cdot)$ represents the tail probability of the standard normal distribution, defined in (2.83), and

$$\begin{aligned} D^{2\,(l)}_{m,m'}(\mu) &= \left|\left|\mathbf{S}^{(m)} - \mathbf{S}^{(m')}\right|\right|^2 \\ &= \sum_{k=0}^{N-1} \left|S_k^{(m)} - S_k^{(m')}\right|^2. \end{aligned} \tag{4.29}$$

Under these conditions, we have

$$P_b \leq \sum_{m=1}^{M^{N_u}} M^{-N_u} \sum_{\mu=1}^{\log_2(M)N_u} \frac{\mu}{\log_2(M)N_u} \sum_{\Phi(\mu,m)} Q\left(\sqrt{\frac{D^{2\,(l)}_{m,m'}(\mu)}{2N_0}}\right). \tag{4.30}$$

As referred before, we are only considering single-bit errors, which is an assumption that lead us to a lower bound of the optimum receiver's performance. In fact, as the Euclidean distance between signals that differ in more than one bit is much higher than the Euclidean distance associated with sequences that differ in one bit, we can take advantage of the decreasing characteristic of the $Q(\cdot)$ function. Therefore, $Q(x) \ll Q(y)$ for $x > y$, which means that $Q(x) + Q(y) \approx Q(\min(x,y))$. Under these conditions, the optimum receiver's BER bound of (4.30) can be approximated by considering only sequences that have $\mu = 1$ bit errors, resulting

$$\begin{aligned} P_b &\approx \sum_{m=1}^{M^{N_u}} \frac{M^{-N_u}}{\log_2(M)N_u} \sum_{\Phi(1,m)} Q\left(\sqrt{\frac{D^{2\,(l)}_{m,m'}(1)}{2N_0}}\right) \\ &\approx \sum_{m=1}^{M^{N_u}} \frac{M^{-N_u}}{\log_2(M)N_u} \sum_{\Phi(1,m)} Q\left(\sqrt{\frac{\left|\left|\mathbf{S}^{(m)} - \mathbf{S}^{(m')}\right|\right|^2}{2N_0}}\right). \end{aligned} \tag{4.31}$$

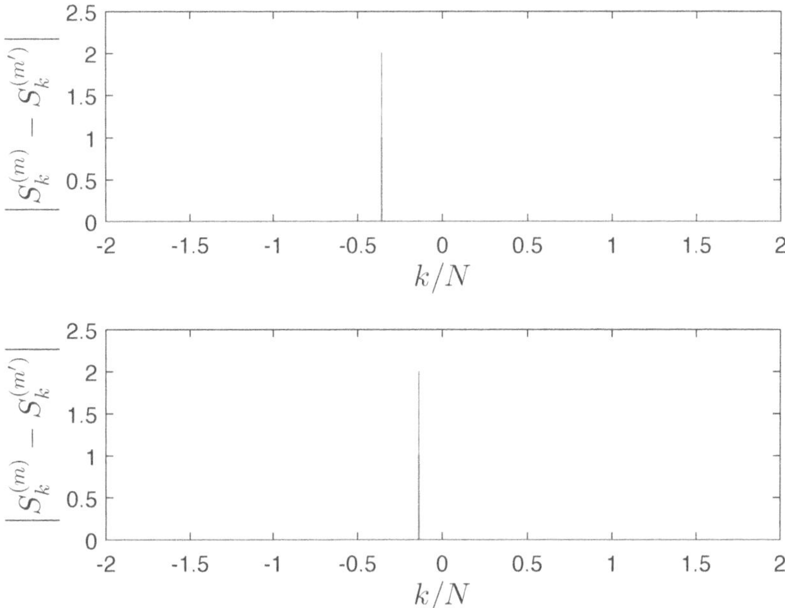

Figure 4.4 Difference between two OFDM sequences that differ in $\mu = 1$ bits.

Clearly, in the asymptotic region, the sequences that differ in one bit are at the minimum Euclidean distance. Note that the squared minimum Euclidean distance D^2_{min} is constant, regardless of the position (the subcarrier's index) of the bit that is in error and regardless of the data sequences $\mathbf{S}^{(m)}$ and $\mathbf{S}^{(m')}$, i.e.,

$$D^{2\ (l)}_{m,m'}(1) = D^{2\ (l)}(1) = D^2_{\text{min}}\ \forall\ m, m'. \tag{4.32}$$

This can be confirmed in Fig. 4.4, which shows the difference between two OFDM sequences generated randomly and their variations of $\mu = 1$ bit. The error positions were also randomly generated. From the figure, it can be noted that the value of the squared Euclidean distance (see (4.29)) is the same in both situations, which means that the squared Euclidean distance is only dependent on the number of bit errors, μ. Additionally, in an OFDM scheme employing M-QAM constellations, the average number of symbols that are at the minimum Euclidean distance relatively to a given sequence $\mathbf{S}^{(m)}$ is $\frac{4\sqrt{M}-4}{\sqrt{M}}$ (note that, for QPSK constellations, this expression yields the exact number of symbols that are at the minimum Euclidean distance). Therefore, the size of the set $\Phi(1,m)$ is $N_u\left(\frac{4\sqrt{M}-4}{\sqrt{M}}\right)$, which makes possible to write

$$
\begin{aligned}
P_b &\approx \sum_{m=1}^{M^{N_u}} \frac{M^{-N_u}}{\log_2(M)N_u} N_u \left(\frac{4\sqrt{M}-4}{\sqrt{M}}\right) Q\left(\sqrt{\frac{D_{\min}^2}{2N_0}}\right) \\
&= \frac{4\sqrt{M}-4}{\log_2(M)\sqrt{M}} Q\left(\sqrt{\frac{D_{\min}^2}{2N_0}}\right),
\end{aligned} \tag{4.33}
$$

since $D_{\min}^2$ is constant (see (4.32)). For an M-QAM constellation, the minimum squared Euclidean distance is given by

$$
\begin{aligned}
D_{\min}^2 &= \frac{6E_s}{M-1} \\
&= \frac{6\log_2(M)E_b}{M-1},
\end{aligned} \tag{4.34}
$$

where $E_s = \log_2(M)E_b$ denotes the average symbol energy. Note that the average bit energy E_b is defined as in (2.84), i.e., it does not depend on the transmitted signal $\mathbf{S}^{(m)}$. Under these conditions, we have

$$
P_b \approx \left(\frac{4\sqrt{M}-4}{\log_2(M)\sqrt{M}}\right) Q\left(\sqrt{\frac{3\log_2(M)E_b}{(M-1)N_0}}\right). \tag{4.35}
$$

For the specific case of QPSK constellations, i.e., $M=4$, (4.34) yields

$$
D_{\min}^2 = 4E_b. \tag{4.36}
$$

Therefore, the optimum performance is given by the well-known expression

$$
P_b = Q\left(\sqrt{\frac{2E_b}{N_0}}\right). \tag{4.37}
$$

Note that the above equation gives the exact BER although, in general, for an M-QAM constellation, (4.35) gives an approximation. This can be explained by the compound effect of the upper and lower bounds that are considered for obtaining (4.35). In fact this compound effect is null for QPSK constellations. Note also that, as $P_s = \log_2(M)P_b$, the symbol error rate (SER) associated with the optimum detection of a linear OFDM system employing QPSK constellations in an ideal AWGN channel is

$$
P_s = 2Q\left(\sqrt{\frac{2E_b}{N_0}}\right). \tag{4.38}
$$

It is worth to mention that the BER represented in (4.37) is identical to the BER associated with the conventional OFDM detection that was presented in Chapter 2 (see (2.82) and (2.85)). This means that the conventional detection of OFDM, made under a subcarrier-by-subcarrier basis, allows to obtain the optimum detection performance. Put differently, the optimum detection reduces to the conventional detection, which is a consequence of the orthogonality of the subcarriers. However, as will be demonstrated in the following, the optimum detection provides substantial potential gains when employed in nonlinear multicarrier schemes.

4.3.2 NONLINEAR MULTICARRIER SCHEMES

In the following, the asymptotic optimum performance for the specific case of OFDM schemes that have a clipping operation on their transmission chain is analyzed by a set of simulation results. However, as will be seen later in this document, the main conclusions presented here are valid for other multicarrier schemes (such as DMT schemes) impaired by either baseband or bandpass nonlinearities. We consider both ideal AWGN and frequency-selective channels.

Ideal AWGN Channels

When a given OFDM signal $\mathbf{S} = [S_0\ S_1\ S_2\ \ldots\ S_{N-1}]^T \in \mathbb{C}^N$ is submitted to an envelope clipping device (see (4.5) and Fig. 4.1), the transmitted signal $\mathbf{Y} = [Y_0\ Y_1\ Y_2\ \ldots\ Y_{N-1}]^T \in \mathbb{C}^N$ can be written as

$$\mathbf{Y} = \alpha_{bp}\mathbf{S} + \mathbf{D}. \tag{4.39}$$

Therefore, the received signal is

$$\begin{aligned}\mathbf{R} &= \mathbf{Y} + \mathbf{N}\\ &= \alpha_{bp}\mathbf{S} + \mathbf{D} + \mathbf{N}.\end{aligned} \tag{4.40}$$

Let us consider two data sequences $\mathbf{S}^{(m)} = [S_0^{(m)}\ S_1^{(m)}\ \ldots\ S_{N-1}^{(m)}]^T \in \mathbb{C}^N$ and $\mathbf{S}^{(m')} = [S_0^{(m')}\ S_1^{(m')}\ \ldots\ S_{N-1}^{(m')}]^T \in \mathbb{C}^N$ that differ in μ bits. As aforementioned, independently of m and m', the squared Euclidean distance between these two signals in AWGN channels $D^{2\ (l)}_{m,m'}(\mu)$ only depends on the number of μ bit differences between them, regardless of their position. However, the situation is different when the squared Euclidean distance between their corresponding nonlinearly distorted versions $\mathbf{Y}^{(m)} = [Y_0^{(m)}\ Y_1^{(m)}\ \ldots\ Y_{N-1}^{(m)}]^T \in \mathbb{C}^N$ and $\mathbf{Y}^{(m')} = [Y_0^{(m')}\ Y_1^{(m')}\ \ldots\ Y_{N-1}^{(m')}]^T \in \mathbb{C}^N$ is taken into account. This is explained by the fact that for a given sequence $\mathbf{S}^{(m)}$, the corresponding nonlinear distortion term $\mathbf{D}^{(m)} = [D_0^{(m)}\ D_1^{(m)}\ D_2^{(m)}\ \ldots\ D_{N-1}^{(m)}]^T \in \mathbb{C}^N$ depends on all elements of $\mathbf{S}^{(m)}$. In fact, when there is a bit modification on $\mathbf{S}^{(m)}$, the entire block of nonlinear distortion terms at the nonlinearity output change. This means that the squared Euclidean distance depends, therefore, on the information spread along all subcarriers. In addition, as the nonlinearity introduces out-of-band radiation, the difference between two nonlinearly distorted sequences spread over the entire band of the signal, i.e., is nonzero for the $N = ON_u$ subcarriers. This can be confirmed in Fig. 4.5, which shows the difference between two different nonlinearly distorted OFDM sequences $\mathbf{Y}^{(m')}$ and $\mathbf{Y}^{(m')}$. In both cases, the sequences differ

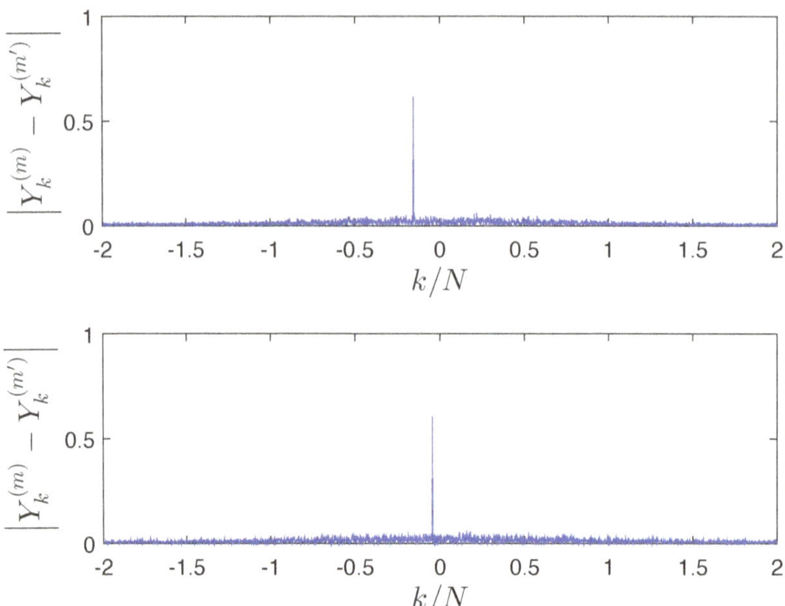

Figure 4.5 Difference between two nonlinearly distorted OFDM sequences that differ in $\mu = 1$ bits.

in $\mu = 1$ bits. The positions of the errors were randomly generated. From the results depicted in the figure, it can be seen that the difference between the nonlinearly distorted signals has components in the entire block, i.e., on the N subcarriers. Additionally, it can also be noted that, differently from the linear transmission case (see Fig. 4.4), the squared Euclidean distance $D^{2\,(nl)}_{m,m'}(\mu)$ (the superscript (nl) is related to the nonlinear nature of the transmission) depends on all energy spread over the signal band, i.e.,

$$\begin{aligned} D^{2\,(nl)}_{m,m'}(\mu) &= \left|\left|\mathbf{Y}^{(m')} - \mathbf{Y}^{(m)}\right|\right|^2 \\ &= \sum_{k=0}^{N-1} \left|Y_k^{(m')} - Y_k^{(m)}\right|^2 \\ &= \sum_{k=0}^{N-1} \left|\alpha_{bp}\left(S_k^{(m')} - S_k^{(m)}\right) + D_k^{(m')} - D_k^{(m)}\right|^2. \end{aligned} \tag{4.41}$$

This means that the BER associated with the optimum detection of nonlinearly distorted OFDM signals can be upper-bounded as

$$P_b \leq \sum_{m=1}^{M^{N_u}} M^{-N_u} \sum_{\mu=1}^{\log_2(M)N_u} \frac{\mu}{\log_2(M)N_u} \sum_{\Phi(\mu,m)} P\left(Y_{m \xrightarrow{\mu} m'}\right), \tag{4.42}$$

where $P\left(Y_{m \xrightarrow{\mu} m'}\right)$ represents the probability of detecting $\mathbf{S}^{(m')}$ given that $\mathbf{S}^{(m)}$ (and correspondingly $\mathbf{Y}^{(m)}$) was transmitted. As this probability can be computed as

$$P\left(Y_{m \xrightarrow{\mu} m'}\right) = Q\left(\sqrt{\frac{D^{2\,(nl)}_{m,m'}(\mu)}{2N_0}}\right), \tag{4.43}$$

we can write

$$P_b \leq \sum_{m=1}^{M^{N_u}} M^{-N_u} \sum_{\mu=1}^{\log_2(M)N_u} \frac{\mu}{\log_2(M)N_u} \sum_{\Phi(\mu,m)} Q\left(\sqrt{\frac{D^{2\,(nl)}_{m,m'}(\mu)}{2N_0}}\right). \tag{4.44}$$

By considering only single-bit errors (as previously seen, an approximation that is valid in the asymptotic region, where the SNR is high and the BER is dominated by the error events associated with 1-bit variations), the approximate BER can be computed as

$$P_b \approx \sum_{m=1}^{M^{N_u}} \frac{M^{-N_u}}{\log_2(M)N_u} \sum_{\Phi(1,m)} Q\left(\sqrt{\frac{D^{2\,(nl)}_{m,m'}(1)}{2N_0}}\right). \tag{4.45}$$

Without loss of generality, let us focus on the case where a QPSK symbol is transmitted on each subcarrier, i.e., the case with $M = 4$. Under these conditions, we can rewrite $D^{2\,(nl)}_{m,m'}(1)$ as

$$D^{2\,(nl)}_{m,m'}(1) = 4G_{m,m'}(1)E^{(nl)}_{b,m}, \tag{4.46}$$

where $G_{m,m'}(1)$ is defined as the asymptotic gain relatively to the linear transmission case when the sequences differ in 1 bit (in general, for sequences differing in μ bits, we will denote this gain as $G_{m,m'}(\mu)$) and $E^{(nl)}_{b,m}$ can be defined in the frequency-domain as

$$E^{(nl)}_{b,m} = \frac{1}{2N_u} \sum_{k=0}^{N-1} \left|Y^{(m)}_k\right|^2. \tag{4.47}$$

In the remaining of this analysis, the average $E^{(nl)}_{b,m}$ is considered, i.e., we consider

$$E^{(nl)}_b = \mathbb{E}_{\mathbf{S}^{(m)}}\left[E^{(nl)}_{b,m}\right], \tag{4.48}$$

i.e., the average $E^{(nl)}_{b,m}$ obtained over different realizations of data symbols $\mathbf{S}^{(m)}$. It should be mentioned that if $G_{m,m'}(1) > 1.0$, then there is a gain relatively to linear OFDM transmissions, where $D^{2\,(l)}_{m,m'}(1) = 4E_b$ and $G_{m,m'}(1) = 1.0$. In the following, it is shown that, although the squared Euclidean distance between sequences that differ in $\mu = 1$ bits, $D^{2\,(nl)}_{m,m'}(1)$, has some fluctuations, we have almost always $G_{m,m'}(1) > 1.0$. This suggests that the optimum detection of nonlinear distorted

OFDM schemes can be even better than the performance of conventional, linear OFDM schemes, at least, in the asymptotic region. However, it is not easy to obtain the approximate optimum receiver's BER represented in (4.45), since it involves the computation of $4^{N_u} \times 2N_u$ Euclidean distances, due to the fact that each one of the 4^{N_u} possible transmitted sequences has $2N_u$ possible variations of $\mu = 1$ bit, which clearly involves a very high complexity, even considering constellations with only $M = 4$ points and/or a small number of subcarriers. For this reason, in order to have an insight on the optimum performance of nonlinearly distorted OFDM signals, we present a method to obtain an approximation of (4.45). This method is based on the histogram of the possible values of $D^{2\,(nl)}_{m,m'}(1)$. Indeed, we can take advantage of the fact that the $4^{N_u} \times 2N_u$ values of squared Euclidean distances may have relatively low fluctuations and, therefore, it may be possible to avoid the computation of all possible values of $D^{2\,(nl)}_{m,m'}(1)$ and still obtain a good approximation of (4.45).

The first step to obtain the histogram of the Euclidean distances is to randomly generate $N_{\text{seq}} \ll 4^{N_u}$ sequences and compute the squared Euclidean distance between them and their $2N_u$ variations of $\mu = 1$ bit[2]. These $N_{\text{seq}}2N_u$ values are divided into a set of intervals where $D_i^{2\,(nl)}(1)$ represents the ith possible value of the squared Euclidean distance. The absolute frequency associated with $D_i^{2\,(nl)}(1)$ is $f_{\text{abs}}\left(D_i^{2\,(nl)}(1)\right)$ and the corresponding relative frequency is $\frac{f_{\text{abs}}\left(D_i^{2\,(nl)}(1)\right)}{N_{\text{seq}}2N_u}$. Given the histogram of the squared Euclidean distances, the approximate BER can be obtained as

$$
\begin{aligned}
P_b &\approx f_{\text{rel}}\left(D_i^{2\,(nl)}(1)\right) Q\left(\sqrt{\frac{D_i^{2\,(nl)}(1)}{2N_0}}\right) \\
&= \frac{f_{\text{abs}}\left(D_i^{2\,(nl)}(1)\right)}{N_{\text{seq}}2N_u} Q\left(\sqrt{\frac{D_i^{2\,(nl)}(1)}{2N_0}}\right).
\end{aligned}
\tag{4.49}
$$

In fact, we can see the squared Euclidean distance as a "random variable", for which we may attach a given distribution computed from the histogram described above. Fig. 4.6 shows the distribution of the squared Euclidean distances between two randomly generated, nonlinearly distorted OFDM sequences that differ in $\mu = 1$ bits. Each OFDM sequence has $O = 4$ and different values of N_u are considered. The normalized clipping level is $s_M/\sigma = 0.5$. From the results depicted in the figure, it can be noted that the distribution of the squared Euclidean distances tends to be Gaussian. In addition, its variance tends to be smaller as the number of subcarriers increases. This effect is clearly illustrated in Fig. 4.7, which shows the average value associated with distribution of the squared Euclidean distances $D^{2\,(nl)}(1)$ (Fig. 4.7.A), as well as its variance (Fig. 4.7.B), considering different values of N_u and $O = 4$. From the

[2]Note that, as the sequences are generated randomly, the variations of the same sequence can be analyzed more than one time. However, as N_u is typically large, this "redundancy" is very unlikely.

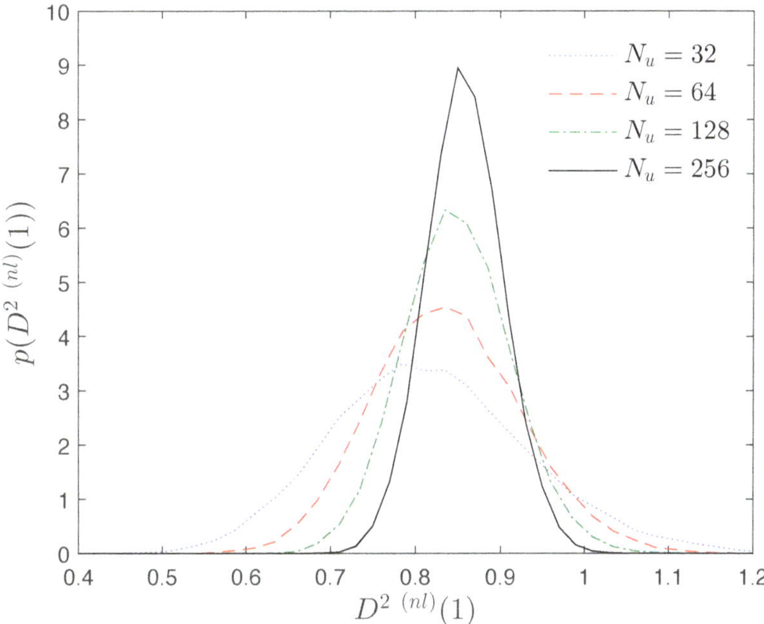

Figure 4.6 Distribution of the squared Euclidean distances between two nonlinearly distorted OFDM sequences that differ in $\mu = 1$ bits.

figure it can be seen that, although the average value of the squared Euclidean distance increases with N_u, it tends to stabilize for large values of N_u (say, for instance, $N_u \geq 512$). Additionally, it can also be seen that the variance tends to zero as N_u increases. This suggests that for very large values of N_u, the squared Euclidean distances tend to be equal to the average value $\mathbb{E}\left[D^{2\,(nl)}(1)\right]$[3]. Under these conditions, the optimum performance may be approximated as

$$P_b \approx Q\left(\sqrt{\frac{D^{2\,(nl)}(1)/2}{N_0}}\right). \tag{4.50}$$

In fact, this approximate BER can be directly compared to the one associated with linear OFDM transmissions (see (4.37)). In addition, by noting that

$$D^{2\,(nl)}(1) = 4G(1)E_b^{(nl)}, \tag{4.51}$$

[3]In the following, we will consider the asymptotic case (where the number of subcarriers N_u is very large). For this reason, we denote $\mathbb{E}\left[D^{2\,(nl)}(1)\right]$ as simply $D^{2\,(nl)}(1)$.

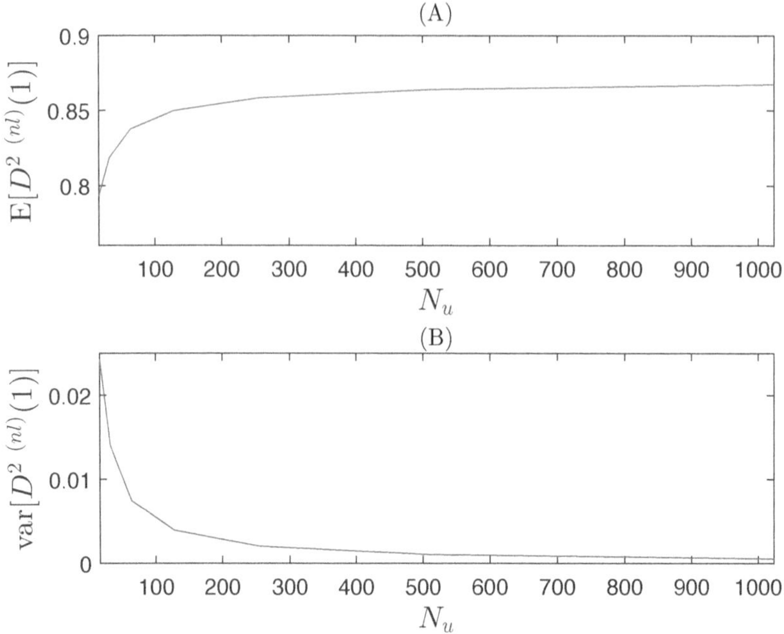

Figure 4.7 Evolution of the average value associated with $p(D^{2\,(nl)}(1))$ (A) as well as its variance (B) considering different values of N_u and $O = 4$.

where

$$G(1) = \frac{D^{2\,(nl)}(1)}{4E_b^{(nl)}}, \tag{4.52}$$

is defined as the average asymptotic gain associated with the optimum detection of nonlinearly distorted OFDM signals, it is clear the existence of an asymptotic gain relatively to the conventional, linear OFDM schemes. Fig. 4.8 shows the asymptotic gain's distribution, $p(G(1))$, obtained through the histogram of the squared Euclidean distances associated with nonlinearly distorted OFDM signals that differ in $\mu = 1$ bits. Those distances are computed between OFDM signals with N_u subcarriers and $O = 4$. The normalized clipping level is $s_M/\sigma = 0.5$. From the figure it can be verified that $G(1) > 1.0$ (this means that, at least for the generated sequences and their variations of 1 bit, we never obtained $G(1) < 0$), which confirms the existence of potential asymptotic gains. Once again, it can be noted that for large values of N_u, the variance of $D^{2\,(nl)}(1)$ (and consequently of $G(1)$) decreases. Under these conditions, one can rewrite the approximate BER of (4.50) as

$$P_b \approx Q\left(\sqrt{2\frac{G(1)E_b^{(nl)}}{N_0}}\right). \tag{4.53}$$

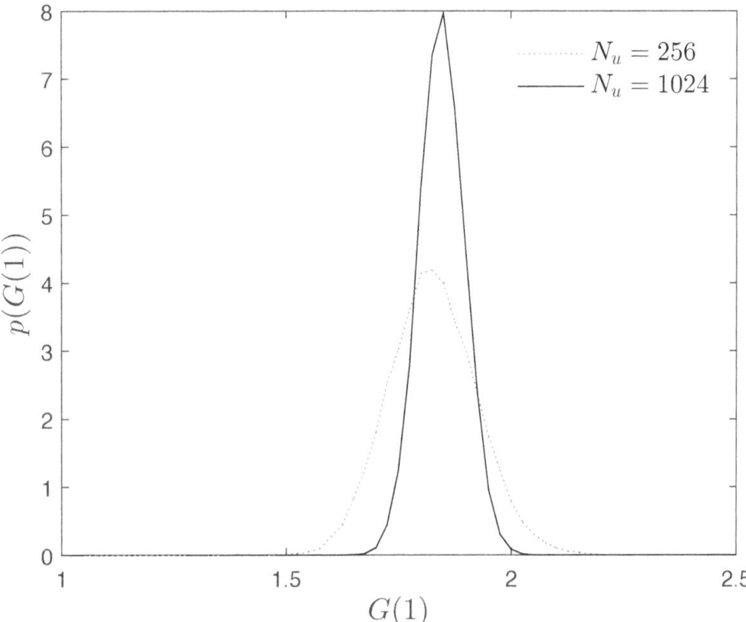

Figure 4.8 Asymptotic gain distribution for nonlinearly distorted OFDM signals that differ in $\mu = 1$ bit.

It should be mentioned, however, that the clipping is commonly followed by a frequency-domain filtering operation to remove the out-of-band radiation and reduce the ACI levels [53,54]. However, in Fig. 4.9, it can be observed that even when part of the nonlinear distortion is removed, there are also potential asymptotic gains, since we almost always have $G(1) > 0$ dB. It should also be noted that the average value of the gain is dependent on the nonlinearity, since the magnitude of the distortion term, that influences the Euclidean distances, depends on the severeness of the nonlinearity. This is illustrated in Fig. 4.10, which shows the distribution of $p(G(1))$ considering $N_u = 1024$ and different normalized clipping levels s_M/σ. As expected, the asymptotic gain decreases as s_M/σ increases. For $s_M/\sigma = +\infty$ (i.e., when the transmission is linear), we cannot observe potential performance gains, since the squared Euclidean distance between two sequences is always $4E_b$ (see (4.32)). On the other hand, as the magnitude of the nonlinear distortion effects increases, the asymptotic gain also increases. More concretely, the average value of the asymptotic gain is approximately 1.05 dB, 1.15 dB, 1.35 dB and 1.85 dB for $s_M/\sigma = 2.0$, 1.5, 1.0 and 0.5, respectively.

Fig. 4.11 shows the BER obtained with (4.49) considering nonlinearly distorted OFDM signals submitted to an envelope clipping operation with different normalized clipping levels s_M/σ, as well as the theoretical BER associated with linear OFDM

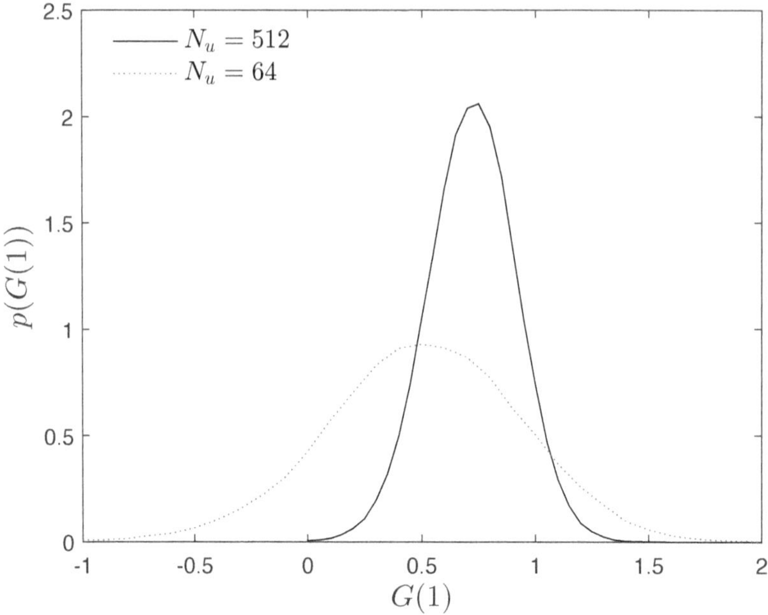

Figure 4.9 Asymptotic gain distribution obtained through the histogram of the squared Euclidean distances considering clipped and filtered OFDM signals that differ in $\mu = 1$ bit.

transmissions (see (4.37)). As expected, the asymptotic gains shown in Fig. 4.10 provide considerable BER improvements. As these gains decrease with the magnitude of the nonlinear distortion effects, the BER obtained from the distribution of the gains for $s_M/\sigma = +\infty$ is equal to the BER associated with linear transmissions. On the other hand, for a target BER of $P_b = 10^{-3}$, the potential asymptotic gain associated with the optimum detection is approximately 2.7 dB, 1.3 dB and 0.6 dB for $s_M/\sigma = 0.5$, $s_M/\sigma = 1.0$ and $s_M/\sigma = 1.0$, respectively.

In order to clearly access the potentialities of the optimum detection, let us look at Fig. 4.12. This figure shows the BER associated with linear transmissions and nonlinear OFDM transmissions, when a normalized clipping level $s_M/\sigma = 1.0$ is considered. For the case of nonlinear transmissions, the figure presents the BER associated with both conventional and ideal BNC receivers. Additionally, it includes the approximate optimum receiver's performance. Clearly, it can be seen that the optimum receiver dealing with nonlinearly distorted signals outperforms all the others. For a target BER of $P_b = 10^{-3}$, the conventional receiver presents a degradation of 6 dB relatively to the linear transmission case. In fact, even if we were able to perfectly estimate and eliminate the nonlinear distortion from the received signal (which is the case of the ideal BNC receivers), we end up with a degradation of about 0.5 dB relatively to the linear transmission scenario. In contrast, regarding the optimum

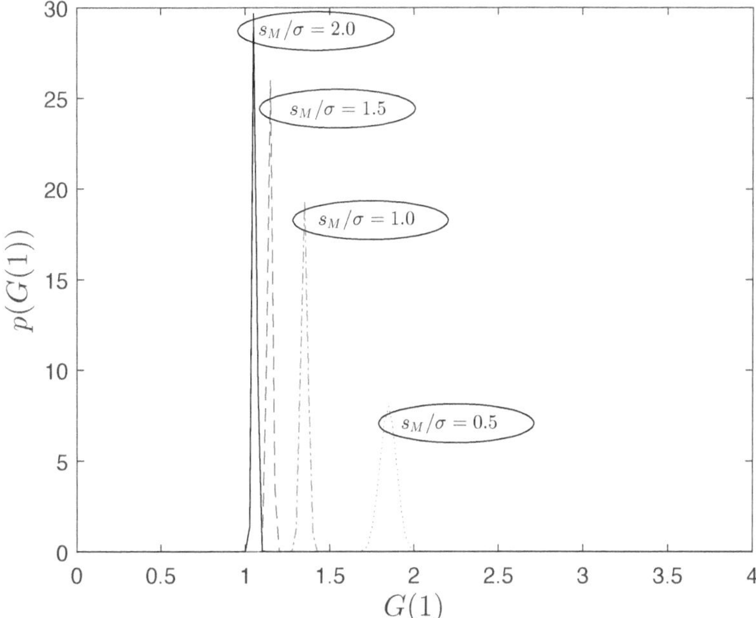

Figure 4.10 Asymptotic gain distribution obtained through the histogram of the squared Euclidean distances between two nonlinearly distorted OFDM signals.

receiver, it can be seen that it not only does not present a degradation relatively to the linear OFDM, but it also has considerable performance gains. For instance, for a target BER of $P_b = 10^{-3}$ or $P_b = 10^{-4}$, the performance gain is around 1.3 dB. This means that the optimum detection of nonlinearly distorted OFDM signals may be even better than the conventional detection of linear, OFDM signals.

Frequency-Selective Channels

Let us know consider a more realistic scenario. Here, we consider a nonlinear OFDM scheme where the channel is assumed to be frequency-selective and composed by L equally-spaced, uncorrelated multipath components. The equivalent, subcarrier-level model for that scenario is depicted in Fig. 4.13. Note that when frequency-selective channels are considered, the transmitted signal associated with the OFDM symbol $\mathbf{S}^{(m)} = [S_0^{(m)}\ S_1^{(m)}\ \ldots\ S_{N-1}^{(m)}]^T \in \mathbb{C}^N$ can be written as

$$\mathbf{B}^{(m)} = \mathbf{H}\mathbf{Y}^{(m)}, \tag{4.54}$$

where $\mathbf{H}$ is the channel matrix associated with a given channel realization (see (2.75)). Regarding the kth subcarrier, we have

$$B_k^{(m)} = H_k\left(\alpha S_k^{(m)} + D_k^{(m)}\right). \tag{4.55}$$

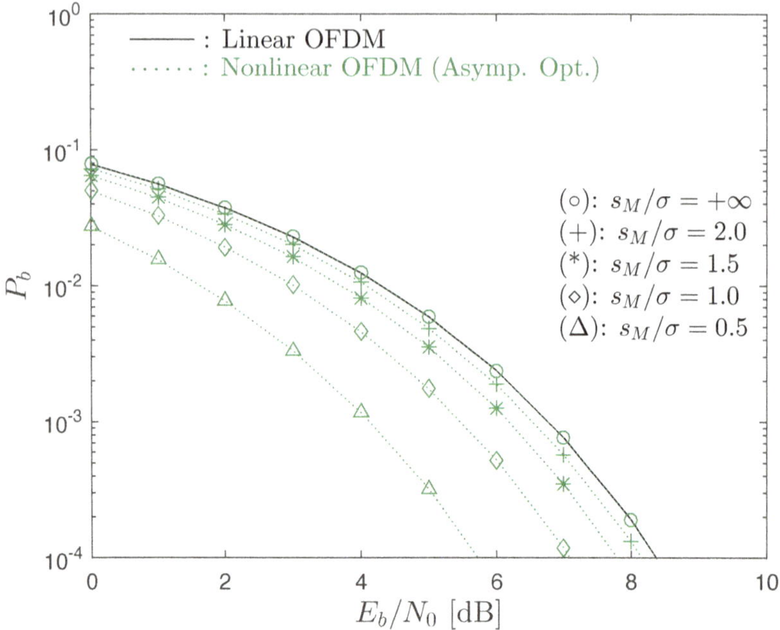

Figure 4.11 Approximate optimum receivers' BER for nonlinearly distorted signals submitted to an envelope clipping operation with normalized clipping level s_M/σ.

Therefore, the squared Euclidean distance between two nonlinearly distorted multicarrier signals $\mathbf{Y}^{(m)} = [Y_0^{(m)}\ Y_1^{(m)}\ \ldots\ Y_{N-1}^{(m)}]^T \in \mathbb{C}^N$ and $\mathbf{Y}^{(m')} = [Y_0^{(m')}\ Y_1^{(m')}\ \ldots\ Y_{N-1}^{(m')}]^T \in \mathbb{C}^N$ that differ in μ bits can be computed as

$$\begin{aligned} D_{m,m'}^{2\ (H,nl)}(\mu) &= \left|\left|\mathbf{B}^{(m')} - \mathbf{B}^{(m)}\right|\right|^2 \\ &= \left|\left|\mathbf{H}\left(\mathbf{Y}^{(m')} - \mathbf{Y}^{(m)}\right)\right|\right|^2 \\ &= \sum_{k=0}^{N-1} |H_k|^2 \left|\alpha\left(S_k^{(m')} - S_k^{(m)}\right) + D_k^{(m')} - D_k^{(m)}\right|^2. \end{aligned} \tag{4.56}$$

Clearly, the squared Euclidean distance is conditioned by the squared magnitudes of the channel frequency responses $|H_k|^2$. Due to their random nature ($|H_k|^2$ is distributed according to (2.86)), we can consider that the asymptotic gain for frequency-selective channels is a random quantity given by

$$G^{(H)}(\mu) = \frac{D^{2\ (H,nl)}(\mu)}{4E_b^{(H,nl)}}. \tag{4.57}$$

Note that for large values of N_u, the quantity $D^{2\ (H,nl)}(\mu)$ is only dependent on the channel frequency responses and, for this reason, the subscript m, m' disappears in

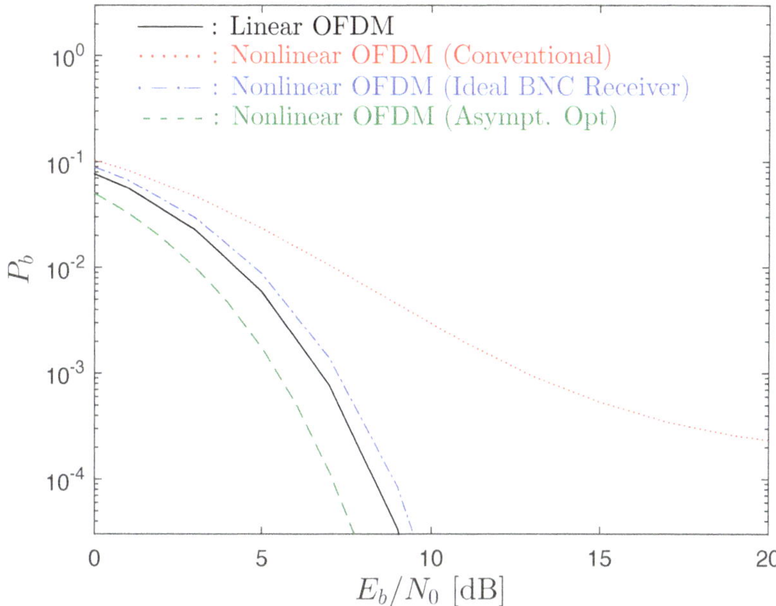

Figure 4.12 Performance of different receivers considering nonlinearly distorted OFDM signals and ideal AWGN channels.

(4.57) (this will be analyzed in more detail in Subsection 4.5). Additionally, as the frequency-selective channel presents $\mathbb{E}\left[|H_k|^2\right] = 1$, it should be noted that we have $\mathbb{E}_H[E_b^{(H,nl)}] = E_b^{(nl)}$, and we can write

$$G^{(H)}(\mu) = \frac{D^{2\ (H,nl)}(\mu)}{4E_b^{(nl)}}. \tag{4.58}$$

Therefore, the asymptotic BER associated with a given channel realization (i.e., associated with a given channel matrix $\mathbf{H}$), considering OFDM sequences that differ in $\mu = 1$ bit, can be computed as

$$P_b(H) \approx Q\left(\sqrt{2\frac{G^{(H)}(1)E_b^{(nl)}}{N_0}}\right). \tag{4.59}$$

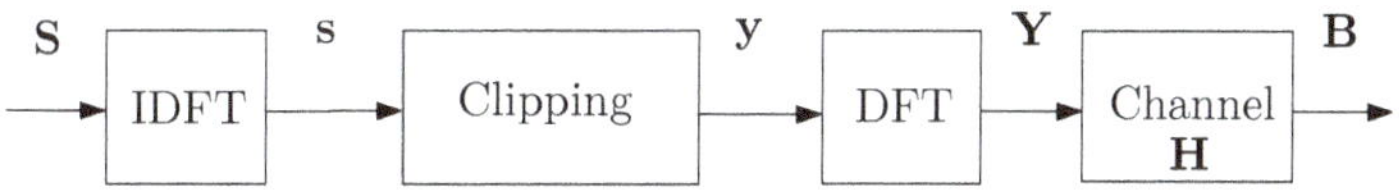

S → IDFT → s → Clipping → y → DFT → Y → Channel **H** → B

Figure 4.13 Equivalent, subcarrier-level model for a nonlinear OFDM transmission in a frequency-selective channel.

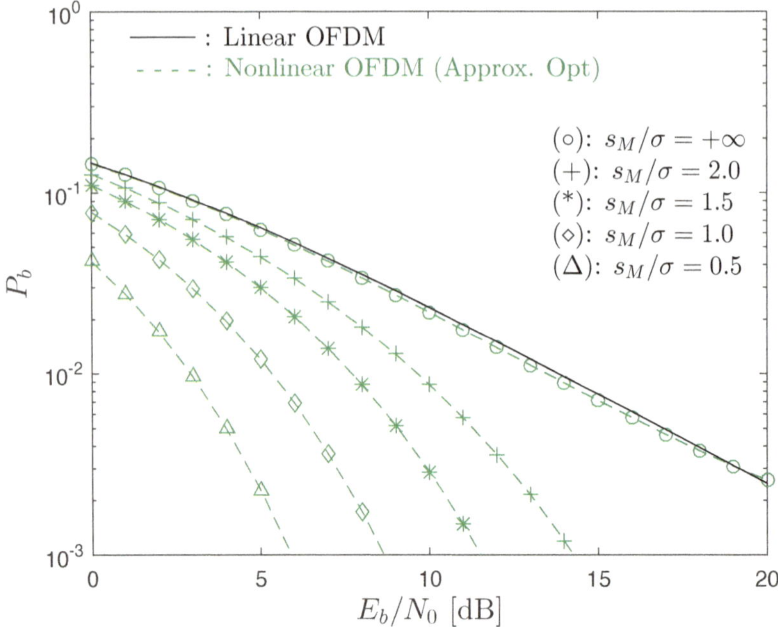

Figure 4.14 Simulated BER associated with the optimum detection of nonlinearly distorted OFDM signals considering frequency-selective channels and an envelope clipping.

Once again, due to the random nature of $|H_k|^2$, the average asymptotic BER associated with the optimum detection is given by

$$P_b \approx \mathbb{E}_H[P_b(H)] = \int_0^{+\infty} P_b(H)p(H)dH. \tag{4.60}$$

Fig. 4.14 shows the simulated average BER associated with the optimum detection of nonlinearly distorted OFDM signals considering $N_u = 512$, $O = 4$, a variable normalized clipping level s_M/σ and frequency-selective channels with $L = 32$ uncorrelated multipath components. From the results depicted in this figure, one can note that there are very large potential gains associated with the optimum detection of nonlinearly distorted OFDM signals, when frequency-selective channels are considered. As in the case of ideal AWGN channels, the potential gains are larger when the magnitude of the nonlinear distortion effects is higher. For instance, for a target BER of $P_b = 10^{-3}$ and $s_M/\sigma = 1.0$, the improvement in the performance relatively to linear OFDM transmissions is around 8.7 dB and may even reach approximately 11 dB when $s_M/\sigma = 0.5$. In fact, the existence of higher gains in frequency-selective channels can be explained by the fact that the nonlinear distortion effects introduce an additional diversity effect. This diversity effect is associated with the correlation

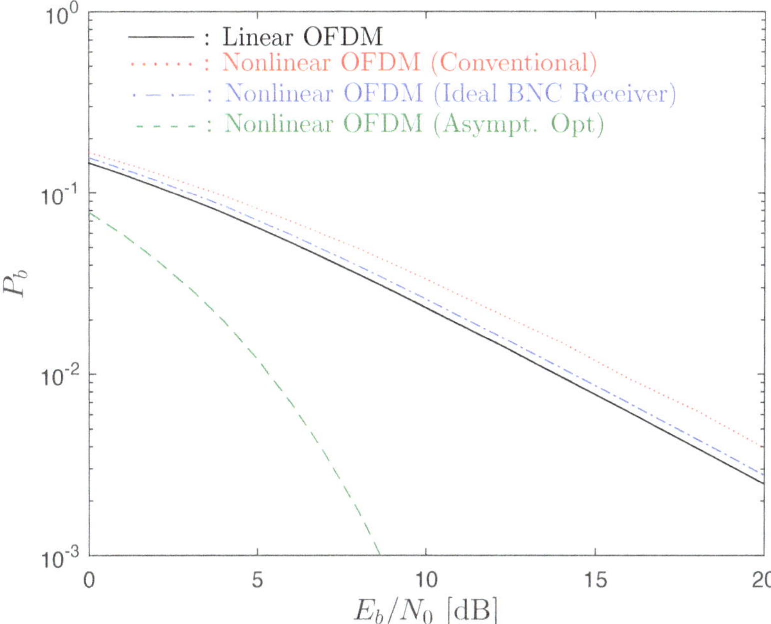

Figure 4.15 Performance of different receivers considering nonlinearly distorted OFDM signals and frequency-selective channels.

between the subcarriers that is introduced by the nonlinearity. Therefore, as the optimum receiver makes the detection in a block-by-block basis, the BER becomes less conditioned on the local deep fades that may exist along the block. Fig. 4.15 shows the BER associated with linear transmissions and nonlinear OFDM transmissions considering a normalized clipping level $s_M/\sigma = 1.0$. The figure presents BERs associated with conventional receivers, ideal BNC receivers and the approximate optimum receiver's performance. For the sake of comparison, the BER associated with a linear OFDM transmission is also shown. Clearly, from the results depicted in the figure, one can note that the approximate optimum receiver's performance is the best performance. For a target BER of $P_b = 10^{-3}$, the performance gain relatively to the conventional detection of linear, OFDM schemes is approximately 8.4 dB.

4.4 THEORETICAL ASYMPTOTIC GAINS IN AWGN CHANNELS

In the previous subsection, it was demonstrated that the optimum receiver presents considerable potential asymptotic gains relatively to conventional receivers that deal with nonlinearly distorted OFDM signals. More surprisingly, it was also shown that the optimum receiver can even outperform the conventional receivers that deal with linear OFDM signals. However, all of these conclusions were drawn based on simulation results, i.e., by obtaining the average squared Euclidean distance between two nonlinearly distorted OFDM signals by simulation and then obtaining the corresponding asymptotic gains in the system's performance.

The main goal of this subsection is to present a theoretical characterization of the average value of these potential asymptotic gains. This theoretical analysis is based on the computation of the average value of squared Euclidean distance between two nonlinearly distorted multicarrier signals and comprises not only OFDM schemes that have an envelope clipping operation on their transmission chain [15,93], but also other multicarrier schemes, impaired by general bandpass nonlinearities (i.e., with AM/AM and AM/PM conversion functions) [94, 95], as well as Cartesian and real-valued nonlinearities. The subsection is divided in two parts: firstly, attention is given to the characterization of the asymptotic gains for nonlinearly distorted, complex-valued multicarrier signals. Then, we present expressions for the average value of the asymptotic gains when baseband, real-valued multicarrier signals are impaired by memoryless nonlinearities. In both cases, the squared Euclidean distance is obtained in the time-domain, considering that the original multicarrier signals differ in μ bits.

4.4.1 COMPLEX-VALUED MULTICARRIER SIGNALS

Let us consider two OFDM data symbols $\mathbf{S}^{(m)} = [S_0^{(m)}\ S_1^{(m)}\ ...\ S_{N-1}^{(m)}]^T \in \mathbb{C}^N$ and $\mathbf{S}^{(m')} = [S_0^{(m')}\ S_1^{(m')}\ ...\ S_{N-1}^{(m')}]^T \in \mathbb{C}^N$, that differ in μ bits. Their corresponding time-domain versions are $\mathbf{s}^{(m)} = \mathbf{F}^{-1}\mathbf{S}^{(m)} = [s_0^{(m)}\ s_1^{(m)}\ ...\ s_{N-1}^{(m)}]^T \in \mathbb{C}^N$ and $\mathbf{s}^{(m')} = \mathbf{F}^{-1}\mathbf{S}^{(m')} = [s_0^{(m')}\ s_1^{(m')}\ ...\ s_{N-1}^{(m')}]^T \in \mathbb{C}^N$, respectively. Let us also define the array $\mathbf{E} = [E_0\ E_1\ ...\ E_{N-1}]^T \in \mathbb{C}^N$ as the error (or the difference) between $\mathbf{S}^{(m)}$ and $\mathbf{S}^{(m')}$. Under these conditions, we have

$$\mathbf{S}^{(m')} = \mathbf{S}^{(m)} + \mathbf{E}. \tag{4.61}$$

The indexes of the nonzero subcarriers of the error term, i.e., the indexes of the subcarriers where there are bit errors are represented by the set $\Upsilon = [\Upsilon_0\ \Upsilon_1\ ...\ \Upsilon_{\mu-1}] \in \mathbb{N}^{\mu}$. Therefore, the kth element of $\mathbf{E}$ is given by

$$E_k = \begin{cases} d_{\text{adj}} \exp(j\upsilon), & k \in \Upsilon \\ 0, & \text{otherwise}, \end{cases} \tag{4.62}$$

where d_{adj} is the absolute value of the difference between two adjacent symbols from a given constellation and υ represents the argument of that difference. In the concrete case of normalized QPSK constellations, we have $S_k = \pm 1 \pm j$, $d_{\text{adj}} = 2$ and $\upsilon \in \{\pm\pi, \pm\pi/2\}$. Regarding the time-domain, the error term is expressed as $\varepsilon = [\varepsilon_0\ \varepsilon_1\ ...\ \varepsilon_{N-1}]^T \in \mathbb{C}^N$, where

$$\varepsilon = \mathbf{F}^{-1}\mathbf{E}. \tag{4.63}$$

According to our IDFT definition (see (2.17), the nth element of ε can be expressed as

$$\begin{aligned}\varepsilon_n &= \sum_{k=0}^{N-1} F_{n,k}^{-1} E_k \\ &= \sum_{k\in\Upsilon} F_{n,k}^{-1} E_k \\ &= \sum_{k\in\Upsilon} \frac{d_{\text{adj}}}{N} \exp\left(\frac{j2\pi nk}{N} + j\upsilon\right) \\ &= \Delta_n \exp(j\vartheta_n),\end{aligned} \tag{4.64}$$

where Δ_n and ϑ_n represent the absolute value and the argument of nth time-domain sample of the error term, ε_n, respectively.

Bandpass Nonlinearities

Here, we are interested in the computation of the squared Euclidean distance between the nonlinearly distorted versions of $\mathbf{s}^{(m)} = [s_0^{(m)}\ s_1^{(m)}\ \ldots\ s_{N-1}^{(m)}]^T \in \mathbb{C}^N$ and $\mathbf{s}^{(m')} = [s_0^{(m')}\ s_1^{(m')}\ \ldots\ s_{N-1}^{(m')}]^T \in \mathbb{C}^N$, which are represented by $\mathbf{y}^{(m)} = [y_0^{(m)}\ y_1^{(m)}\ \ldots\ y_{N-1}^{(m)}]^T \in \mathbb{C}^N$ and $\mathbf{y}^{(m')} = [y_0^{(m')}\ y_1^{(m')}\ \ldots\ y_{N-1}^{(m')}]^T \in \mathbb{C}^N$, respectively. These nonlinearly distorted signals are obtained at the output of a given bandpass memoryless nonlinearity $f_{bp}(\cdot)$ (see Subsection 3.1.2). To obtain the average value of the squared Euclidean distance between two OFDM signals submitted to bandpass nonlinearities, we focus, firstly, on the difference between these two signals at the nth time instant. To obtain that difference, it is important to notice that the nth time-domain sample of a given OFDM signal can be written in its polar form as

$$s_n = r_n \exp(j\theta_n), \tag{4.65}$$

where θ_n is the phase associated with the nth time-domain sample s_n. Under these conditions, by applying the IDFT to (4.61) and considering (4.64), we have, for the nth sample of $\mathbf{s}^{(m')}$,

$$\begin{aligned} r_n^{(m')} \exp\left(j\theta_n^{(m')}\right) &= s_n^{(m)} + \varepsilon_n \\ &= r_n^{(m)} \exp\left(j\theta_n^{(m)}\right) + \Delta_n \exp\left(j\vartheta_n\right). \end{aligned} \tag{4.66}$$

However, due to the circular nature of $s_n^{(m)}$ and ε_n, we can assume, without loss of generality, that $\theta_n^{(m)} = 0$. Therefore,

$$\begin{aligned} r^{(m')} \exp\left(j\theta^{(m')}\right) &= r_n^{(m)} + \Delta_n \exp\left(j\vartheta_n\right) \\ &= \left(r_n^{(m)} + \Delta_n \cos(\vartheta_n)\right) + j\Delta_n \sin(\vartheta_n), \end{aligned} \tag{4.67}$$

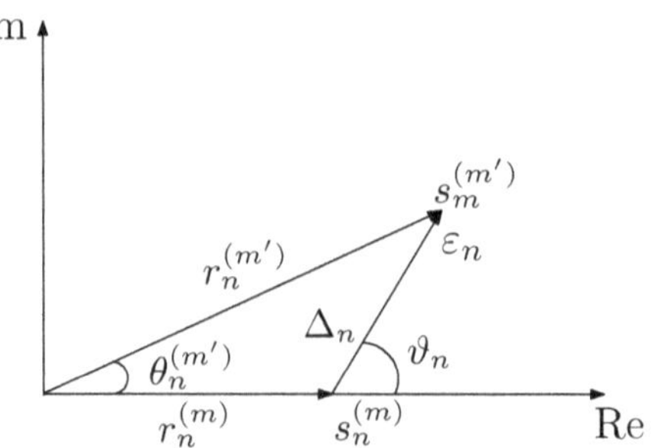

Figure 4.16 Vectorial representation of the samples $s_n^{(m)}$ and $s_n^{(m')}$ when $\theta_n^{(m)} = 0$.

as can be observed in Fig. 4.16. Defining $\varsigma_n = \theta_n^{(m')} - \theta_n^{(m)}$, we have

$$\begin{aligned}
\varsigma_n &= \arg\left(r_n^{(m)} + \Delta_n \exp(j\vartheta_n)\right) \\
&= \arctan\left(\frac{\Delta_n \sin(\vartheta_n)}{r_n^{(m)} + \Delta_n \cos(\vartheta_n)}\right) \\
&\overset{(a)}{\approx} \frac{\Delta_n \sin(\vartheta_n)}{r_n^{(m)}},
\end{aligned} \tag{4.68}$$

where approximation (a) is valid if $r_n^{(m)} \gg \Delta_n$. Similarly, $r_n^{(m')}$ can be approximated by

$$\begin{aligned}
r_n^{(m')} &= \left|r_n^{(m)} + \Delta_n \exp(j\vartheta_n)\right| \\
&= \sqrt{(r_n^{(m)} + \Delta_n \cos(\vartheta_n))^2 + \Delta_n^2 \sin^2(\vartheta_n)} \\
&\approx r_n^{(m)} + \Delta_n \cos(\vartheta_n).
\end{aligned} \tag{4.69}$$

Under these conditions, regarding the input sample $s_n^{(m)}$, the bandpass nonlinearity output is

$$\begin{aligned}
y_n^{(m)} &= A\left(r_n^{(m)}\right) \exp\left(j\left(\theta_n^{(m)} + \Theta\left(r_n^{(m)}\right)\right)\right) \\
&= A\left(r_n^{(m)}\right) \exp\left(j\Theta\left(r_n^{(m)}\right)\right).
\end{aligned} \tag{4.70}$$

On the other hand, when the input is $s_n^{(m')}$, the nonlinearity yields

$$\begin{aligned}
y_n^{(m')} &= A\left(r_n^{(m')}\right) \exp\left(j\left(\theta_n^{(m')} + \Theta\left(r^{(m')}\right)\right)\right) \\
&= A\left(r^{(m)} + \Delta_n \cos(\vartheta_n)\right) \exp(j\varsigma_n) \exp\left(j\Theta\left(r_n^{(m)} + \Delta_n \cos(\vartheta_n)\right)\right) \\
&= A\left(r_n^{(m)} + \Delta_n \cos(\vartheta_n)\right) \exp\left(j\frac{\Delta_n \sin(\vartheta_n)}{r_n^{(m)}}\right) \\
&\quad . \exp\left(j\Theta\left(r_n^{(m)} + \Delta_n \cos(\vartheta_n)\right)\right).
\end{aligned} \tag{4.71}$$

Let us now consider the Cartesian form of the bandpass nonlinearity

$$f_{bp}(r) = A(r)\cos(\Theta(r)) + jA(r)\sin(\Theta(r)) \\ = f_{bp,I}(r) + jf_{bp,Q}(r). \tag{4.72}$$

Under these conditions, when the input is $s_n^{(m)}$, the nonlinearity outputs

$$y_n^{(m)} = f_{bp,I}(r_n) + jf_{bp,Q}(r_n). \tag{4.73}$$

Additionally, by considering a Taylor approximation, we can expand $f_{bp,I}(r_n + \Delta_n\cos(\vartheta_n))$ and $f_{bp,Q}(r_n + \Delta_n\cos(\vartheta_n))$ around r_n, resulting,

$$f_{bp,I}(r_n + \Delta_n\cos(\vartheta_n)) = f_{bp,I}(r_n) + f'_{bp,I}(r_n)\Delta_n\cos(\vartheta_n), \tag{4.74}$$

and,

$$f_{bp,Q}(r_n + \Delta_n\cos(\vartheta_n)) = f_{bp,Q}(r_n) + f'_{bp,Q}(r_n)\Delta_n\cos(\vartheta_n), \tag{4.75}$$

where $f'_{bp,I}(r_n)$ and $f'_{bp,Q}(r_n)$ are the derivatives of $f_{bp,I}(r_n)$ and $f_{bp,Q}(r_n)$, respectively. With this approximation, we can rewrite the nonlinearity output for $s_n^{(m')}$ as[4]

$$y_n^{(m')} \approx \left(f_{bp,I}(r_n) + f'_{bp,I}(r_n)\Delta_n\cos(\vartheta_n) + j\left(f_{bp,Q}(r_n)\right.\right. \\ \left.\left. + f'_{bp,Q}(r_n)\Delta_n\cos(\vartheta_n)\right)\right)\exp(j\varsigma_n). \tag{4.76}$$

Considering (4.73) , (4.73) and $z'_{r_n} = f'_{bp,I}(r_n) + jf'_{bp,Q}(r_n)$ we can write the difference between the two nonlinearly distorted signals at the nth time-domain sample as

$$\begin{aligned} y_n^{(m')} - y_n^{(m)} &\approx \left(f_{bp,I}(r_n) + jf_{bp,Q}(r_n) + \Delta_n\cos(\vartheta_n)z'_{r_n}\right) \\ &.\exp(j\varsigma_n) - f_{bp,I}(r_n) + jf_{bp,Q}(r_n) \\ &= \left(\left(f_{bp,I}(r_n) + jf_{bp,Q}(r_n)\right)(1 - \exp(-j\varsigma_n)) + \Delta\cos(\vartheta_n)z'_{r_n}\right) \\ &\exp(j\varsigma_n) \\ &= \left(A(r_n)\exp(j\Theta(r_n))(1 - \exp(-j\varsigma_n)) + \Delta_n\cos(\vartheta_n)z'_{r_n}\right) \\ &\exp(j\varsigma_n). \end{aligned} \tag{4.77}$$

Furthermore, by taking into account that $\sin(\varsigma_n) \approx \varsigma_n$ and $\cos(\varsigma_n) \approx 1$ for low values of ς_n (say $\varsigma_n \ll 1$) and using (4.68), we can write

$$\exp(-j\varsigma_n) \approx 1 - j\varsigma_n \\ = 1 - j\frac{\Delta\sin(\vartheta_n)}{r_n}. \tag{4.78}$$

[4]For the sake of notation simplicity, the superscript (m) is omitted in the following computations.

Under these conditions, we can rewrite (4.77) as

$$\begin{aligned} y_n^{(m')} - y_n^{(m)} &\approx \left(A(r_n) \exp\left(j\Theta(r_n)\right) j \left(\frac{\Delta_n \sin(\vartheta_n)}{r_n} \right) + \Delta_n \cos(\vartheta_n) z'_{r_n} \right) \\ &\quad \exp(j\varsigma_n) \\ &\approx \Delta_n \left(A(r_n) \exp\left(j\Theta(r_n)\right) j \left(\frac{\sin(\vartheta_n)}{r_n} \right) + \cos(\vartheta_n) z'_{r_n} \right) \\ &\quad \exp(j\varsigma_n). \end{aligned} \tag{4.79}$$

Since

$$f'_{bp,Q}(r_n) = A'(r_n)\sin(\Theta(r_n)) + A(r_n)\cos(\Theta(r_n))\Theta'(r_n), \tag{4.80}$$

and $f'_{bp,I}(r_n) = f'_{bp,Q}(r_n + \frac{\pi}{2})$, we have

$$\begin{aligned} f'_{bp,I}(r_n) + jf'_Q(r_n) &= \exp\left(j\Theta(r_n)\right) \left(A'(r_n) + A(r)\Theta'(r_n) \exp\left(j\frac{\pi}{2} \right) \right) \\ &= \exp\left(j\Theta\left(r_n\right)\right) \left(A'(r_n) + jA(r_n)\Theta'(r_n) \right). \end{aligned} \tag{4.81}$$

Using this result, (4.79) turns to

$$\begin{aligned} y_n^{(m')} - y_n^{(m)} \approx \Delta_n \exp(j\varsigma_n) \exp\left(j\Theta(r_n)\right) \Bigg(& j\frac{A(r_n)\sin(\vartheta_n)}{r_n} \\ & + \cos(\vartheta_n) \left(A'(r_n) + jA(r_n)\Theta'(r_n) \right) \Bigg). \end{aligned} \tag{4.82}$$

Therefore, the squared absolute value of (4.82) is

$$\begin{aligned} &\left| y_n^{(m')} - y_n^{(m)} \right|^2 \approx \\ &\Delta_n^2 \left(A'^2(r_n)\cos^2(\vartheta_n) + \frac{A^2(r_n)\sin^2(\vartheta_n)}{r_n^2} + \Theta'^2(r_n)A^2(r_n)\cos^2(\vartheta_n) \right) \\ &+ \Delta_n^2 \left(\frac{2\Theta'(r_n)A^2(r_n)\cos(\vartheta_n)\sin(\vartheta_n)}{r_n} \right) \\ &= \Delta_n^2 \left(A'^2(r_n)\cos^2(\vartheta_n) + \frac{A^2(r_n)\sin^2(\vartheta_n)}{r_n^2} + \Theta'^2(r_n)A^2(r_n)\cos^2(\vartheta_n) \right) \\ &+ \Delta_n^2 \left(\frac{\Theta'(r_n)A^2(r_n)\sin(2\vartheta_n)}{r_n} \right) \\ &= \Delta_n^2 \left(A'^2(r_n)\frac{1+\cos(2\vartheta_n)}{2} + \frac{A^2(r_n)}{r_n^2}\frac{1-\cos(2\vartheta_n)}{2} \right) \\ &+ \Delta_n^2 \left(\Theta'^2(r_n)A^2(r_n)\frac{1+\cos(2\vartheta_n)}{2} + \frac{\Theta'(r_n)A^2(r_n)\sin(2\vartheta_n)}{r_n} \right) \\ &= \frac{\Delta_n^2}{2} \left(A'^2(r_n)(1+\cos(2\vartheta_n)) + \frac{A^2(r_n)}{r_n^2}(1-\cos(2\vartheta_n)) \right) \\ &+ \frac{\Delta_n^2}{2} \left(\Theta'^2(r_n)A^2(r_n)(1+\cos(2\vartheta_n)) + \frac{2\Theta'(r_n)A^2(r_n)\sin(2\vartheta_n)}{r_n} \right). \end{aligned} \tag{4.83}$$

On the other hand, the squared Euclidean distance between two nonlinearly distorted OFDM signals can be computed as

$$D^{2\,(nl)}_{m,m'}(\mu) = \sum_{n=0}^{N-1} \left| y_n^{(m')} - y_n^{(m)} \right|^2. \tag{4.84}$$

As we are interested in the average value of $D^{2\,(nl)}_{m,m'}(\mu)$ we make use of the fact that the absolute values of the time-domain samples of the error term, $|\varepsilon_n| = \Delta_n$, can be modeled by the random variable Δ, and their phases, ϑ_n, can be modeled by the random variable ϑ. In fact, it is important to note that, although we do not have the distribution of Δ, $p(\Delta)$, it can be easily demonstrated that

$$\mathbb{E}[\Delta^2] = \frac{\mu}{N^2} d^2_{\text{adj}}. \tag{4.85}$$

Moreover, the random variable ϑ has uniform distribution in $[0, 2\pi[$ and, consequently, its PDF is given by

$$p(\vartheta) = \begin{cases} \frac{1}{2\pi}, & \vartheta \in [0, 2\pi[, \\ 0, & \text{otherwise}, \end{cases} \tag{4.86}$$

whereas the samples r_n are modeled by the Rayleigh random variable r (see (3.75)). Under these conditions, the average value of the squared Euclidean distance between two OFDM signals submitted to a bandpass nonlinearity is

$$\begin{aligned} D^{2\,(nl)}(\mu) &\approx N\mathbb{E}\left[\left| y_n^{(m')} - y_n^{(m)} \right|^2\right] \\ &\approx N\mathbb{E}\left[\frac{\Delta^2}{2}\left(A'^2(r)(1+\cos(2\vartheta)) + \frac{A^2(r)}{r^2}(1-\cos(2\vartheta))\right)\right] \\ &+ N\mathbb{E}\left[\frac{\Delta^2}{2}\left(\Theta'^2(r)A^2(r)(1+\cos(2\vartheta)) + \frac{2\Theta'(r)A^2(r)\sin(2\vartheta)}{r}\right)\right]. \end{aligned} \tag{4.87}$$

However, we can take advantage of the fact that the random variables ϑ, Δ and r can be considered to be independent from each other. In addition, we can take advantage of the fact that

$$\mathbb{E}[\cos(2\vartheta)] = \int_0^{2\pi} \cos(2\vartheta) p(\vartheta) d\vartheta = 0, \tag{4.88}$$

and

$$\mathbb{E}[\sin(2\vartheta)] = \int_0^{2\pi} \sin(2\vartheta) p(\vartheta) d\vartheta = 0. \tag{4.89}$$

Therefore, by considering (4.85), the average value of the squared Euclidean distance can be approximated as

$$
\begin{aligned}
D^{2\,(nl)}(\mu) &\approx \frac{N}{2}\left(\mathbb{E}\left[\Delta^2\right]\mathbb{E}\left[A'^2(r)+\frac{A^2(r)}{r^2}+\Theta'^2(r)A^2(r)\right]\right)\\
&=\frac{\mu}{2N}d_{\mathrm{adj}}^2\int_0^{+\infty}\left(A'^2(r)+\frac{A^2(r)}{r^2}+\Theta'^2(r)A^2(r)\right)p(r)dr. \qquad (4.90)
\end{aligned}
$$

Furthermore, in addition to (4.47) and (4.48), we may define the average bit energy theoretically regarding the time-domain as

$$
\begin{aligned}
E_b^{(nl)} &= \frac{N}{2N_u}\mathbb{E}\left[|y_n|^2\right]=\frac{N}{2N_u}\mathbb{E}\left[\left|f_{bp}(r)\right|^2\right]\\
&=\frac{O}{2}\int_0^{+\infty}|f_{bp}(r)|^2p(r)dr\\
&=\frac{O}{2}\int_0^{+\infty}A^2(r)p(r)dr. \qquad (4.91)
\end{aligned}
$$

Therefore, by using (4.90), (4.91) and (2.25), we can express the theoretical average asymptotic gain associated with the optimum detection of nonlinearly distorted OFDM signals submitted to bandpass nonlinearities as

$$
\begin{aligned}
G(\mu) &\approx \frac{D^{2\,(nl)}(\mu)}{4E_b^{(nl)}}\\
&=\frac{\frac{\mu}{2N}d_{\mathrm{adj}}^2\int_0^{+\infty}\left(A'^2(r)+\frac{A^2(r)}{r^2}+\Theta'^2(r)A^2(r)\right)p(r)dr}{4\frac{O}{2}\int_0^{+\infty}A^2(r)p(r)dr}\\
&=\frac{\mu d_{\mathrm{adj}}^2\int_0^{+\infty}\left(A'^2(r)+\frac{A^2(r)}{r^2}+\Theta'^2(r)A^2(r)\right)p(r)dr}{4N_uO^2\int_0^{+\infty}A^2(r)p(r)dr}\\
&=\frac{\mu(d_{\mathrm{adj}}\sigma)^2}{4}\frac{\int_0^{+\infty}\left(A'^2(r)+\frac{A^2(r)}{r^2}+\Theta'^2(r)A^2(r)\right)p(r)dr}{\int_0^{+\infty}A^2(r)p(r)dr}. \qquad (4.92)
\end{aligned}
$$

Cartesian Nonlinearities

Here, we are interested in the computation of the squared Euclidean distance between the nonlinearly distorted versions of $\mathbf{s}^{(m)} = [s_0^{(m)}\ s_1^{(m)}\ \ldots\ s_{N-1}^{(m)}]^T \in \mathbb{C}^N$ and $\mathbf{s}^{(m')} = [s_0^{(m')}\ s_1^{(m')}\ \ldots\ s_{N-1}^{(m')}]^T \in \mathbb{C}^N$, which are represented by $\mathbf{y}^{(m)} = [y_0^{(m)}\ y_1^{(m)}\ \ldots\ y_{N-1}^{(m)}]^T \in \mathbb{C}^N$ and $\mathbf{y}^{(m')} = [y_0^{(m')}\ y_1^{(m')}\ \ldots\ y_{N-1}^{(m')}]^T \in \mathbb{C}^N$, respectively. These nonlinearly distorted signals are obtained by submitting $\mathbf{s}^{(m)}$ and $\mathbf{s}^{(m')}$ to a Cartesian nonlinearity. In order to do that, we start by obtaining the difference between these two signals at the nth time instant. Due to the Cartesian nature of the nonlinearity, it is useful to express the nth time-domain sample of a multicarrier signal in its Cartesian form (as was made in (2.24)),

$$s_n = s_{n,I} + js_{n,Q}. \tag{4.93}$$

Moreover, the nth time-domain sample of error term represented in (4.64), can also be written in its Cartesian form, resulting

$$\begin{aligned} \varepsilon_n &= \varepsilon_{n,I} + j\varepsilon_{n,Q} \\ &= \Delta_n \cos(\vartheta_n) + j\Delta_n \sin(\vartheta_n). \end{aligned} \tag{4.94}$$

Applying the IDFT to (4.61) and considering (4.94) we have, for the nth time-domain sample,

$$\begin{aligned} s_n^{(m')} &= s_n^{(m)} + \varepsilon_n \\ &= \left(s_{n,I}^{(m)} + \varepsilon_{n,I}\right) + j\left(s_{n,Q}^{(m)} + \varepsilon_{n,Q}\right), \end{aligned} \tag{4.95}$$

as is geometrically represented in Fig. 4.17. Considering (3.4), the nonlinearity output for the input $s_n^{(m)}$ is

$$y_n^{(m)} = f_{bb,I}\left(s_{n,I}^{(m)}\right) + jf_{bb,Q}\left(s_{n,Q}^{(m)}\right). \tag{4.96}$$

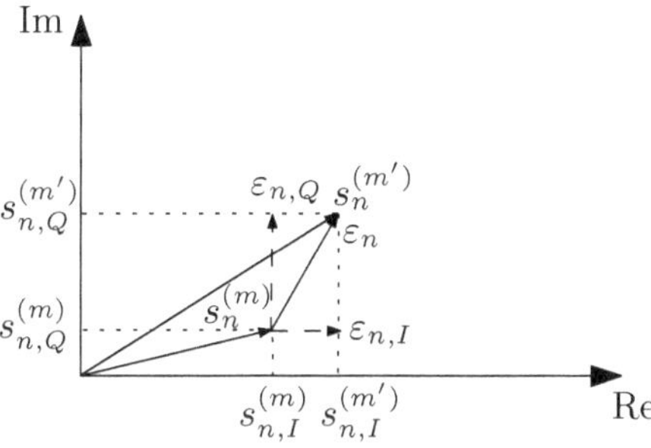

Figure 4.17 Vectorial representation of $s_n^{(m)}$, $s_n^{(m')}$ and their corresponding Cartesian components.

On the other hand, for the input $s_n^{(m')}$, the nonlinearity yields

$$y_n^{(m')} = f_{bb,I}\left(s_{n,I}^{(m)} + \varepsilon_I\right) + j f_{bb,Q}\left(s_{n,Q}^{(m)} + \varepsilon_{n,Q}\right). \tag{4.97}$$

In the following, we consider that the nonlinearity that operates on the real and imaginary parts of the signal are equal, i.e., $f_{bb,I}(\cdot) = f_{bb,Q}(\cdot) = f(\cdot)$. By considering a Taylor approximation of $f(\cdot)$ around $s_{n,I}$ and $s_{n,Q}$, we can approximate $y_n^{(m')}$ as

$$y_n^{(m')} \approx \left(f\left(s_{n,I}^{(m)}\right) + \varepsilon_{n,I} f'\left(s_{n,I}^{(m)}\right)\right) + j\left(f\left(s_{n,Q}^{(m)}\right) + \varepsilon_{n,Q} f'\left(s_{n,Q}^{(m)}\right)\right), \tag{4.98}$$

where $f'(\cdot)$ represents the derivative of the nonlinear function $f(\cdot)$. Under these conditions, the difference between the two samples $y_n^{(m)}$ and $y_n^{(m')}$ is given by

$$y_n^{(m')} - y_n^{(m)} \approx \varepsilon_{n,I} f'\left(s_{n,I}^{(m)}\right) + j\varepsilon_{n,Q} f'\left(s_{n,Q}^{(m)}\right). \tag{4.99}$$

Furthermore, the squared absolute value of that difference is approximately

$$\begin{aligned}
\left|y_n^{(m')} - y_n^{(m)}\right|^2 &\approx \left(\varepsilon_{n,I} f'\left(s_{n,I}^{(m)}\right)\right)^2 + \left(\varepsilon_{n,Q} f'\left(s_{n,Q}^{(m)}\right)\right)^2 \\
&= \varepsilon_{n,I}^2 f'^2\left(s_{n,I}^{(m)}\right) + \varepsilon_{n,Q}^2 f'^2\left(s_{n,Q}^{(m)}\right) \\
&= \Delta_n^2\left(\cos^2(\vartheta_n) f'^2\left(s_{n,I}^{(m)}\right) + \sin^2(\vartheta_n) f'^2\left(s_{n,Q}^{(m)}\right)\right) \\
&= \frac{\Delta_n^2}{2}\left((1+\cos(2\vartheta_n)) f'^2\left(s_{n,I}^{(m)}\right)\right) + \frac{\Delta_n^2}{2}\left((1-\cos(2\vartheta_n)) f'^2\left(s_{n,Q}^{(m)}\right)\right).
\end{aligned} \tag{4.100}$$

To obtain the average squared Euclidean distance between two nonlinearly distorted OFDM signals submitted to a Cartesian nonlinearity, we make use of the fact that both $s_{n,I}$ and $s_{n,Q}$ can be modeled with Gaussian random variable s, whose the corresponding PDF is given in (2.26). In addition, under the assumption of a large number of subcarriers, each sample of the difference between the two signals can be replaced by its average value, i.e.,

$$\begin{aligned}
D^{2\,(nl)}(\mu) &\approx N\mathbb{E}\left[\left|y_n^{(m')} - y_n^{(m)}\right|^2\right] \\
&= N\mathbb{E}\left[\frac{\Delta^2}{2} f'^2(s)\left((1+\cos(2\vartheta)) + (1-\cos(2\vartheta))\right)\right] \\
&= N\mathbb{E}\left[\Delta^2 f'^2(s)\right].
\end{aligned} \tag{4.101}$$

Once again, as Δ, s and ϑ can be considered to be independent, the average value of the squared Euclidean distance can be written as

$$\begin{aligned}
D^{2\,(nl)}(\mu) &\approx N\left(\frac{\mu d_{\mathrm{adj}}^2}{N^2}\mathbb{E}\left[f'^2(s)\right]\right) \\
&= \frac{\mu d_{\mathrm{adj}}^2}{N}\int_{-\infty}^{+\infty} f'^2(s) p(s) ds,
\end{aligned} \tag{4.102}$$

given that $\mathbb{E}_\phi[\cos(2\vartheta)] = 0$ and considering (4.85). On the other hand, the average bit energy can be computed theoretically as

$$E_b^{(nl)} = \frac{N}{2N_u}\mathbb{E}\left[|y_n|^2\right] = \frac{2N}{2N_u}\int_{-\infty}^{+\infty} f^2(s)p(s)ds$$
$$= O\int_{-\infty}^{+\infty} f^2(s)p(s)ds. \tag{4.103}$$

Thus, by considering (4.102) and (4.103), the average value of the asymptotic gain is given by

$$G(\mu) = \frac{D^{2\,(nl)}(\mu)}{4E_b^{(nl)}} = \frac{\mu d_{\text{adj}}^2 \int_{-\infty}^{+\infty} f'^2(s)\,p(s)ds}{4N_u O^2 \int_{-\infty}^{+\infty} f^2(s)p(s)ds}$$
$$= \frac{\mu(d_{\text{adj}}\sigma)^2}{4}\frac{\int_{-\infty}^{+\infty} f'^2(s)\,p(s)ds}{\int_{-\infty}^{+\infty} f^2(s)p(s)ds}. \tag{4.104}$$

4.4.2 REAL-VALUED MULTICARRIER SIGNALS

Here, we study the asymptotic optimum performance for scenarios where we have nonlinearities operating in real-valued multicarrier signals. This happens, for instance, when carrier-less UWB-OFDM systems [99] are considered. Let us consider two real-valued multicarrier signals $\mathbf{s}^{(m)} = [s_0^{(m)}\ s_1^{(m)}\ \ldots\ s_{N-1}^{(m)}]^T \in \mathbb{C}^N$ and $\mathbf{s}^{(m')} = [s_0^{(m')}\ s_1^{(m')}\ \ldots\ s_{N-1}^{(m')}]^T \in \mathbb{C}^N$. Note that when these two data sequences differ in μ bits, their corresponding frequency-domain data versions $\mathbf{S}^{(m)} = [S_0^{(m)}\ S_1^{(m)}\ \ldots\ S_{N-1}^{(m)}]^T \in \mathbb{C}^N$ and $\mathbf{S}^{(m')} = [S_0^{(m')}\ S_1^{(m')}\ \ldots\ S_{N-1}^{(m')}]^T \in \mathbb{C}^N$ differ in 2μ subcarriers. This is explained by the Hermitian symmetry of the frequency-domain signal, necessary to assure that the multicarrier signals are real-valued at the IDFT output. Fig. 4.18 depicts the difference between $\mathbf{S}^{(m')}$ and $\mathbf{S}^{(m)}$. These two frequency-domain symbols have $N_u = 256$, $O = 4$ and $\mu = 1$ bit differences. As expected, it can be seen that the signals differ in two subcarriers.

Let us start by defining $\mathbf{S}^{(m')}$ through $\mathbf{S}^{(m)}$ and the error term $\mathbf{E}$, i.e.,

$$\mathbf{S}^{(m')} = \mathbf{S}^{(m)} + \mathbf{E}. \tag{4.105}$$

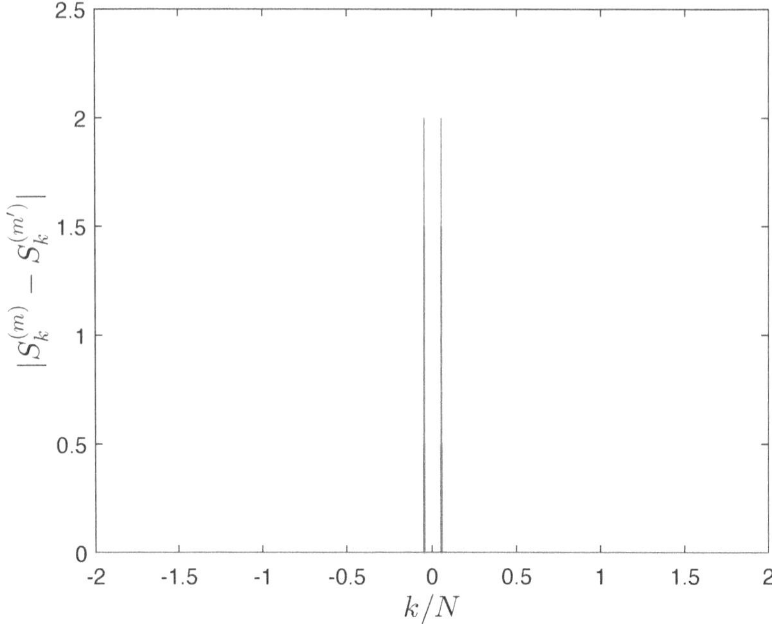

Figure 4.18 Difference between two frequency-domain OFDM symbols with Hermitian symmetry considering $\mu = 1$.

The indexes associated with the subcarriers where there are bit errors are represented by the set $\Upsilon = [\Upsilon_0\ \Upsilon_1\ ...\ \Upsilon_{2\mu-1}] \in \mathbb{N}^{2\mu}$. Therefore, the kth element of $\mathbf{E}$ is given by

$$E_k = \begin{cases} d_{\text{adj}} \exp(j\upsilon), & k \in \Upsilon \\ 0, & \text{otherwise.} \end{cases} \tag{4.106}$$

Note that the time-domain samples of the error term $\varepsilon = \mathbf{F}^{-1}\mathbf{E} = [\varepsilon_0\ \varepsilon_1\ ...\ \varepsilon_{N-1}]^T \in \mathbb{C}^N$ are real-valued, which means that $\varepsilon_n = \Delta_n$. In addition, it can be shown that

$$\mathbb{E}[\Delta^2] = \frac{2\mu}{N} d_{\text{adj}}^2. \tag{4.107}$$

In the following, we are interested in obtaining the squared Euclidean distance between two real-valued, nonlinearly distorted signals $\mathbf{y}^{(m)}$ and $\mathbf{y}^{(m')}$, obtained by submitting $\mathbf{s}^{(m)}$ and $\mathbf{s}^{(m')}$ to a real-valued, baseband nonlinearity $f(\cdot)$ (see (3.1)). Let us start by considering the difference between these two signals at the nth time instant. One one hand, we can express the nonlinearly distorted version of $s_n^{(m)}$ as

$$y_n^{(m)} = f\left(s_n^{(m)}\right). \tag{4.108}$$

On the other hand, for $s_n^{(m')} = s_n^{(m)} + \varepsilon_n$, the nonlinearity yields

$$y_n^{(m')} = f\left(s_n^{(m)} + \varepsilon_n\right). \tag{4.109}$$

As ε_n is real-valued, we have $\varepsilon_n = \Delta_n$. Additionally, by employing the Taylor series, we can approximate $y_n^{(m')}$ as

$$y_n^{(m')} \approx f\left(s_n^{(m)}\right) + f'\left(s_n^{(m)}\right)\Delta_n. \tag{4.110}$$

Under these conditions, the difference between the two nonlinearly distorted time-domain samples is

$$y_n^{(m')} - y_n^{(m)} \approx f'\left(s_n^{(m)}\right)\Delta_n, \tag{4.111}$$

and the squared absolute value of their difference is

$$\left|y_n^{(m')} - y_n^{(m)}\right|^2 \approx \Delta_n^2\left|f'\left(s_n^{(m)}\right)\right|^2. \tag{4.112}$$

This means that the average value of the squared Euclidean distance between the two nonlinearly distorted signals can be computed as

$$\begin{aligned} D^{2\,(nl)}(\mu) &\approx N\mathbb{E}_{\Delta,s}\left[\Delta^2\left|f'(s)\right|^2\right] \\ &= \frac{2\mu d_{\text{adj}}^2}{N}\int_{-\infty}^{+\infty}\left|f'(s)\right|^2 p(s)ds, \end{aligned} \tag{4.113}$$

where $p(s)$ is given by (2.26). On the other hand, the average bit energy is theoretically given by

$$E_b^{(nl)} = \frac{N}{N_u-2}\mathbb{E}\left[|y_n|^2\right] = \frac{N}{N_u-2}\int_{-\infty}^{+\infty}|f(s)|^2 p(s)ds. \tag{4.114}$$

Using (4.113) and (4.114), the average asymptotic gain associated with the optimum detection can be expressed as

$$\begin{aligned} G(\mu) = \frac{D^{2\,(nl)}(\mu)}{4E_b^{(nl)}} &= \frac{\frac{2\mu d_{\text{adj}}^2}{N}\int_{-\infty}^{+\infty}\left|f'(s)\right|^2 p(s)ds}{\frac{4N}{N_u-2}\int_{-\infty}^{+\infty}|f(s)|^2 p(s)ds} \\ &= \frac{\mu d_{\text{adj}}^2\int_{-\infty}^{+\infty}\left|f'(s)\right|^2 p(s)ds}{2\frac{N^2}{N_u-2}\int_{-\infty}^{+\infty}|f(s)|^2 p(s)ds} = \frac{\mu(d_{\text{adj}}\sigma)^2}{4}\frac{\int_{-\infty}^{+\infty}\left|f'(s)\right|^2 p(s)ds}{\int_{-\infty}^{+\infty}|f(s)|^2 p(s)ds}, \end{aligned} \tag{4.115}$$

where it should be noted that the variance of the real-valued multicarrier signal is given by $\sigma^2 = 2(N_u - 2)/N^2$.

4.5 THEORETICAL ASYMPTOTIC GAINS IN FREQUENCY-SELECTIVE CHANNELS

In this subsection, we are interested in deriving the theoretical asymptotic optimum performance associated with multicarrier transmissions under frequency-selective channels. This theoretical analysis makes use of the results of the previous section and considers the statistical characterization of the frequency-selective channels made in Subsection 2.2.2.

Note that, in addition to (4.56), the squared Euclidean distance between two nonlinearly distorted multicarrier signals $\mathbf{Y}^{(m)} = [Y_0^{(m)}\ Y_1^{(m)}\ \ldots\ Y_{N-1}^{(m)}]^T \in \mathbb{C}^N$ and $\mathbf{Y}^{(m')} = [Y_0^{(m')}\ Y_1^{(m')}\ \ldots\ Y_{N-1}^{(m')}]^T \in \mathbb{C}^N$ that differ in μ bits and are transmitted through a frequency-selective can be written as

$$\begin{aligned}
D_{m,m'}^{2\ (H,nl)}(\mu) &= \left|\left|\mathbf{B}^{(m')} - \mathbf{B}^{(m)}\right|\right|^2 \\
&= \sum_{k=0}^{N-1} \left|B_k^{(m')} - B_k^{(m)}\right|^2 \\
&= \sum_{k=0}^{N-1} |H_k|^2 \underbrace{\left|\alpha\left(S_k^{(m')} - S_k^{(m)}\right) + D_k^{(m')} - D_k^{(m)}\right|^2}_{D_{m,m'}^{2\ (nl)}(\mu)_k} \\
&= \sum_{k=0}^{N-1} |H_k|^2 D_{m,m'}^{2\ (nl)}(\mu)_k,
\end{aligned} \tag{4.116}$$

where $D_{m,m'}^{2\ (nl)}(\mu)_k$ represents the kth component of the squared Euclidean distance between $\mathbf{Y}^{(m)}$ and $\mathbf{Y}^{(m')}$. In fact, by summing up all these components, we have the squared Euclidean distance between $\mathbf{Y}^{(m)}$ and $\mathbf{Y}^{(m')}$ in an ideal AWGN channel, i.e.,

$$D_{m,m'}^{2\ (nl)}(\mu) = \sum_{k=0}^{N-1} D_{m,m'}^{2\ (nl)}(\mu)_k. \tag{4.117}$$

From (4.116), it can be noted that the squared difference between the kth subcarrier of the two signals is weighted by the squared absolute value of the channel frequency response associated with that subcarrier, $|H_k|^2$. Additionally, as mentioned in the previous subsection, under the assumption of a large number of subcarriers, the squared Euclidean distance between two sequences, $D_{m,m'}^{2\ (nl)}(\mu)$, tends to a fixed value that does not depend on $\mathbf{S}^{(m)}$ and $\mathbf{S}^{(m')}$. Under these conditions, it is possible to approximate this value by the sum of two components: one component associated with the difference between the μ subcarriers that have bit errors, $D_d^{2\ (nl)}(\mu)$, and other component associated with the difference between the nonlinear distortion terms, $D_c^{2\,(nl)}(\mu)$. This means that

$$
\begin{aligned}
D_{m,m'}^{2\,(nl)}(\mu) &= \sum_{k=0}^{N-1}\left(\left|\alpha\left(S_k^{(m')}-S_k^{(m)}\right)+D_k^{(m')}-D_k^{(m)}\right|^2\right)\\
&\overset{(a)}{\approx} \sum_{k\in\Upsilon}\left|\alpha\left(S_k^{(m')}-S_k^{(m)}\right)\right|^2+\sum_{k=0}^{N-1}\left|D_k^{(m')}-D_k^{(m)}\right|^2\\
&\overset{(b)}{\approx} \underbrace{\sum_{k\in\Upsilon}\left|\alpha\left(S_k^{(m')}-S_k^{(m)}\right)\right|^2}_{D_d^{2\,(nl)}(\mu)}+\underbrace{\sum_{k=0}^{N-1}\mathbb{E}\left[\left|D_k^{(m')}-D_k^{(m)}\right|^2\right]}_{D_c^{2\,(nl)}(\mu)}\\
&= D_d^{2\,(nl)}(\mu)+D_c^{2\,(nl)}(\mu).
\end{aligned}
\tag{4.118}
$$

The approximation (a) is valid since, although the size of the first term of the sum is comparable to the size of the second term (associated with the nonlinear distortion components), the second term is formed by many more elements[5] (see Fig. 4.5 and note that the size of Υ is typically $\mu \ll N$). Moreover, from the Bussgang's theorem [10], we known that the terms $D_k^{(i)}$ are uncorrelated with $S_k^{(i)}$ and that the cross terms between $S_k^{(m)}$ and $D_k^{(m')}$ or $D_k^{(m)}$ and $S_k^{(m')}$ are also uncorrelated with zero mean. The approximation (b) is related to the fact that for $D_k^{(i)}$ we have, when $N \gg 1$,

$$
\sum_{k=0}^{N-1}\left|D_k^{(i)}\right|^2 \approx \sum_{k=0}^{N-1}\mathbb{E}\left[\left|D_k^{(i)}\right|^2\right], \tag{4.119}
$$

since the different $D_k^{(i)}$ are almost uncorrelated and the expected values change slowly with k. Note that the decomposition of (4.118) holds independent of type of nonlinearity. Under these conditions, we can separate the asymptotic gain associated with the optimum detection in ideal AWGN channels as the sum of two components, i.e.,

$$
\begin{aligned}
G(\mu) &\approx \frac{D_d^{2\,(nl)}(\mu)}{4E_b^{(nl)}}+\frac{D_c^{2\,(nl)}(\mu)}{4E_b^{(nl)}}\\
&= G_d(\mu)+G_c(\mu).
\end{aligned}
\tag{4.120}
$$

Additionally, we can also decompose the asymptotic gain associated with the optimum detection in frequency-selective channels in two components, i.e,

$$
G^{(H)}(\mu) \approx G_d^{(H)}(\mu)+G_c^{(H)}(\mu), \tag{4.121}
$$

[5] Due to distortion term in the μ subcarriers, this decomposition has an error of order μ/N that becomes negligible when $N \gg \mu$.

where the first component, $G_d^{(H)}(\mu)$, is related to the μ subcarriers that have bit errors and the second term, $G_c^{(H)}(\mu)$, is related two the difference between the nonlinear distortion components. In fact, from (4.116), we can write that

$$G_d^{(H)}(\mu) \approx \frac{\frac{D_d^{2\ (nl)}(\mu)}{\mu} \sum_{k\in\Upsilon} |H_k|^2}{4E_b^{(nl)}} = \frac{G_d(\mu)\gamma_d}{\mu}, \tag{4.122}$$

where γ_d is a random variable defined as

$$\gamma_d = \sum_{k\in\Upsilon} |H_k|^2. \tag{4.123}$$

The computation of $G_c^{(H)}(\mu)$ is more difficult, since it has a term of type

$$G_c^{(H)}(\mu) = \frac{\sum_{k=0}^{N-1} |H_k|^2 \underbrace{\left|D_k^{(m')} - D_k^{(m)}\right|^2}_{D_c^{2\ (nl)}{}_k}}{4E_b^{(nl)}}. \tag{4.124}$$

To obtain it, we should have in mind that the variances of the terms $D_c^{2\ (nl)}{}_k$ vary slowly in the frequency (i.e., with k), and their mean is constant and equal to 0. Therefore, one can admit that $D_c^{2\ (nl)}{}_k$ is approximately stationary in a neighborhood around k (the length of this neighborhood can be on the order of tens of subcarriers). Moreover, we will assume that the statistical means in that neighborhood are equal to the ensemble average, which means that $D_c^{2\ (nl)}{}_k$ can be considered to be approximately ergodic. Indeed, one should note that:

(a) $D_c^{2\ (nl)}{}_k$ changes much faster than $|H_k|^2$ (see Fig. 4.19);
(b) $D_c^{2\ (nl)}{}_k$ has a huge number of oscillations over the sum (see Fig. 4.19);
(c) The statistical properties $D_c^{2\ (nl)}{}_k$ change very slowly over the sum, namely when compared with $|H_k|^2$;
(d) $|H_k|^2$ has a large number of oscillations;
(e) The sum of $|H_k|^2$ over a sliding window with length R, with $1 \ll R \ll N$ is almost constant and equal to $\frac{R}{N}\sum_k |H_k|^2$ [6].

[6] Indeed, this is not necessarily equal to $\sum_k \mathbb{E}\left[|H_k|^2\right] = N$, due to statistical fluctuations in the amplitudes of the corresponding multipath components associated with that channel realization.

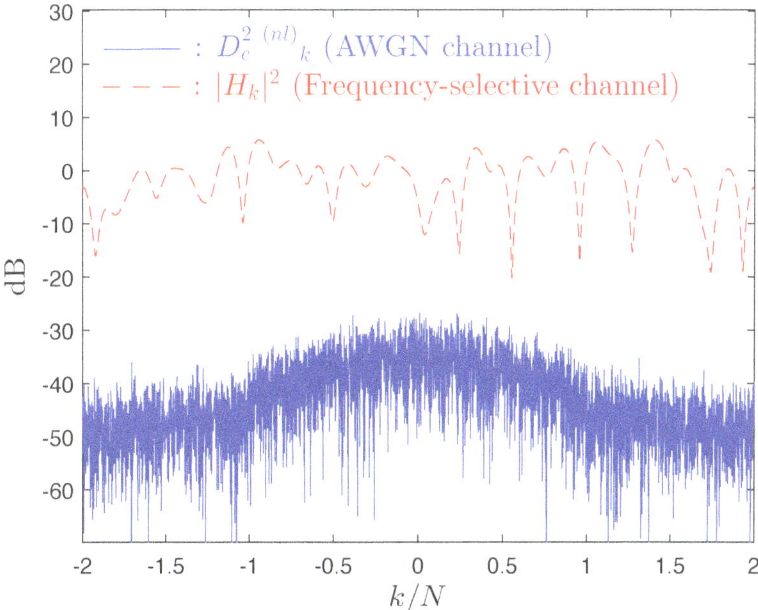

Figure 4.19 Evolution of $D_c^{2\ (nl)}{}_k$ in an ideal AWGN channel and the squared value of the channel frequency responses in a frequency-selective channel.

Therefore, we have

$$\begin{aligned}\sum_{k=0}^{N-1} |H_k|^2 D_c^{2\ (nl)}{}_k &\approx \sum_{k=0}^{N-1} |H_k|^2 \mathbb{E}\left[D_c^{2\ (nl)}{}_k\right] \\ &\approx \left(\frac{1}{N}\sum_{k=0}^{N-1} |H_k|^2\right) \times \left(\sum_{k=0}^{N-1} \mathbb{E}[D_c^{2\ (nl)}{}_k]\right) \\ &= \frac{D_c^{2\ (nl)}(\mu)}{N} \sum_{k=0}^{N-1} |H_k|^2,\end{aligned} \tag{4.125}$$

where the first approximation comes from (a), (b) and (c) and the second approximation comes from (c), (d) and (e). Under these conditions

$$\begin{aligned}G_c^{(H)}(\mu) &\approx \frac{\frac{D_c^{2\ (nl)}(\mu)}{N} \sum_{k=0}^{N-1} |H_k|^2}{4E_b^{(nl)}} \\ &= \frac{G_c(\mu)\gamma_c}{N},\end{aligned} \tag{4.126}$$

where γ_c is a random variable defined as

$$\gamma_c = \sum_{k=0}^{N-1} |H_k|^2. \tag{4.127}$$

In fact, the asymptotic gain associated with the optimum detection of nonlinear multicarrier schemes in frequency-selective channels, $G^{(H)}(\mu)$, is a random variable due to the random nature of its two components (see (4.121)). In order to obtain its distribution, we must have the corresponding asymptotic gain in ideal AWGN channels, since we need both $G_d(\mu)$ and $G_c(\mu)$, as well as the statistical behavior of the frequency-selective channel, since we also need to characterize the random variables γ_d and γ_c. In the following, we derive the distribution of $G^{(H)}(\mu)$ by taking advantage of the statistical analysis of the channel frequency responses made in Subsection 2.2.2. In that subsection, it was shown that the absolute value of the channel frequency response associated with the kth subcarrier has Rayleigh distribution with scale parameter $\sqrt{1/2}$, i.e., $|H_k| \sim \text{Rayleigh}(1/\sqrt{2})$. This means that the different $|H_k|^2$ have a Gamma distribution with unitary scale and shape parameters, i.e., $|H_k|^2 \sim \Gamma(1,1)$ (see Fig. 2.7).

Let us start by characterizing the distribution of γ_d, that is the random variable related to the frequency-responses that weight $G_d(\mu)$, as well as the distribution of γ_c, associated with $G_c(\mu)$. With these two distributions, we are able to derive the distribution of the asymptotic gain $G^{(H)}(\mu)$, as well as obtain its basic statistical properties. Regarding the definition of γ_c in (4.127), we have

$$\gamma_c = \sum_{k=0}^{N-1} |H_k|^2 \approx N \sum_{l=0}^{L-1} |\alpha_l|^2, \tag{4.128}$$

where α_l is the complex amplitude of the lth multipath component (see (2.42)). Therefore, the random variable γ_c has a Gamma distribution with shape parameter L and scale parameter $\frac{N}{L}$, i.e., $\gamma_c \sim \Gamma\left(L, \frac{N}{L}\right)$. Similarly, from (4.123), we have $\gamma_d \sim \Gamma(\mu, 1)$. Fig. 4.20 shows the distribution of γ_c obtained both theoretically and by simulation. The figure confirms the high accuracy of the theoretical expression for $p(\gamma_c)$. In fact, the random variable $G^{(H)}(\mu)$ is given by the sum of two independent gamma variables with different parameters, i.e., $G_d^{(H)}(\mu) \sim \Gamma(\mu, \frac{G_d(\mu)}{\mu})$ and $G_c^{(H)}(\mu) \sim \Gamma(L, \frac{G_c(\mu)}{L})$. Fig. 4.21 shows the distribution $G_d^{(H)}(\mu)$ and $G_c^{(H)}(\mu)$ obtained both theoretically and by simulation considering $L = 32$, $N_u = 512$, $O = 4$, $\mu = 1$ and an envelope clipping with normalized clipping level $s_M/\sigma = 1.0$. From the results depicted in the figure, it can be noted that distribution of both gain components can be obtained with accuracy. In Appendix B, by making use of these distributions, we derive the distribution of $G^{(H)}(\mu)$, which is shown to be

$$p(G^{(H)}(\mu)) = \frac{\left(\frac{\mu}{G_d(\mu)}\right)^{\mu}\left(\frac{L}{G_c(\mu)}\right)^{L}}{\Gamma(L+\mu)} G^{(H)}(\mu)^{L+\mu-1} \exp\left(-\frac{G^{(H)}(\mu)L}{G_c(\mu)}\right)$$
$$\times M\left(\mu, L+\mu, G^{(H)}(\mu)\left(\frac{L}{G_c(\mu)} - \frac{\mu}{G_d(\mu)}\right)\right). \tag{4.129}$$

With this distribution, we are able to derive the asymptotic optimum performance in frequency-selective channels for a variety of multicarrier schemes and nonlinearities.

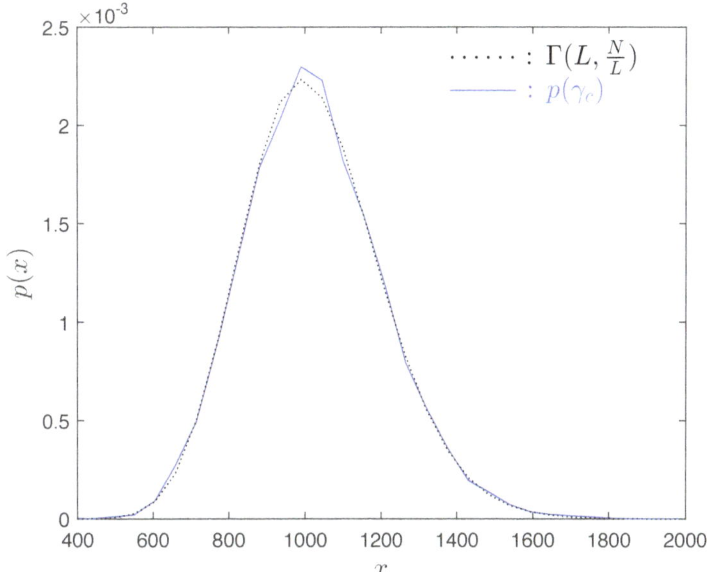

Figure 4.20 Distribution of γ_c obtained both theoretically and by simulation.

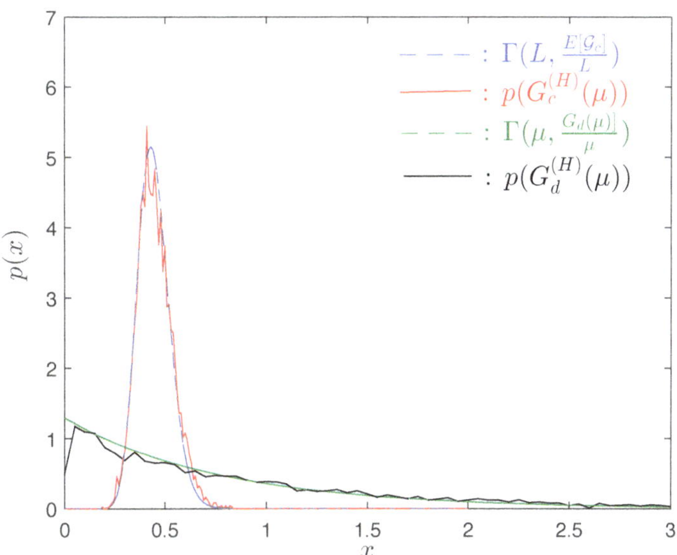

Figure 4.21 Distribution of $G_d^{(H)}(\mu)$ and $G_c^{(H)}(\mu)$ obtained both theoretically and by simulation for $\mu = 1$.

5 Applications

This chapter presents several applications where the theoretical analysis of the previous chapters can be employed to provide an accurate characterization of nonlinearly distorted multicarrier schemes. These applications consist of different multicarrier schemes with nonlinearities on their transceivers. For each application, we provide an accurate theoretical characterization of the corresponding nonlinearly distorted signals. Additionally, for most of applications, the potential optimum performance considering both ideal AWGN and frequency-selective channels is also studied. This optimum performance is accessed through the potential asymptotic gains, as well as by a set of performance results associated to practical sub-optimum receivers.

This chapter is divided as follows: in Section 5.1, we analyze the nonlinear distortion effects and the optimum performance of nonlinear bandpass OFDM schemes. In that section, results associated with transmitter-side nonlinearities such as the ones associated with nonlinear power amplifiers and envelope clipping techniques are presented. In Section 5.2, OFDM systems with LINC transmitter structures are considered. In Section 5.3 the analytical characterization and the optimum performance of constant envelope OFDM schemes are presented. Finally, Section 5.4 concerns with the analytical characterization of nonlinearly distorted signals associated with an amplify-and-forward, relay-based OFDM system.

In all results presented in this chapter, it is assumed the existence of a CP adequate to remove both the ISI and IBI. In addition, perfect channel estimation and synchronization are also considered.

5.1 CONVENTIONAL OFDM

Due to their large PAPR, it is very common that conventional OFDM systems are impaired by bandpass nonlinearities (see Subsection 3.1.2). These nonlinearities are typically associated with a nonlinear amplification process, but can also be related to the simplest solutions that avoid this nonlinear amplification (e.g., envelope clipping techniques for PAPR reduction [51,52]), which were discussed in Subsection 2.3.2. In this section, results regarding both the analytical characterization and the optimum performance of nonlinear, bandpass OFDM systems are shown.

5.1.1 LOW-COMPLEXITY ANALYTICAL SIGNAL'S CHARACTERIZATION

Although bandpass nonlinearities act "physically" on the bandpass OFDM signal, we consider the equivalent baseband scenario presented in Subsection 3.1.2, which takes into account the AM/AM and AM/PM models for the characterization of such nonlinearities. Therefore, a given bandpass nonlinearity is described by the complex nonlinear function $f_{bp}(r) = A(r)\exp(j\Theta(r))$. Fig. 5.1 shows the equivalent, sub-carrier level for a nonlinear OFDM transmission. Although conventional OFDM systems

DOI: 10.1201/9781032708744-5

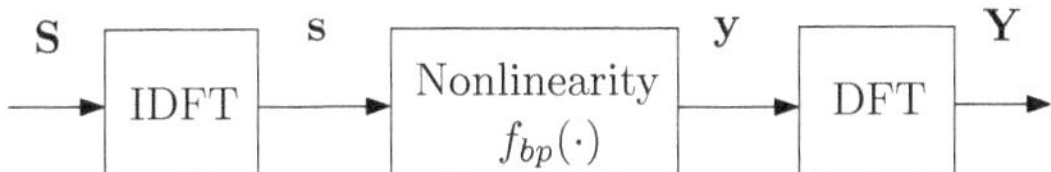

Figure 5.1 Equivalent, sub-carrier level model for a nonlinear OFDM transmission considering an envelope clipping.

are described in detail in Subsection 2.2.1, a brief characterization of such systems is reproduced here. Each OFDM symbol $\mathbf{S} = [S_0\ S_1\ ...\ S_{N-1}]^T \in \mathbb{C}^N$ is composed by N_u useful subcarriers that carry M-QAM symbols (otherwise stated, we consider that $M = 4$, i.e., each useful subcarrier employs a QPSK symbol and $S_k = \pm 1 \pm j$) and an oversampling factor of O, which means that the total number of subcarriers is $N = ON_u$. Under these conditions, at the input of the IDFT block, we have a rectangular signal, since $\mathbb{E}[|S_k|^2] = 2$ for the in-band subcarriers and $\mathbb{E}[|S_k|^2] = 0$ for the out-of-band subcarriers. A rectangular, frequency-domain OFDM signal with N_u subcarriers and $O = 4$ is depicted in Fig. 5.2. After the IDFT block, we have the set of time-domain samples $\mathbf{s} = [s_0\ s_1\ ...\ s_{N-1}]^T \in \mathbb{C}^N$. As mentioned before, when the number of subcarriers is large, the time-domain samples of a given OFDM signal can be seen as samples of a complex stationary, Gaussian random process.

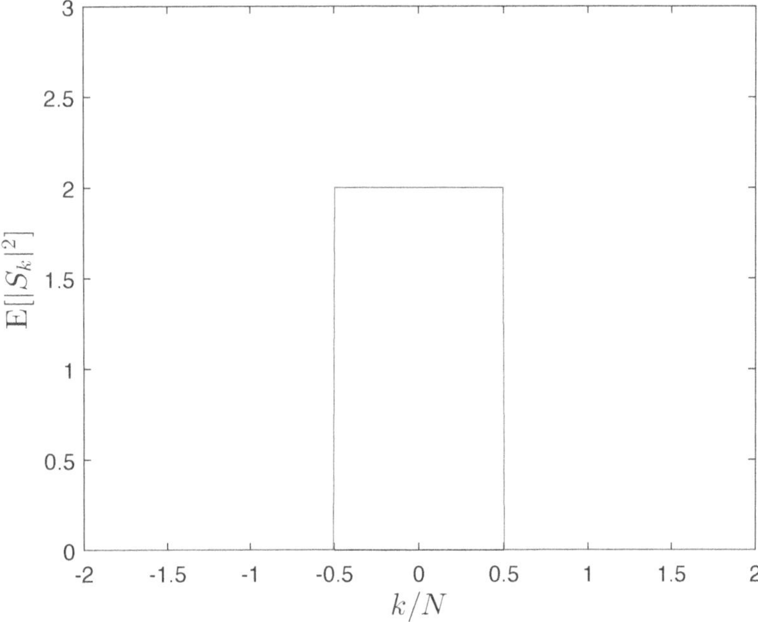

Figure 5.2 Rectangular PSD at the nonlinearity input considering N_u QPSK symbols and $O = 4$.

Therefore, the real and imaginary parts of each sample are Gaussian distributed with variance $\sigma^2 = 2/(N_u O^2)$ (see (2.25)) and PDF, $p(s)$, given by (2.26). Additionally, it is important to note that $\mathbb{E}[s_n] = 0$ and that the input signal autocorrelation is

$$R_{\tilde{s},bp}\left(n'\right) = \mathbb{E}\left[s_n s_{n-n'}^*\right], \tag{5.1}$$

Therefore, we have

$$R_{\tilde{s},bp}\left(n-n'\right) = \mathbb{E}\left[s_n s_{n-(n-n')}^*\right] = \mathbb{E}\left[s_n s_{n'}^*\right]. \tag{5.2}$$

By considering (2.17) and (2.4), the autocorrelation can be written as

$$\begin{aligned} R_{\tilde{s},bp}\left(n-n'\right) &= \mathbb{E}[s_n s_{n'}^*] \\ &= \frac{1}{N^2}\sum_{k=0}^{N-1}\sum_{k'=0}^{N-1}\mathbb{E}[S_k S_{k'}^*]\exp\left(-j2\pi\frac{kn-k'n'}{N}\right) \\ &= \frac{1}{N^2}\sum_{k=0}^{N-1}\mathbb{E}\left[|S_k|^2\right]\exp\left(-j2\pi\frac{k(n-n')}{N}\right). \end{aligned} \tag{5.3}$$

As expected, when $n' = n$, the autocorrelation gives the average power associated with the complex envelope of the bandpass OFDM signal, i.e.,

$$\begin{aligned} R_{\tilde{s},bp}(0) &= \mathbb{E}\left[|s_n|^2\right] \\ &= \frac{1}{N^2}\sum_{k=0}^{N-1}\mathbb{E}\left[|S_k|^2\right] \\ &= \frac{2}{N_u O^2} = 2\sigma^2. \end{aligned} \tag{5.4}$$

It is also important to note that

$$\begin{aligned} \mathbb{E}[S_k S_{k'}^*] &= \sum_{n=0}^{N-1}\sum_{n'=0}^{N-1}\mathbb{E}\left[s_n s_{n'}^*\right]\exp\left(-j2\pi\frac{kn-k'n'}{N}\right) \\ &= \sum_{n=0}^{N-1}\sum_{n'=0}^{N-1}R_{\tilde{s},bp}\left(n-n'\right)\exp\left(-j2\pi\frac{kn-k'n'}{N}\right) \\ &= \sum_{n'=0}^{N-1}\left(\sum_{n=0}^{N-1}R_{\tilde{s},bp}\left(n-n'\right)\exp\left(-j2\pi\frac{kn}{N}\right)\right)\exp\left(j2\pi\frac{k'n'}{N}\right). \\ &= \sum_{n'=0}^{N-1}G_{\tilde{s},bp}(k)\exp\left(-j2\pi\frac{(k-k')n'}{N}\right), \end{aligned} \tag{5.5}$$

where $G_{\tilde{s},bp}(k) \triangleq G_{\tilde{s},bp}(k/T)$ represents the kth sample associated with the PSD of the OFDM signal $G_{\tilde{s},bp}(f)$, that is the DFT of (5.1). Note that if $k = k'$, we obtain that

$$\mathbb{E}\left[|S_k|^2\right] = N G_{\tilde{s},bp}(k), \tag{5.6}$$

which means that the signal represented in Fig. 5.2 is just a scaled version of the PSD of the OFDM signal. As was mentioned in the previous chapter, the Bussgang's theorem [10] can be employed to characterize nonlinearly distorted OFDM signals, since OFDM samples present a Gaussian distribution when the number of subcarriers is large. Under these conditions, the samples of a nonlinearly distorted signal at the output of a bandpass nonlinearity $\mathbf{y} = [y_0\ y_1\ y_2\ ...\ y_{N-1}]^T \in \mathbb{C}^N$ can be written as the sum of two uncorrelated terms, i.e.,

$$\mathbf{y} = \alpha_{bp}\mathbf{s} + \mathbf{d}, \tag{5.7}$$

where $\mathbf{d} = [d_0\ d_1\ d_2\ ...\ d_{N-1}]^T \in \mathbb{C}^N$ represents the nonlinear distortion components and α_{bp} is given by (3.78). By focusing on the nth time-domain sample, we have

$$y_n = \alpha_{bp}s_n + d_n, \tag{5.8}$$

where d_n, that is uncorrelated with the input sample s_n, i.e., $\mathbb{E}\left[s_n d_{n'}^*\right] = 0$, represents the nonlinear distortion term associated with the nth time-domain sample.

To quantify the performance degradation associated with the conventional OFDM schemes impaired by bandpass nonlinearities it is useful to obtain the PSD of the transmitted signals. In fact, by making use of the Gaussian nature of OFDM signals, one can obtain the analytical characterization of the nonlinearly distorted OFDM signals. In Subsection 3.2.2, it was shown that the autocorrelation of a nonlinearly distorted Gaussian signal submitted to a bandpass nonlinearity is given by (3.104), i.e.,

$$R_{\tilde{y},bp}(\tau) = 2\sum_{\gamma=0}^{+\infty} P_{2\gamma+1}^{bp} \frac{R_{\tilde{s},bp}(\tau)^{\gamma+1} R_{\tilde{s},bp}^*(\tau)^{\gamma}}{R_{\tilde{s},bp}(0)^{2\gamma+1}}, \tag{5.9}$$

where $P_{2\gamma+1}^{bp}$ represents the power associated with the IMP of order $2\gamma+1$, defined in (3.102). Therefore, we have

$$\mathbb{E}\left[y_n y_{n'}^*\right] = R_{\tilde{y},bp}(n-n') \overset{(a)}{\approx} 2\sum_{\gamma=0}^{n_\gamma} P_{2\gamma+1}^{bp} \frac{R_{\tilde{s},bp}(n-n')^{\gamma+1} R_{\tilde{s},bp}^*(n-n')^{\gamma}}{\sigma^{2(2\gamma+1)}}, \tag{5.10}$$

where the approximation (a) is related to the truncated IMP approach, i.e., only n_γ IMPs are considered for the computation of the autocorrelation. In addition, as

$$R_{\tilde{y},bp}(n-n') = |\alpha_{bp}|^2 R_{\tilde{s},bp}(n-n') + R_{\tilde{d},bp}(n-n'), \tag{5.11}$$

and, since the IMP of order 1 is associated with the useful signal at the nonlinearity output, we have

$$\mathbb{E}\left[d_n d_{n'}^*\right] = R_{\tilde{d},bp}(n-n') \approx 2\sum_{\gamma=1}^{n_\gamma} P_{2\gamma+1}^{bp} \frac{R_{\tilde{s},bp}(n-n')^{\gamma+1} R_{\tilde{s},bp}^*(n-n')^{\gamma}}{\sigma^{2(2\gamma+1)}}. \tag{5.12}$$

By recalling (4.39), we have that frequency-domain samples of the nonlinearly distorted signal $\mathbf{Y} = [Y_0\ Y_1\ Y_2\ ...\ Y_{N-1}]^T \in \mathbb{C}^N$ can be divided as follows

$$\mathbf{Y} = \alpha_{bp}\mathbf{S} + \mathbf{D}, \tag{5.13}$$

where $\mathbf{D} = [D_0\ D_1\ D_2\ ...\ D_{N-1}]^T \in \mathbb{C}^N$ is a block with the frequency-domain version of the nonlinear distortion terms. Under these conditions,

$$Y_k = \alpha_{bp} S_k + D_k. \tag{5.14}$$

As in (5.5), we have that

$$\mathbb{E}\left[|Y_k|^2\right] = N G_{\tilde{y},bp}(k), \tag{5.15}$$

and

$$\mathbb{E}\left[|D_k|^2\right] = N G_{\tilde{d},bp}(k), \tag{5.16}$$

where $G_{\tilde{y},bp}(k)$ and $G_{\tilde{d},bp}(k)$ represent the DFT of (5.10) and (5.12), respectively. Additionally, by applying the DFT to (5.11), we have

$$G_{\tilde{y},bp}(k) = |\alpha_{bp}|^2 G_{\tilde{s},bp}(k) + G_{\tilde{d},bp}(k). \tag{5.17}$$

In the following, a set of results regarding the PSD of nonlinearly distorted signals obtained by simulation and theoretically are presented. Fig. 5.3 shows the PSD of a clipped OFDM signal obtained both by simulation (i.e., by considering $G_{\tilde{y},bp}(k) = \mathbb{E}\left[|Y_k|^2\right]/N$) and theoretically through the truncated IMP approach. Each OFDM

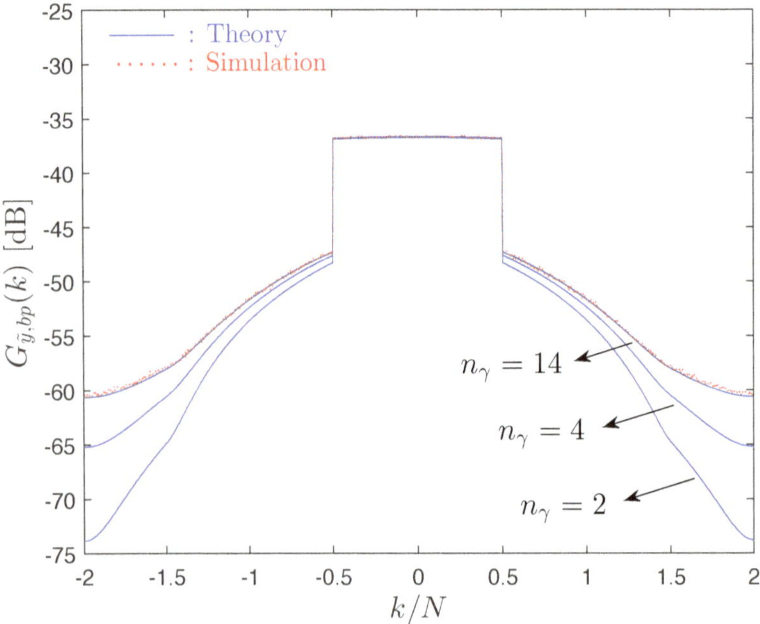

Figure 5.3 PSD of a clipped OFDM signal considering $s_M/\sigma = 0.5$ and different values of n_γ.

signal has $N_u = 256$ subcarriers and $O = 4$. The envelope clipping has a normalized clipping level $s_M/\sigma = 0.5$.

From the figure it can be seen that the PSD obtained through the truncated IMP approach can be very accurate. However, to obtain that degree of accuracy, and a maximum error relatively to the PSD obtained by simulation lower than 0.2 dB, a value of $n_\gamma = 14$ must be considered, which means that a large number of IMPs must be computed. In fact, when $n_\gamma = 4$, the error can reach a value around 0.5 dB, but it can even be approximately 13 dB when only 3 IMPs are considered, i.e., when $n_\gamma = 2$. Clearly, the accuracy of the truncated IMP approach is related to the number of IMPs.

To obtain the degradation associated with the nonlinear distortion for a given subcarrier, it is common to compute the corresponding signal-to-interference ratio (SIR) for that subcarrier. The SIR is defined as the ratio between the PSDs associated with the useful part and the distortion parts of the nonlinearly distorted signal, i.e.,

$$\begin{aligned} \text{SIR} &= \frac{|\alpha_{bp}|^2 G_{\tilde{s},bp}(k)}{G_{\tilde{d},bp}(k)} \\ &= \frac{|\alpha_{bp}|^2 \mathbb{E}\left[|S_k|^2\right]}{\mathbb{E}\left[|D_k|^2\right]}. \end{aligned} \tag{5.18}$$

Fig. 5.4 shows the SIR obtained by simulation and theoretically considering the truncated IMP approach and $n_\gamma = 14$. Each OFDM signal has $N_u = 512$ subcarriers, $O = 4$ and different clipping levels are considered. From the figure, it can be seen that accurate estimates of the SIR can be obtained. As expected, when the clipping level is lower, the SIR levels decrease and, for $s_M/\sigma = 0.5$, the SIR can even reach a minimum of around 8.6 dB at the middle of the band, since the distortion is maximum at that frequency (note that the PSD of the input signal is constant and the nonlinear distortion terms are approximately Gaussian distributed at the subcarrier level). On the other hand, for $s_M/\sigma = 1.5$, the SIR increases substantially since the distortion levels are lower. In that case, it can be observed that the minimum SIR is approximately 14.2 dB.

Let us now study the impact of the "severeness" of the nonlinearity in the number of required IMPs to obtain accurate PSDs. Fig. 5.5 shows the PSD of an OFDM signal amplified through a TWTA. The PSD was obtained both by simulation considering $G_{\tilde{y},bp}(k) = \mathbb{E}\left[|Y_k|^2\right]/N$ and theoretically with the truncated IMP approach and two values of n_γ. The OFDM signals have $N_u = 256$ useful subcarriers and $O = 4$. The TWTA has a normalized clipping level $s_M/\sigma = 0.5$ and a phase rotation at the saturation region of $\theta_M = \pi/4$ (see the corresponding AM/AM and AM/PM conversion functions of the Saleh's model represented in (3.27) and (3.28), respectively).

When such TWTA is employed, it can be noted that even if $n_\gamma = 14$, which is actually a high value, the error between the simulated and the theoretical PSD can reach approximately 1.6 dB. On the other hand, to obtain an error of approximately 0.2 dB, one should consider $n_\gamma = 20$. As expected, this confirms that the higher the magnitude of the nonlinear distortion effects, i.e., the severeness of the nonlinearity, the higher the number of IMPs required to obtain an accurate estimate of the PSD,

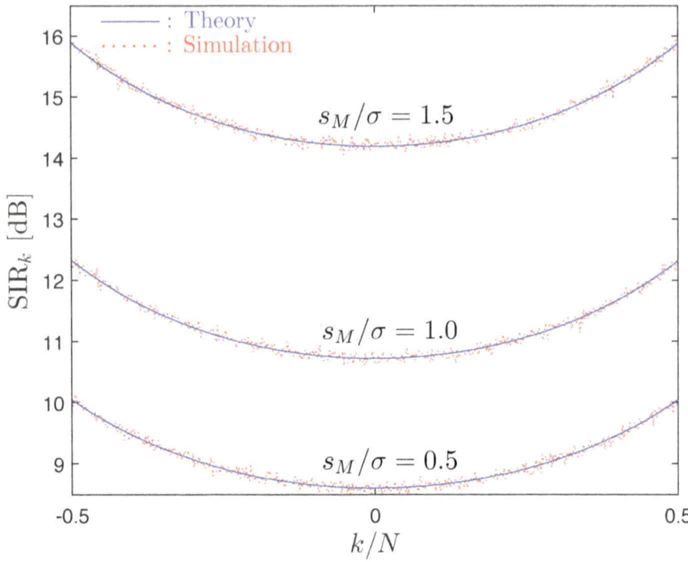

Figure 5.4 Simulated and theoretical SIR of a clipped OFDM signal considering different normalized clipping levels.

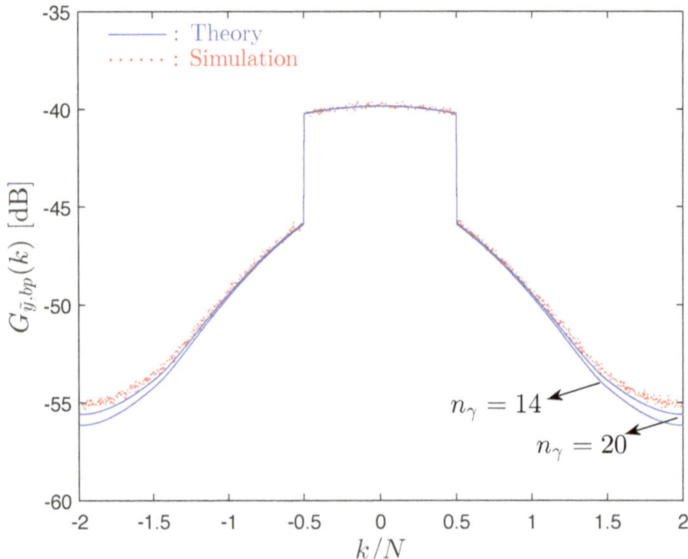

Figure 5.5 PSD associated an OFDM signal amplified with a TWTA with $s_M/\sigma = 0.5$ and $\theta_M = \pi/4$.

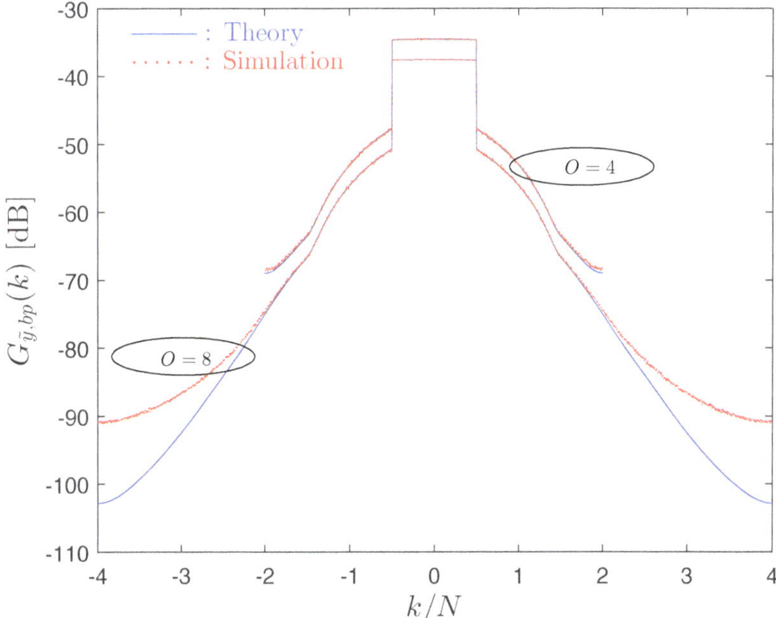

Figure 5.6 PSD associated an OFDM signal amplified by an SSPA with $s_M/\sigma = 1.0$ and $p = 3$ considering different oversampling factors.

since this number is intimately related to the number of polynomial terms employed to approximate the nonlinear characteristic.

Let us now assess the impact of the oversampling factor on the number of required IMP to obtain accurate PSDs. Fig. 5.6 presents the PSD of an OFDM signal amplified by an SSPA obtained both by simulation and theoretically through the truncated IMP approach. Each OFDM signal has $N_u = 512$ useful subcarriers and different oversampling factors are considered. The SSPA has a normalized clipping level $s_M/\sigma = 1.0$ and sharpness factor of $p = 3$ (see the corresponding AM/AM function of Rapp's model for SSPAs in (3.31)). In fact, although when $O = 4$ the difference between the simulated and theoretical PSD has a maximum value of approximately 0.4 dB, when the oversampling factor increases to $O = 8$, this difference increases substantially and can even reach a value around 10 dB. This means that when higher oversampling factors are considered, the number of required IMPs to obtain an accurate spectral characterization of the nonlinearly distorted signal is also higher.

Besides the high complexity associated with high n_γ values, convergence problems can also take place, i.e., increasing n_γ may not guarantee substantial lower errors between the simulated and the theoretical PSDs. This fact is illustrated in Fig. 5.7, that shows the PSD of a clipped OFDM signal considering $s_M/\sigma = 0.5$, $N_u = 512$, $O = 8$ and two values of n_γ. From the results depicted in the figure, one

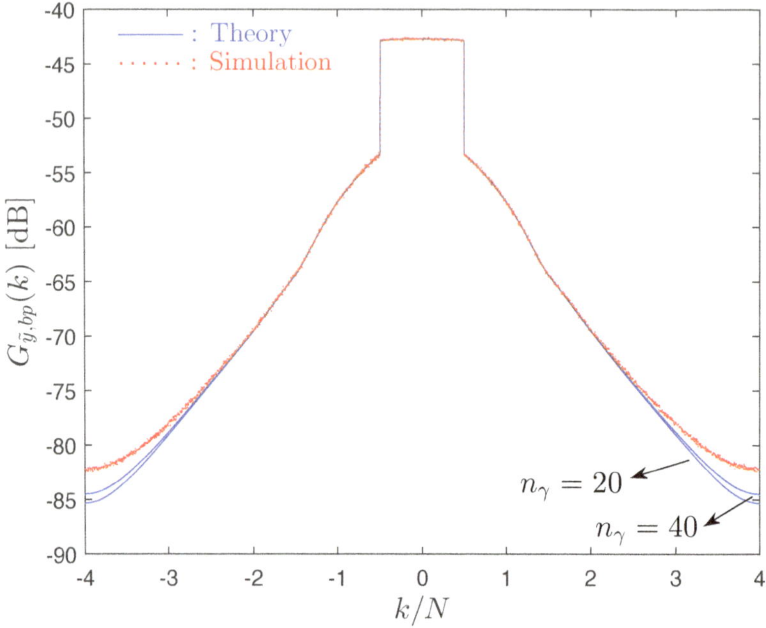

Figure 5.7 PSD of a clipped OFDM signal considering $s_M/\sigma = 0.5$ and $O = 8$.

can clearly note that even when the number of IMPs is doubled from $n_\gamma = 20$ to $n_\gamma = 40$, the maximum error between the simulated and the theoretical PSDs decreases only approximately 0.8 dB. This means that we can have unavoidable errors even when substantially increase n_γ. On the other hand, n_γ should not be large to avoid high complexity. Fig. 5.8 shows the maximum error between the PSDs of a nonlinearly distorted OFDM signal obtained by simulation and theoretically considering the truncated IMP approach, different nonlinearities and different oversampling factors. These figures summarize the conclusions taken above. In fact, it confirms that the number of IMP required to obtain good accuracy varies with the sharpness of the nonlinearity and with the oversampling factor and can be very large in some scenarios (i.e., when there are high oversampling factors and/or severe nonlinearities), not to mention the existence of convergence problems that are expressed as "error floors". In fact, the convergence and complexity problems are the key limitations of the truncated IMP approach.

In [91, 92] it was shown that the use of equivalent nonlinearities to substitute the conventional non-smooth, nonlinear characteristics can reduce the convergence and complexity problems associated with the truncated IMP approach, providing a simple and accurate spectral characterization of sampled, nonlinearly distorted signals. In the following, results for the PSD associated with a nonlinearly distorted, bandpass OFDM signal are given. These results are obtained by simulation and theoretically

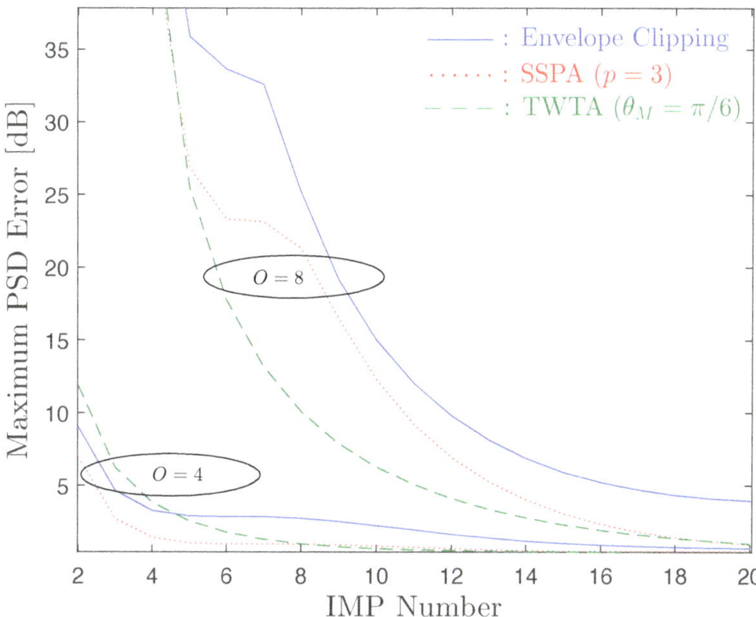

Figure 5.8 Maximum error between the simulated and the theoretical version of the PSD of a nonlinearly distorted OFDM signal considering the truncated IMP approach, different nonlinearities and different oversampling factors.

and the accuracy of the truncated IMP approach with n_γ IMPs is compared to the accuracy of the equivalent nonlinearity approach (see Subsection 3.3.2 for theoretical details of that approach) with $\gamma_{\max}$ IMPs.

Fig. 5.9 shows the simulated and the theoretical PSD of a clipped OFDM signal with $N_u = 512$ and $O = 4$ when $s_M/\sigma = 0.5$.

From the figure, it can be noted that the theoretical PSD obtained with the equivalent nonlinearity is more accurate than the one associated with the conventional truncated IMP approach. In fact, although they are obtained with the same number of IMPs, i.e., $n_\gamma = \gamma_{\max} = 8$, the theoretical PSD associated with the equivalent nonlinearity has a maximum error of approximately 0.2 dB when compared to the simulated PSD, while this value increases to approximately 1.3 dB when the truncated IMP approach is considered. This "accuracy" gain is more visible when a severe nonlinearity is considered, as is shown in Fig. 5.10. This figure shows the simulated and the theoretical PSD associated with an OFDM signal amplified by a TWTA. Note that although the signal that goes through the TWTA is continuous, we can have an insight on the impact of the nonlinear distortion by considering its sampled version, provided that the oversampling factor is high enough. Each OFDM signal has $N_u = 512$ and $O = 4$. The TWTA has a saturation output of $s_M/\sigma = 0.5$ and a phase rotation of $\theta_M = \pi/3$.

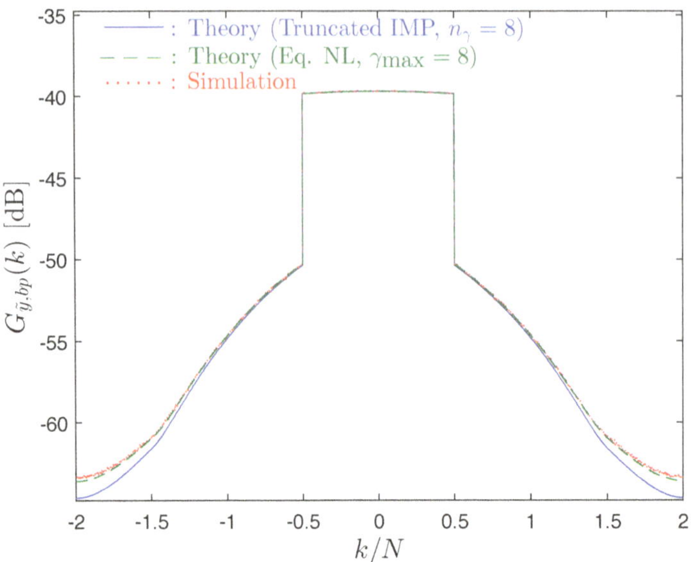

Figure 5.9 PSD of a clipped OFDM signal considering $s_M/\sigma = 0.5$, $O = 4$ and $n_\gamma = \gamma_{\max} = 8$.

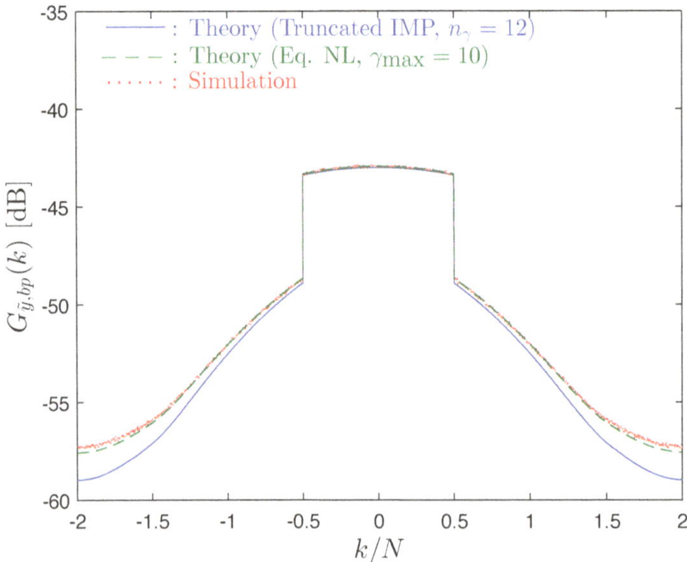

Figure 5.10 PSD of a nonlinearly amplified OFDM signal considering both the truncated IMP approach and the equivalent nonlinearity.

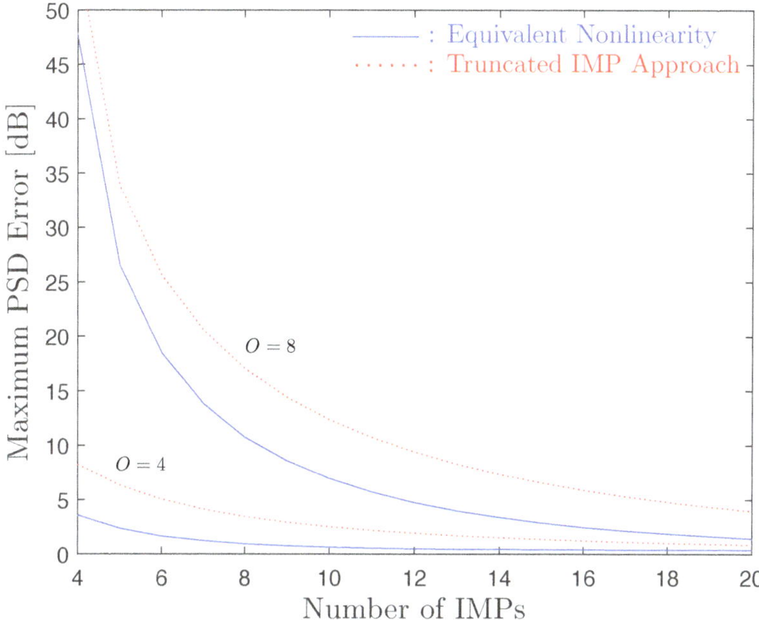

Figure 5.11 Maximum error between the simulated and theoretical PSD of a nonlinearly amplified OFDM signal considering the truncated IMP approach and the equivalent nonlinearity.

As previously mentioned, when a TWTA is considered, the truncated IMP approach requires a large number of IMPs to provide an accurate PSD. However, even when a very high number of IMPs is considered, i.e., $n_\gamma = 12$, the truncated IMP approach presents a maximum error of approximately 1.8 dB. In contrast, when the equivalent nonlinearity is employed, there is an almost negligible error, even when a lower number of IMPs is considered, i.e., for $\gamma_{\max} = 10$.

Fig. 5.11 shows the maximum error between the PSDs of a nonlinearly amplified OFDM signal obtained by simulation and theoretically, considering both the truncated IMP and the equivalent nonlinearity approaches, was well as different oversampling factors. These two approaches are compared with the same number of IMPs, i.e., $n_\gamma = \gamma_{\max}$. The nonlinear amplifier is a TWTA with normalized saturation level $s_M/\sigma = 0.5$ and $\theta_M = \pi/3$. Clearly, the use of the equivalent nonlinearity approach leads to accuracy gains relatively to the truncated IMP approach. As expected, these gains are higher when higher oversampling factors are employed. For instance, when $O = 4$ and 6 IMPs are considered, the accuracy gain is around 6 dB. However, when $O = 8$ and 8 IMPs are taken into account, this accuracy gain increases to approximately 7 dB. Besides these considerable accuracy gains that are more noticeable in the out-of-band region, it can be observed that when the equivalent nonlinearity is

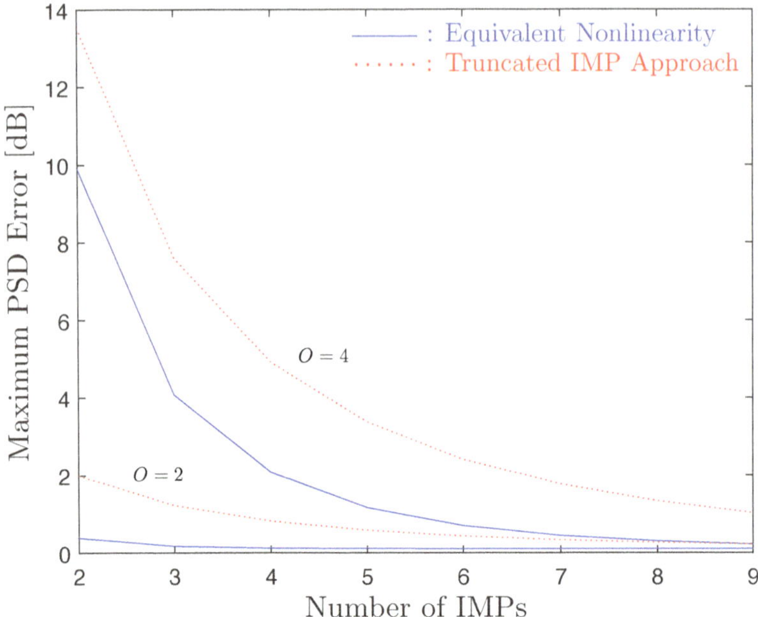

Figure 5.12 Maximum error between the simulated and theoretical PSD of a clipped OFDM signal considering the truncated IMP approach and the equivalent nonlinearity.

considered, the error relatively to the simulated PSD converges to zero at a higher "rate". This effect can also be seen in Fig. 5.12, which shows the maximum error between the simulated and theoretical PSDs considering an envelope clipping with $s_M/\sigma = 0.5$ and oversampling factors $O = 2$ and $O = 4$. For these reasons, the use of equivalent nonlinearities constitutes an efficient, low-complex method for obtaining the analytical spectral characterization of nonlinearly distorted, bandpass OFDM signals [91,92].

5.1.2 POTENTIAL OPTIMUM PERFORMANCE

Here, we present results regarding the optimum performance of nonlinearly distorted OFDM schemes whose equivalent model is depicted in Fig. 5.13. Some of these results, namely the asymptotic optimum performance, derive from the theory presented in Section 4.2 .

As seen in Subsection 4.1, the existence of nonlinear distortion effects can severely prejudice the performance of conventional OFDM systems. Let us recall the unwanted effects associated with the nonlinear distortion. In fact, if the nonlinear distortion is considered as an additional noise component, the BER can be substantially higher comparatively to the one of linear, OFDM systems. As verified in (5.14), the

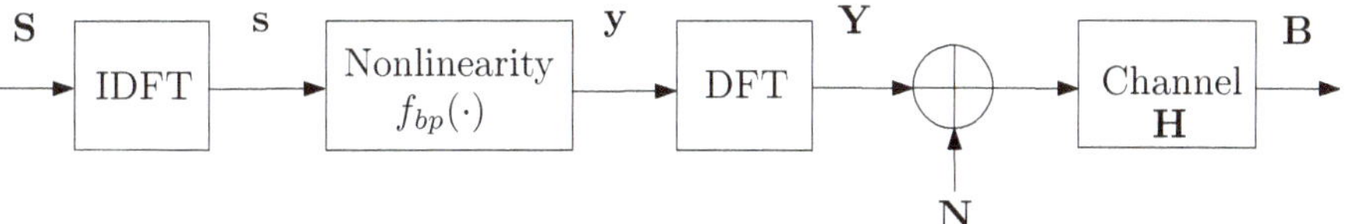

Figure 5.13 Equivalent, subcarrier-level model for a nonlinearly distorted OFDM scheme.

transmitted signal for a given subcarrier k, Y_k, can be divided into uncorrelated useful and distortion components [10], and each subcarrier may present a low SIR, if the nonlinearity is severe. In addition, depending on the nature of the scale factor α_{bp}, which can be either real-valued or complex-valued, the constellation of the transmitted signals can be rotated. Fig. 5.14 shows the constellation of the transmitted symbols when a TWTA with normalized saturation level $s_M/\sigma = 3.0$ and $\theta_M = \pi/4$ is considered for the amplification process. Clearly, due to the complex nature of the scale factor α_{bp}, a rotation of the constellation can be observed from the figure (this rotation effect, however, is not observed in Fig. 4.2 since, in this case, the scale factor α_{bp} is real-valued). In addition, due to the existence of nonlinear distortion terms D_k, exists a "cloud" of points around the shrunken and rotated version of the

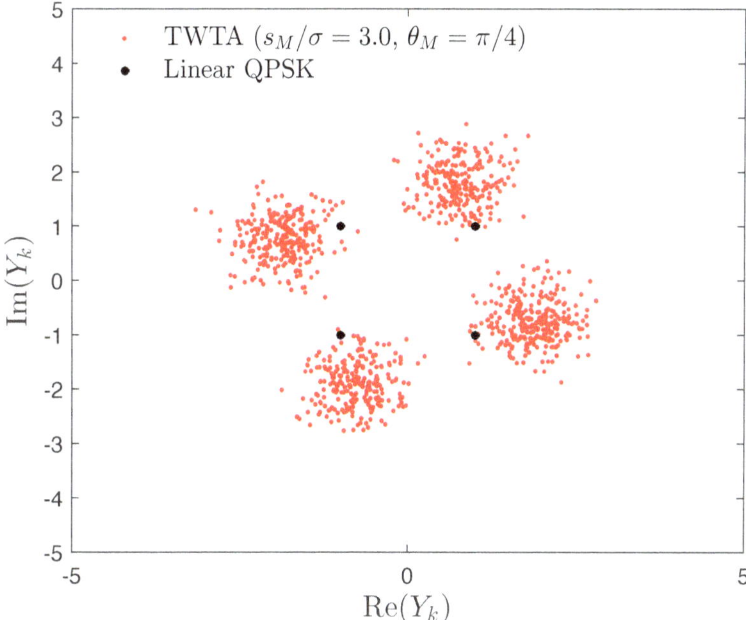

Figure 5.14 Constellation of the transmitted symbols when a TWTA with $s_M/\sigma = 3.0$ and $\theta_M = \pi/4$ is employed.

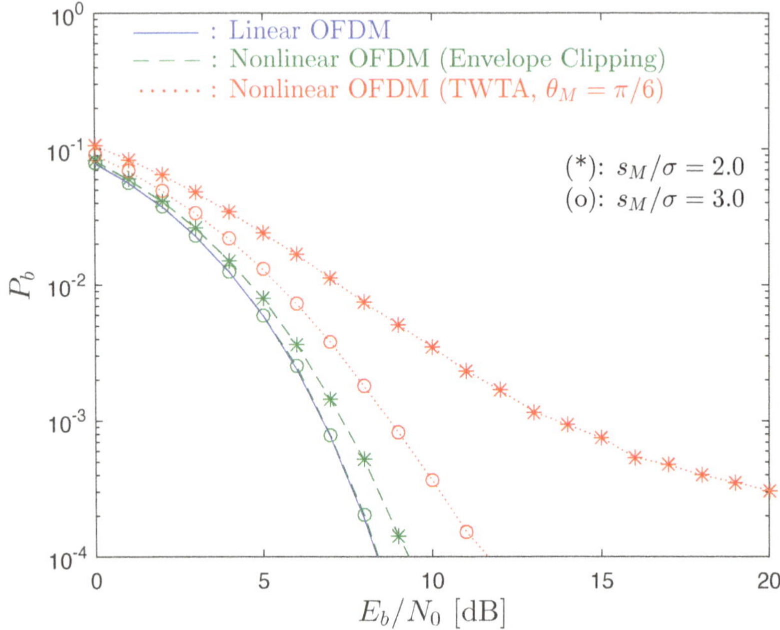

Figure 5.15 Simulated BER of a conventional OFDM receiver considering both linear and a nonlinear OFDM transmissions.

constellation. As aforementioned, the existence of these nonlinear distortion effects degrades the performance of OFDM systems. This is confirmed in Fig. 5.15, which shows the simulated BER of conventional receivers considering both linear and nonlinear OFDM transmissions with $N_u = 512$ and $O = 4$. Results for both an envelope clipping and TWTA with $\theta_M = \pi/6$ are shown. It should be also mentioned that a perfect channel estimation implicitly involves a perfect estimation of α_{bp}[1]. From the figure, it can be noted that the performance degradation associated to the nonlinear distortion can be severe and we can even have an irreducible error floor. Although when $s_M/\sigma \geq 2.0$ this degradation is relatively low when an envelope clipping is employed (the impact on the performance is noticeable for lower clipping levels as shown in Fig. 4.3), this is not the situation when a nonlinear power amplifier such as a TWTA is considered, where considerable performance penalties can be observed.

As demonstrated in the previous chapter, the optimum detection can be a good alternative to conventional detection of nonlinearly distorted multicarrier signals. In the following, results associated with the theoretical asymptotic optimum

[1] Although for QPSK constellations and bandpass nonlinearities with only AM/AM conversion function the estimation of α_{bp} is not necessary, the estimation of α_{bp} for bandpass nonlinearities with AM/PM conversion function should be made since, in those cases, the constellation is rotated. However, in practice the estimation of this factor can implicitly be done by conventional channel estimation procedures [100].

performance of conventional OFDM schemes that have bandpass nonlinearities on their transmission chain (see Fig. 5.1) are presented [15, 93–95]. As shown in Subsection 4.4, the theoretical average asymptotic gain associated with the optimum performance of OFDM signals impaired by bandpass nonlinearities is given by

$$G(1) \approx \frac{(d_{\text{adj}}\sigma)^2}{4} \frac{\int_0^{+\infty} \left(A'^2(r) + \frac{A^2(r)}{r^2} + \Theta'^2(r)A^2(r)\right) p(r)dr}{\int_0^{+\infty} A^2(r)p(r)dr}. \tag{5.19}$$

Therefore, the BER in the asymptotic region can be expressed as

$$P_b \approx Q\left(\sqrt{\frac{2G(1)E_b^{(nl)}}{N_0}}\right). \tag{5.20}$$

This approximation involves two assumptions: (i) a large number of subcarriers N_u since, in this case, the squared Euclidean distance between two nonlinearly distorted OFDM signals tends to assume a unique value and (ii) the error events associated with sequences that differ in more than $\mu = 1$ bits are neglected (this assumes that the squared Euclidean distance between two nonlinearly distorted signals that differ in $\mu > 1$ bits is higher than for the cases where $\mu = 1$ with very high probability). In the concrete case of an ideal envelope clipping, whose the AM/AM conversion function is given by (4.1), (5.19) turns into

$$G(\mu) = \frac{\mu(d_{\text{adj}}\sigma)^2}{4} \frac{\int_0^{+\infty} \left(A'^2(r) + \frac{A^2(r)}{r^2}\right) p(r)dr}{\int_0^{+\infty} A^2(r)p(r)dr}, \tag{5.21}$$

since $\Theta(R) = 0$. Fig. 5.16 shows the evolution of the average asymptotic gain considering an ideal envelope clipping with different normalized clipping levels s_M/σ, $\mu = 1$, $O = 4$ and different values of N_u. Clearly, (4.92) constitutes an accurate theoretical expression to obtain the average value of the asymptotic gain. As expected, this accuracy increases with N_u, since the Taylor approximations of (4.92) (for instance, the one made in (4.76)) become more precise when the number of subcarriers is large. When $N_u = 256$, it can be seen that the simulated and theoretical average values of the asymptotic gain are almost equal. In fact, even when a lower number of subcarriers is considered, the theoretical results present a considerable accuracy and only for small clipping levels deviate from the simulated results. From the figure, it can also be noted that when s_M/σ increases, the gain tends to 0 dB, which is in agreement with the results of the simulated PDFs of Fig. 4.10. Indeed, when $s_M/\sigma \to \infty$,

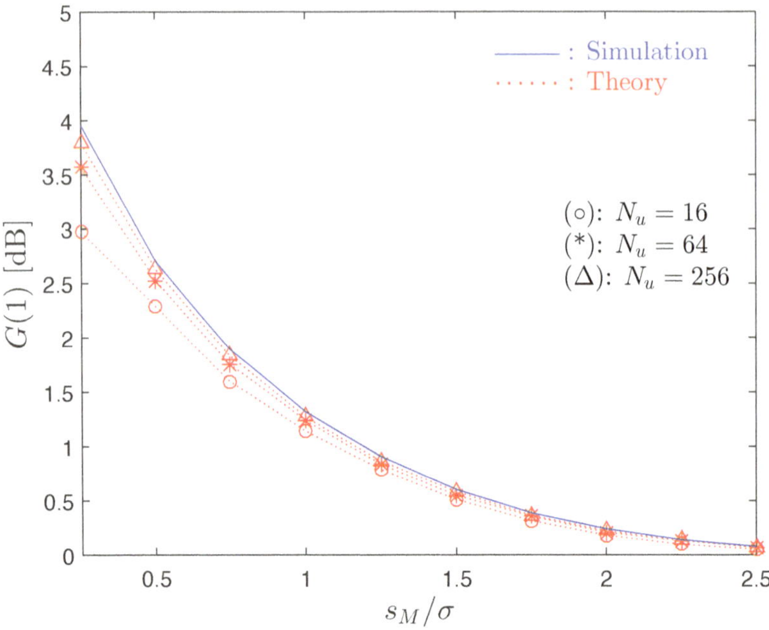

Figure 5.16 Evolution of the average asymptotic gain considering different normalized clipping levels, $\mu = 1$ and different values of N_u.

the transmission is linear and there is no asymptotic gain. On the other hand, it can be seen that the asymptotic gain increases when the clipping level decreases, i.e., increases with the severeness of the clipping. This can be clearly observed in Fig. 5.17 that shows the simulated and theoretical average asymptotic gain associated with the optimum detection of nonlinearly amplified signals considering different amplifiers and different values of N_u. The TWTA has $\theta_M = \pi/3$ and the SSPA has $p = 1$. From the results depicted in the figure, it can be noted that when a severe nonlinearity such as a TWTA is considered, the average asymptotic gain can be very large, which might mean higher performance improvements, comparatively to the conventional detection of linear, OFDM signals. Regardless of the type of the amplifier, it can also be noted that (5.19) yields very accurate estimates of the asymptotic gain. Fig. 5.18 shows the asymptotic BER of linear and nonlinear bandpass OFDM transmissions, obtained through the distribution of the asymptotic gain $G(\mu)$, for $\mu = 1$, $N_u = 256$, $O = 4$ and an envelope clipping with normalized clipping level s_M/σ.

From this figure, asymptotic gains relatively to linear OFDM transmissions can be clearly observed. For instance, for a target BER of $P_b = 10^{-4}$, the performance gain relatively to linear OFDM transmissions is around 1.3 dB when $s_M/\sigma = 1.0$, and increases to approximately 2.5 dB, when $s_M/\sigma = 0.5$.

As seen in the previous chapter, the optimum detection presents also potential asymptotic gains when frequency-selective channels with L uncorrelated Rayleigh

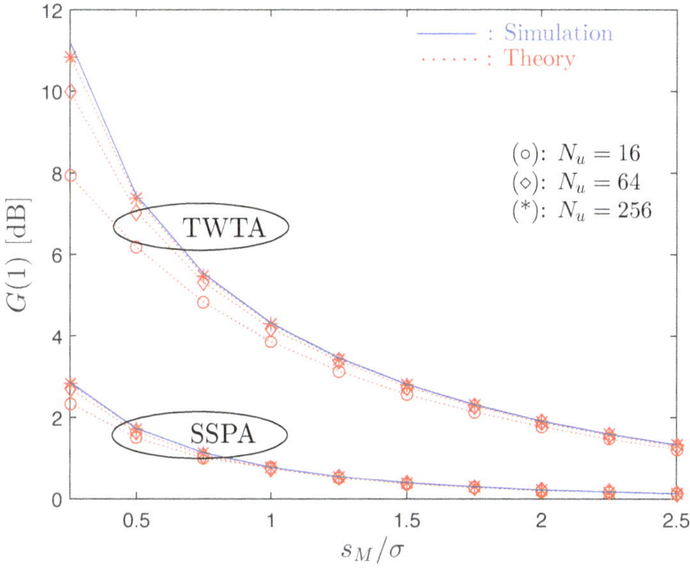

Figure 5.17 Evolution of the average asymptotic gain considering nonlinearly amplified OFDM signals, $\mu = 1$ and different values of N_u.

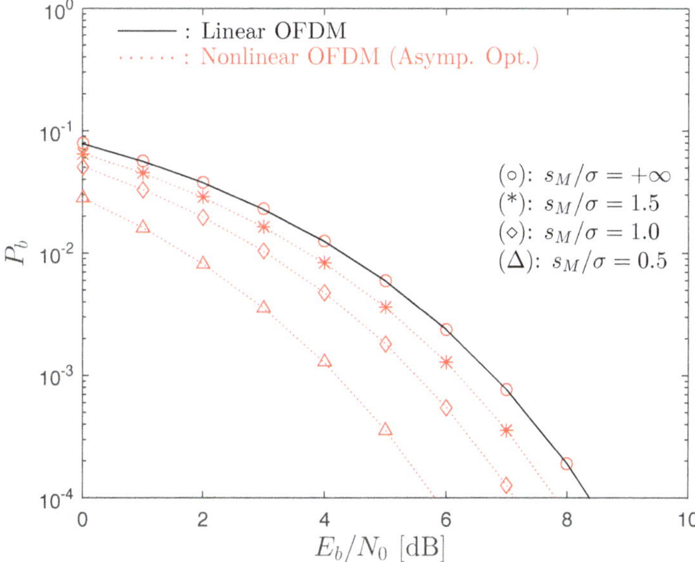

Figure 5.18 Asymptotic optimum receiver BER considering nonlinearly distorted signals submitted to an envelope clipping operation with normalized clipping level s_M/σ.

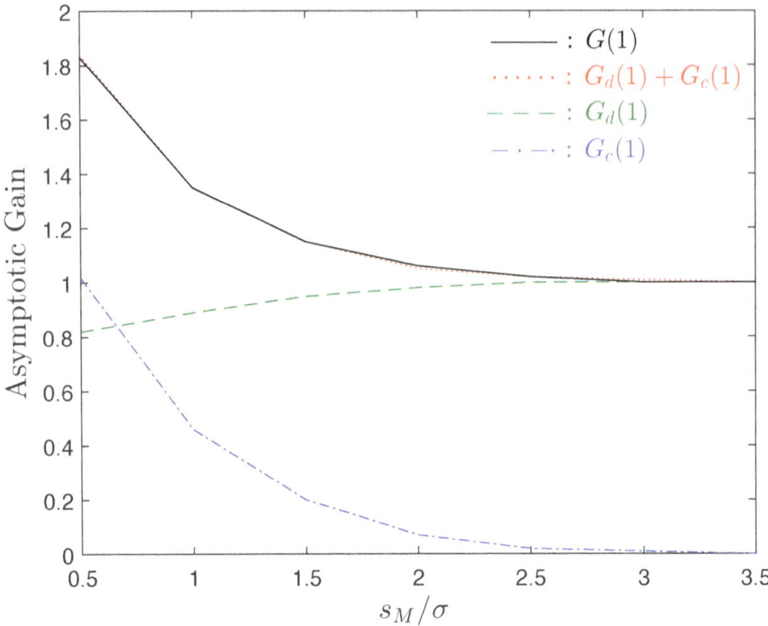

Figure 5.19 Evolution of the asymptotic gain components with the clipping level considering ideal AWGN channels.

components are considered. In addition, it was demonstrated that the asymptotic gain's distribution associated with the optimum detection in those scenarios is given by

$$p(G^{(H)}(\mu)) = \frac{\left(\frac{\mu}{G_d(\mu)}\right)^{\mu}\left(\frac{L}{G_c(\mu)}\right)^{L}}{\Gamma(L+\mu)} G^{(H)}(\mu)^{L+\mu-1}\exp\left(-\frac{G^{(H)}(\mu)L}{G_c(\mu)}\right)$$
$$\times M\left(\mu, L+\mu, G^{(H)}(\mu)\left(\frac{L}{G_c(\mu)}-\frac{\mu}{G_d(\mu)}\right)\right). \quad (5.22)$$

In fact, from the above equation, one can note that the asymptotic gains in frequency-selective channels are function of the asymptotic gain in ideal AWGN channels, $G(\mu)$, which can be decomposed as

$$G(\mu) \approx G_d(\mu) + G_c(\mu). \quad (5.23)$$

Fig. 5.19 shows the asymptotic gains associated with the optimum detection of OFDM signals in AWGN channels, as well as its two components, $G_d(\mu)$ and $G_c(\mu)$, considering $N_u = 512$, $O = 4$, $\mu = 1$ and an envelope clipping with different clipping levels s_M/σ.

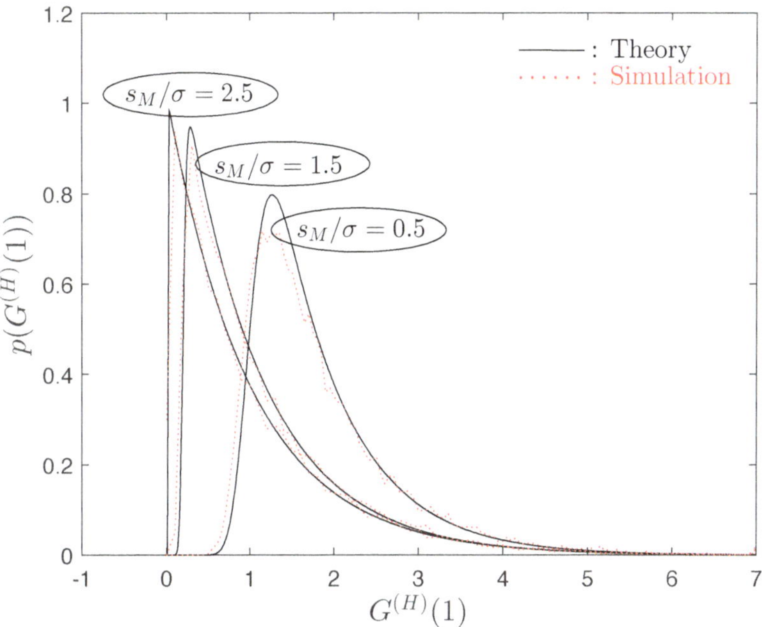

Figure 5.20 Distribution of $G^{(H)}(1)$ considering an envelope clipping.

From the figure, it can be noted that the division of the asymptotic gain into two components (the approximation of (5.23)) can be considered with accuracy. In fact, it can be seen that the asymptotic gain in ideal AWGN channels (given theoretically by (5.19)) can be obtained by the sum of $G_d(\mu)$ and $G_c(\mu)$. On the other hand, these two gain components can be employed to obtain (5.22). As expected, when s_M/σ increases, the gain component of the nonlinear distortion terms, $G_c(1)$, tends to 0, since the magnitude of the distortion effects decreases. In those scenarios, the asymptotic gain tends to 1 (0 dB) and is almost only associated with subcarriers that have bit errors, i.e., only the gain component $G_d(1)$ contributes for the "total gain" $G(1)$.

Fig. 5.20 shows asymptotic gain's distribution considering a frequency-selective channel with $L = 32$, $N_u = 512$, $O = 4$ and an envelope clipping. In this figure, it can be seen that (5.22) constitutes an accurate theoretical expression for obtaining the distribution of the gain when the channel presents frequency selectivity. It should be mentioned that although the gain can be lower than 1, this is also the case when linear OFDM transmissions are considered, as can be noted when the clipping level increases and the magnitude of the nonlinear distortion effects is small. However, when the signals are nonlinearly distorted, the event of having an asymptotic gain lower than 1 is more rare, which leads to performance gains over the conventional detection of linear, OFDM schemes.

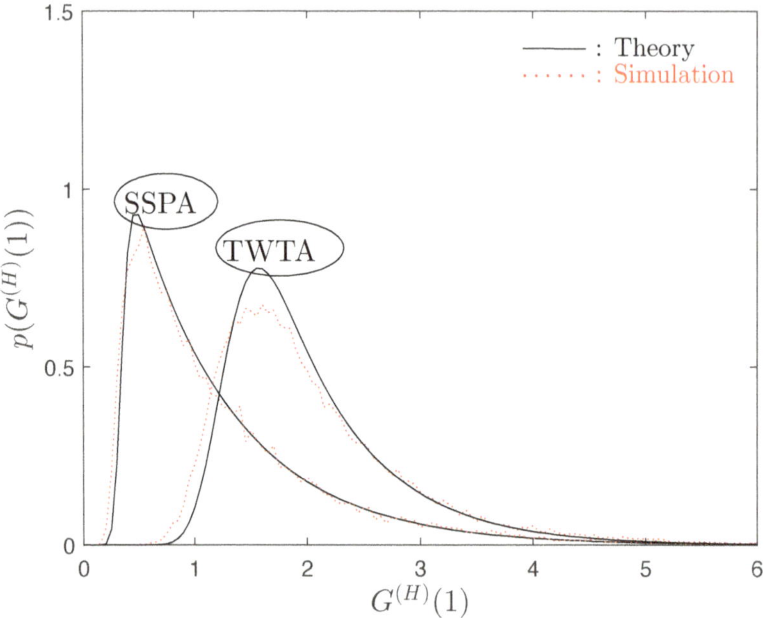

Figure 5.21 Distribution of $G^{(H)}(1)$ considering an OFDM system impaired by different bandpass nonlinearities.

Fig. 5.21 shows the distribution of the asymptotic gain considering $N_u = 1024$, $O = 4$, and $s_M/\sigma = 1.0$ and two different amplifiers: an SSPA with $p = 3$ and a TWTA with $\theta_M = \pi/6$. The results depicted in this figure reveal the accuracy of (5.22) for these different nonlinearities. Furthermore, it should be mentioned that when an SSPA with $p = 3$ and $s_M/\sigma = 1.0$ is employed, the gains are lower than when a TWTA with the same clipping level and $\theta_M = \pi/6$ is considered, since the latter nonlinearity is more severe and leads to the existence of stronger nonlinear distortion effects. Fig. 5.22 presents the asymptotic BER of linear and nonlinear bandpass OFDM transmissions in frequency-selective channels calculated through (5.22), considering $\mu = 1$, $L = 32$, $N_u = 256$, $O = 4$, and an envelope clipping with normalized clipping level s_M/σ. From the results depicted in the figure, it can be seen that the optimum receiver presents potential asymptotic gains relatively to linear OFDM transmissions in frequency-selective channels. For instance, for a target BER of $P_b = 10^{-2}$, the performance gain is around 6.3 dB when $s_M/\sigma = 1.5$, but can reach a value around 11 dB when $s_M/\sigma = 0.5$. As expected, when $s_M/\sigma \to \infty$, the optimum receiver's performance coincides with the performance of conventional, linear OFDM receivers.

It is widely known that the performance of OFDM in frequency-selective channels can be very poor and it is strongly conditioned by the subcarriers that are in deepest

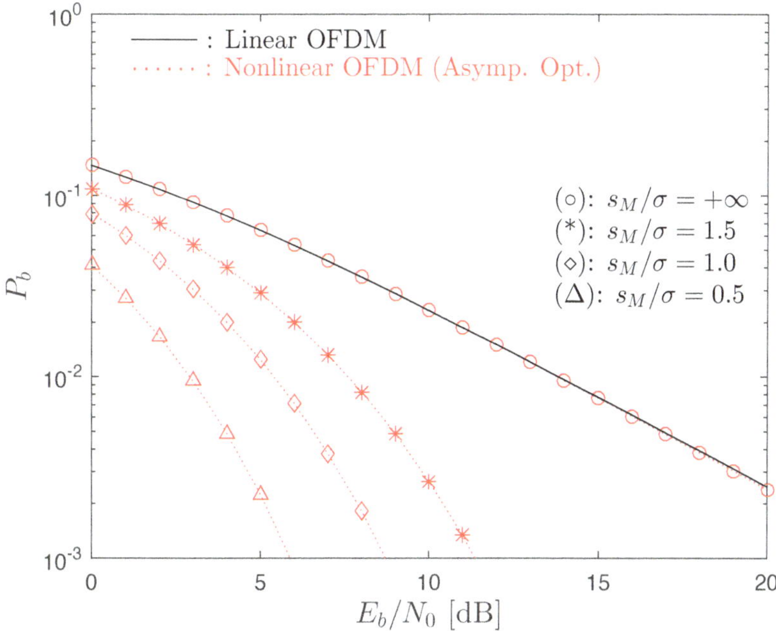

Figure 5.22 Approximate optimum performance of a nonlinear distorted OFDM scheme considering an envelope clipping operation and frequency-selective channels.

fades. To reduce the performance degradation, it is common to employ channel coding schemes. Although the performance of COFDM with nonlinear transmitters and conventional receivers is worse than with linear transmitters, the degradation can be negligible due to the high coding gains of COFDM. Therefore, one might ask what is the optimum asymptotic performance of COFDM with strong nonlinear effects, and how it compares with the optimum performance of linear OFDM. In COFDM schemes, the encoded data sequences differ in many bits. For this reason, it is of interest to study a scenario where the data sequences differ in $\mu > 1$ bit, since this is the situation of COFDM signals. Note that a scenario where the sequences differ in $\mu > 1$ bit can be seen as a scenario where we employ channel coding with minimum Hamming distance equal to μ, combined with an appropriate interleaving (to assure that the bits where the sequences differ are in uncorrelated positions). Let us start by taking into account the scenario of COFDM in ideal AWGN channels. From (5.21), it is clear that the asymptotic gain increases linearly with μ, i.e., which suggests that the optimum detection of nonlinear COFDM schemes present asymptotic gains relatively to linear COFDM schemes. Fig. 5.23 presents the theoretical and simulated asymptotic gain obtained with $N_u = 512$, $O = 4$, different values of μ and an SSPA with $p = 1$. In fact, it can be noted that the gain increases linearly with μ (a similar effect was observed for other values of μ). For instance, when $s_M/\sigma = 1.0$

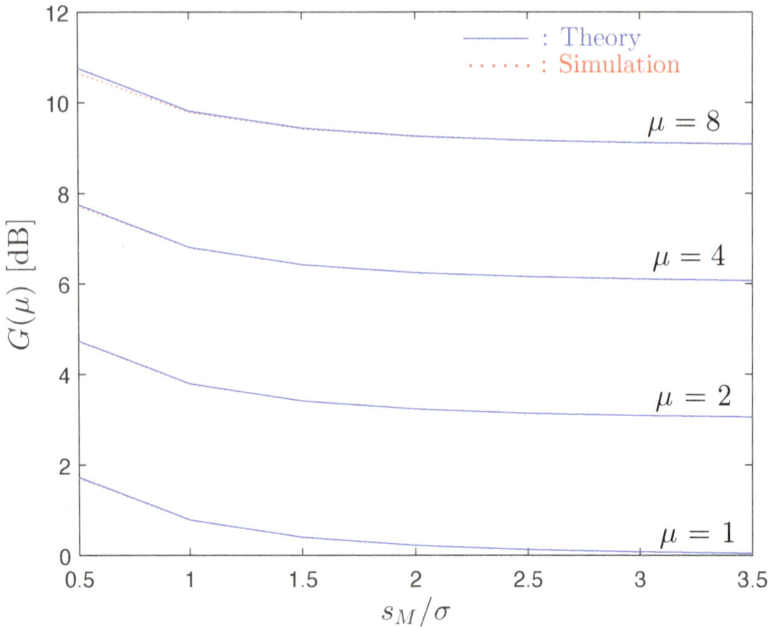

Figure 5.23 Evolution of the average asymptotic gain considering an SSPA with $p = 1$ and different values of μ.

and $\mu = 1$, the asymptotic gain is approximately 0.8 dB. However, it increases to approximately 3.8 dB, 6.8 dB and 9.8 dB for $\mu = 2$, $\mu = 4$ and $\mu = 8$, respectively. Fig. 5.24 shows the asymptotic optimum performance associated nonlinear OFDM and COFDM schemes in AWGN channels. A 64-state convolutional code with rate $C_r = 1/2$ and minimum Hamming distance $D_{\text{Hamming}} = 10$ is considered. The nonlinearity is an envelope clipping with clipping level s_M/σ. Clearly, it can be noted from the figure that besides the natural performance improvements that COFDM presents relatively to uncoded OFDM due to the coding gains, the optimum performance of COFDM schemes present potential asymptotic gains[2] relatively to linear COFDM schemes [101]. In fact, it can be observed that for a target BER of $P_b = 10^{-4}$, the use of an envelope clipping with $s_M/\sigma = 1.0$ leads to a performance gain of approximately 0.4 dB relatively to the linear COFDM. This performance gain increases to a value around 0.85 dB for $s_M/\sigma = 0.5$.

Let us now focus on the case of frequency-selective channels. Fig. 5.25 shows the asymptotic gain's distribution considering $N_u = 1024$, $O = 4$, $s_M/\sigma = 1.0$ and different values of μ. From the results depicted in the figure, it can be noted that

[2]Note that to obtain the asymptotic gain relatively to linear COFDM schemes, the squared Euclidean distance between nonlinearly distorted COFDM signals should be compared to $4C_rD_{\text{Hamming}}E_b$.

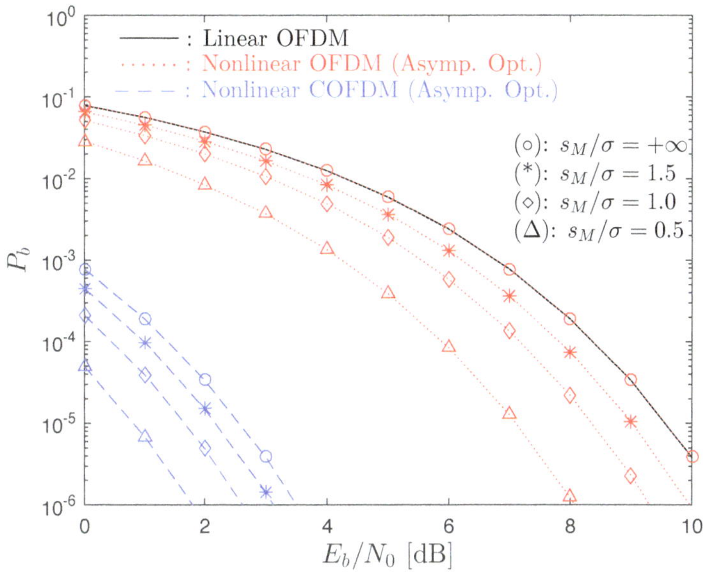

Figure 5.24 Asymptotic optimum performance of COFDM signals submitted to an envelope clipping operation in AWGN channels.

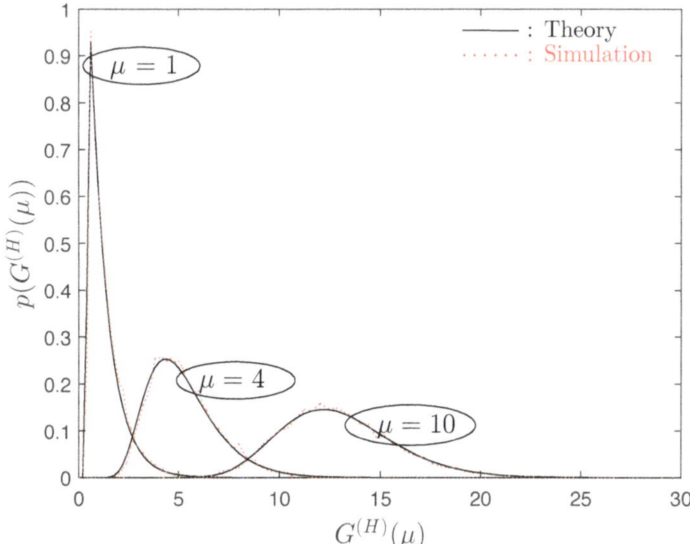

Figure 5.25 Distribution of $G^{(H)}(\mu)$ considering an envelope clipping with $s_M/\sigma = 1.0$ and different values of μ.

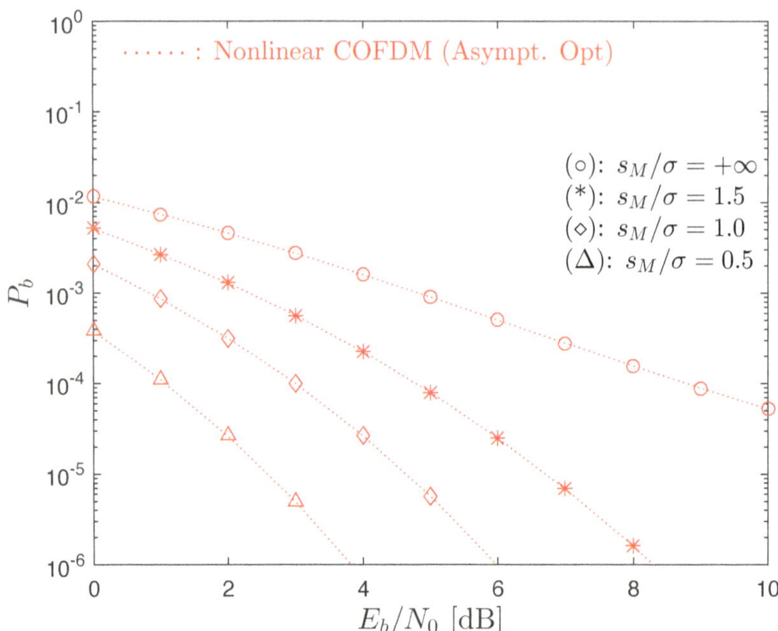

Figure 5.26 Asymptotic optimum performance of nonlinear COFDM schemes in frequency-selective channels.

(5.22) yields an accurate distribution of the asymptotic gains when $\mu \geq 1$. As expected, these asymptotic gains increase with the number of different bits between the sequences, which suggests that the use of channel coding schemes might lead to performance improvements, namely when the optimum detection of nonlinear COFDM is compared to the conventional detection of linear COFDM. Fig. 5.26 shows the asymptotic optimum performance associated nonlinear COFDM schemes in frequency-selective channels. Once again, a 64-state convolutional code with rate $C_r = 1/2$ and minimum Hamming distance $D_{\text{Hamming}} = 10$ is used. The nonlinearity is an envelope clipping. In this figure, it can be noted that in frequency-selective channels there are also potential asymptotic improvements associated with the optimum detection of nonlinearly distorted COFDM signals. In fact, to obtain a target BER of $P_b = 10^{-4}$, one only needs an E_b/N_0 of around 8.8 dB, which means that the optimum performance of linear COFDM is substantially better than the performance of linear OFDM schemes, which is an expected result. However, when nonlinear transmissions are considered, the required E_b/N_0 to obtain that BER decreases to approximately 4.8 dB when $s_M/\sigma = 1.5$ and to only 1 dB when $s_M/\sigma = 0.5$.

5.1.3 SUB-OPTIMUM DETECTION

In fact, although these large asymptotic gains point out to considerable performance improvements, the complexity of the optimum receiver constitutes an important problem that can compromise its practical implementation, even when the number of subcarriers and/or the constellation size is small. Nevertheless, it was already demonstrated that even sub-optimum, less complex receivers present performance gains relatively to the conventional detection of linear OFDM transmissions [102, 103]. This was observed not only for the specific scenarios associated with the results of the previous section, but also for other scenarios such as for OFDM systems employing clipping and filtering or iterative clipping and filtering techniques [104], receive diversity or M-QAM constellations (with $M > 4$) [105]. The main idea behind the sub-optimum detection is to reduce the complexity relatively to the optimum detection, but at the same time try to achieve its potential asymptotic gains. This can be accomplished by testing a reduced number of possible transmitted sequences, instead of analyzing all of them. In [102], different sub-optimum receivers were proposed. For instance, one can consider sub-optimum receivers that:

(a) Compute all the possible combinations of the q less reliable bits and choose the one that presents the lower Euclidean distance relatively to the received signal. Note that this receiver tests 2^q sequences for making a decision.
(b) Test all the 1 bit variations of the hard-decision sequence and verify what is the one that presents the lower Euclidean distance relatively to the received signal. This sub-optimum receiver tests $\log_2(M)N_u$ sequences before making a decision.

Although these sub-optimum receivers allow us to obtain better performance than the conventional receivers that deal with nonlinearly distorted OFDM signals, we will consider the following another sub-optimum receiver, which was also proposed in [102], since it presents better performance than the ones identified above (i.e., than the sub-optimum receivers (a) and (b)). Its algorithm is described as follows:

$\tilde{\mathbf{S}} \leftarrow$ hard-decision sequence
$D^2_{\text{min}} \leftarrow$ get Euclidean distance between the received signal and the nonlinearly distorted version of the hard-decision $\tilde{\mathbf{S}}$
$\tilde{\mathbf{S}}_{\mathbf{bits}} \leftarrow$ get the bits from the hard-decision $\tilde{\mathbf{S}}$
$\tilde{\mathbf{S}}^{\mathbf{sopt}}_{\mathbf{bits}} \leftarrow \tilde{\mathbf{S}}_{\mathbf{bits}}$
for $v = 1 \rightarrow V$ **do**
  for $b = 1 \rightarrow \log_2(M)N_u$ **do**
    Modify the bth bit of $\tilde{\mathbf{S}}^{\mathbf{sopt}}_{\mathbf{bits}}$
    $D^2 \leftarrow$ get Euclidean distance between the nonlinearly distorted version of $\tilde{\mathbf{S}}^{\mathbf{sopt}}$ and the received signal
    if $D^2 \leq D^2_{\text{min}}$ **then**
      $D^2_{\text{min}} \leftarrow D^2$
    else
      return the bth bit to its original value
    end if
  end for
end for

Basically, its decision process starts by taking the hard-decision sequence at the conventional receiver's output. After that, the receiver changes the first bit and the corresponding modulated signal is passed through the same nonlinearity associated with the transmitted signal. Then, the squared Euclidean distance relatively to the received signal is computed. If the squared Euclidean distance reduce, the bit is effectively changed, if not, it returns to its original value. This process is repeated in all the $\log_2(M)N_u$ bits of the hard-decision data sequence. As the sub-optimum sequence can be modified during the process, the cycle of bit modifications can be repeated V times. Note that this sub-optimum receiver computes the Euclidean distance between the received signal and a total number of $\log_2(M)N_uV \ll M^{N_u}$ variations of it, which means that the complexity relatively to the "full" optimum detection is substantially lower.

Approximate Optimum Performance

As pointed out before, it is very difficult to obtain the optimum performance. However, it will be interesting if we can compare the sub-optimum receiver's performance with the one obtained by the "full" optimum receiver. In fact, as in the simulation environment we have "access" to the transmitted sequence, an approximation of the optimum receiver's performance can be obtained by considering that the sub-optimum receiver makes bit modifications on the transmitted sequence, instead of considering the hard-decision sequence at the conventional receiver's output. We denote the performance of this variation of the sub-optimum receiver as "approximate optimum performance".

In the following, performance results associated with the sub-optimum receiver described above are shown. All these results were obtained through Monte Carlo simulations. Fig. 5.27 presents the BER of the sub-optimum receiver for ideal AWGN channels. Each OFDM signal has $N_u = 128$, $O = 4$, M-QAM constellations. The sub-optimum receiver performs $V = 2$ cycles of bit modifications. It should be noted that different normalized clipping levels are adopted since, the higher the size of the constellation, the higher the sensitivity to nonlinear distortion effects. For this reason, the normalized clipping level is $s_M/\sigma = 1.0$ for QPSK, but increases to $s_M/\sigma = 1.6$ and $s_M/\sigma = 2.0$ for 16-QAM and 64-QAM constellations, respectively. In fact, we adopt a higher s_M/σ when M is higher to approximately obtain the same distortion level for the different constellations. From the results depicted in the figure it can be noted that regardless of the constellation's size, the sub-optimum detection presents performance gains relatively to conventional, linear OFDM schemes. Another important aspect is that the sub-optimum receiver with even only $V = 2$ cycles of bit modifications has almost the "approximate optimum performance", especially for large values of E_b/N_0. This behavior reveals that the optimum sequence may typically differ in few bits relatively to the received signal and there is no need to test the M^{N_u} possible transmitted sequences for obtaining the full-optimum receiver's performance. Considering QPSK constellations and a target BER of $P_b = 10^{-3}$, it can be seen that the conventional receiver dealing with nonlinearly distorted signals presents

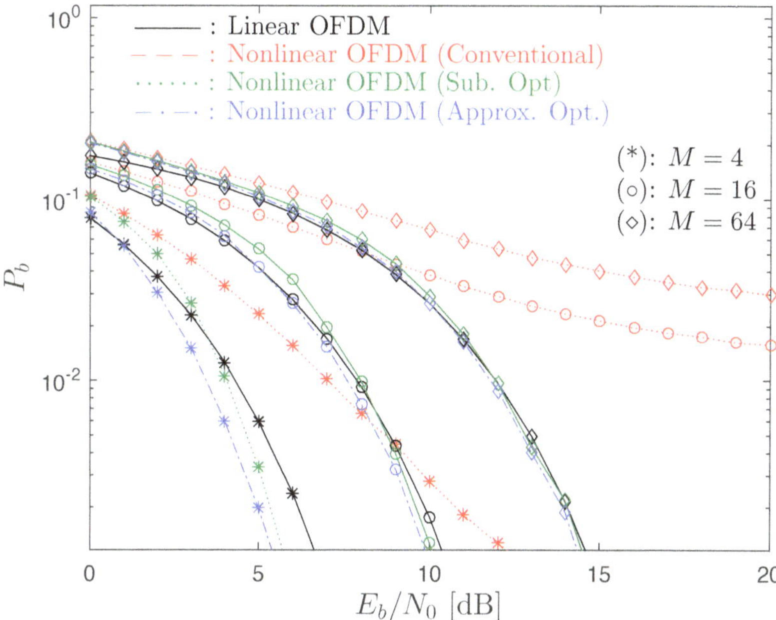

Figure 5.27 Sub-optimum receiver's BER considering ideal AWGN channels and different constellations.

a degradation of approximately 6 dB when compared to the linear OFDM. However, the sub-optimum receiver even outperforms the linear OFDM by presenting a performance gain of approximately 1 dB. When $M > 4$, the gain for the sub-optimum receiver is even higher, since the conventional nonlinear OFDM receiver presents an error floor at $P_b = 1.6 \times 10^{-2}$ and it is not able to obtain $P_b = 10^{-3}$. Fig. 5.28 presents the BER under the same conditions of the previous figure but considering frequency-selective channels with $L = 32$ uncorrelated multipath components. In the figure, it can be seen that the sub-optimum receiver outperforms not only the conventional receivers that deal with nonlinear OFDM signals, but even the linear OFDM. At $P_b = 10^{-3}$, these gains are approximately 8 dB, 9 dB and 13 dB for $M = 4$, $M = 16$ and $M = 64$, respectively. In fact, the performance gains in frequency-selective channels are even higher than the ones obtained in ideal AWGN channels. This can be explained by the additional diversity effect that is introduced by the nonlinearity. This additional diversity effect is related to the correlation between the subcarriers which, in its turn, is associated with the nonlinear distortion terms. Indeed, as the information of each subcarrier is spread along the entire block, even when a subcarrier is in a deep fade, the fact that other subcarriers may have good SNRs means that the optimum receiver may still be able to correctly detect the transmitted signal, since

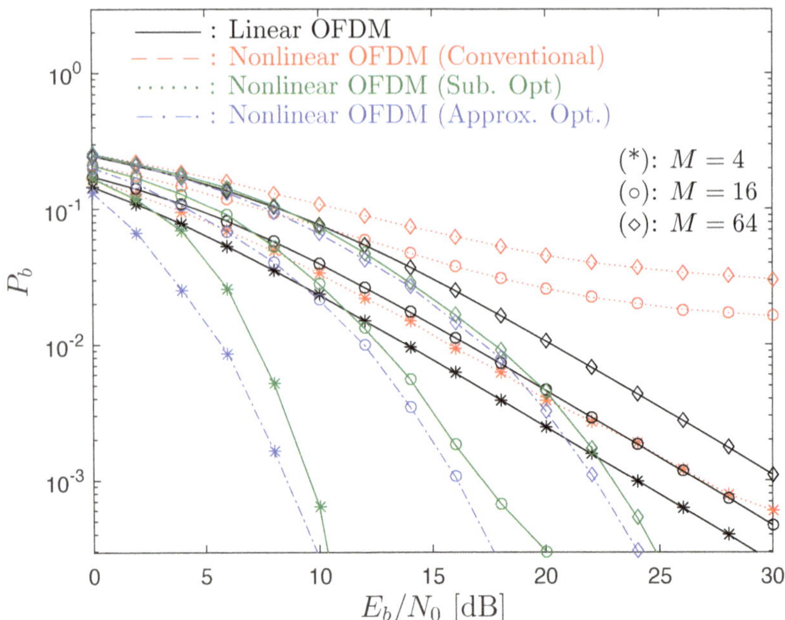

Figure 5.28 Sub-optimum receiver's BER considering frequency-selective channels and different constellations.

it detects the signals in a block-by-block basis. Fig. 5.29 shows the simulated BER in frequency-selective channels with $L = 64$ when $N_u = 128$ and $O = 4$. A receive diversity of order $R = 2$ is considered (R represents the number of received antennas). The nonlinearity is an envelope clipping with $s_M/\sigma = 1.0$. In fact, although is expected that the performance improves when more receive antennas are considered, due to the reduction of the deep fade probability, it can also be noted from the figure that the sub-optimum detection of nonlinear OFDM schemes presents performance gains even relatively to the linear OFDM when receive diversity is considered. For a target BER of $P_b = 10^{-2}$ and $R = 2$, the performance gain for the sub-optimum receiver is approximately 4.3 dB, but it can reach a value around 9.5 dB for $P_b = 10^{-3}$.

5.2 LINC TRANSMITTERS FOR OFDM SIGNALS

Even when simple and efficient PAPR reducing techniques such as clipping techniques are employed, the use of nonlinear, low-cost power amplifiers is precluded since these amplifiers require input signals with a PAPR of 0 dB or almost 0 dB to avoid severe nonlinear distortion effects at the transmitter output [44]. A highly efficient, nonlinear amplification technique is the LINC technique [106], which can be employed in OFDM transmissions [107, 108]. However, although LINC techniques have a good amplification efficiency, they may introduce nonlinear distortion

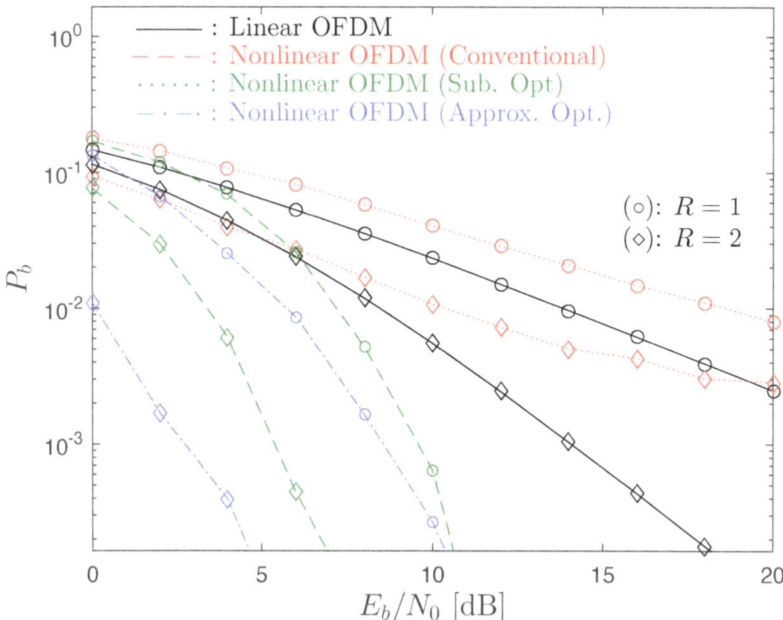

Figure 5.29 Sub-optimum receiver's BER considering frequency-selective channels and receive diversity.

effects in the transmitted signals due to the existence of phase and/or gain imbalances between the power amplifiers that compose the LINC structure. The constant envelope paired burst (CEPB)-OFDM techniques [109], based on LINC techniques, try to solve the problems associated with those imbalances between the amplifiers. In fact, as in CEPB-OFDM only one amplifier is needed and the two LINC components are transmitted one after another, the gain and phase imbalances do not take place. On the other hand, the use of CEPB-OFDM techniques leads to a loss of 50% in the spectral efficiency. Clearly, such a spectral degradation cannot be accommodated due to the spectrum scarcity and cost, which precludes its use in the majority of the applications. This justifies the use of traditional LINC techniques, even knowing that the amplifiers might be unbalanced. For this reason, it is important to accurately characterize the nonlinear distortion effects associated with the gain and phase imbalances inherent to those techniques, in order to assess their corresponding impact on the system's performance.

In this section, we present an accurate spectral characterization as well as the optimum performance of OFDM schemes with LINC transmitter structures. Due to the high severeness of the nonlinearities associated with LINC techniques, we consider the use of equivalent nonlinearities for the characterization of the corresponding nonlinearly distorted signals [110]. This allows us to avoid the very high complexity and the convergence problems of conventional truncated IMP approach.

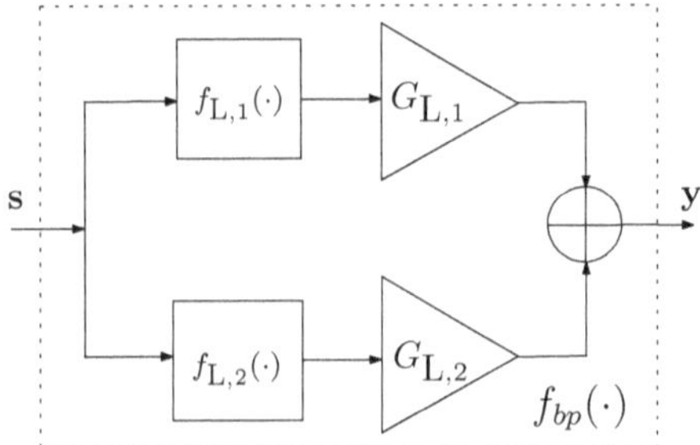

Figure 5.30 LINC transmitter structure.

The model for the LINC transmitter structure is depicted in Fig. 5.30. In fact, the LINC transmitter can be modeled as a bandpass memoryless function. To perform the LINC decomposition, the absolute value of the time-domain OFDM signal $\mathbf{r} = |\mathbf{s}| = [r_0\ r_1\ \dots\ r_{N-1}]^T \in \mathbb{C}^N$ is submitted in parallel to the nonlinear functions $f_{\mathrm{L},1}(r_n)$ and $f_{\mathrm{L},2}(r_n)$. The nonlinear function for the ith branch is defined as

$$f_{\mathrm{L},i}(r_n) = f_{\mathrm{L},c}(r_n) + j(-1)^{i+1} f_{\mathrm{L},e}(r_n), \qquad i = 1,2, \tag{5.24}$$

where $f_{\mathrm{L},c}(r_n)$ represents and envelope clipping function with clipping level s_M, i.e.,

$$f_{\mathrm{L},c}(r_n) = \begin{cases} \frac{1}{2} r_n, & r_n \leq s_M \\ \frac{1}{2} s_M, & r_n > s_M, \end{cases} \tag{5.25}$$

and $f_{\mathrm{L},e}(r_n)$ is defined as

$$f_{\mathrm{L},e}(r_n) = \begin{cases} \frac{1}{2}\sqrt{s_M^2 - r_n^2}, & r_n \leq s_M \\ 0, & r_n > s_M. \end{cases} \tag{5.26}$$

At the output of these nonlinearities, the signals in both branches have a constant envelope since $|f_{\mathrm{L},1}(r_n)| = |f_{\mathrm{L},2}(r_n)| = s_M/2$. Therefore, they can be submitted to nonlinear power amplifiers that are solely characterized by the complex gain coefficients $G_{\mathrm{L},1}$ and $G_{\mathrm{L},2}$, respectively. The two branches are then converted to the RF band and combined to form the transmitted signal. As mentioned before, the LINC transmitter can be characterized by a single bandpass nonlinearity. This nonlinearity can be expressed by the function

$$\begin{aligned} f_{bp}(r_n) &= G_{\mathrm{L},1} f_{\mathrm{L},1}(r_n) + G_{\mathrm{L},2} f_{\mathrm{L},2}(r_n) \\ &= (G_{\mathrm{L},1} + G_{\mathrm{L},2}) f_{\mathrm{L},c}(r_n) + j(G_{\mathrm{L},1} - G_{\mathrm{L},2}) f_{\mathrm{L},e}(r_n). \end{aligned} \tag{5.27}$$

Ideally, the complex amplification coefficients associated with each amplifier should be equal. In those conditions, we have $G_{\mathrm{L},1} = G_{\mathrm{L},2}$ and (5.27) yields

$$f_{bp}(r_n) = 2 f_{\mathrm{L},c}(r_n), \tag{5.28}$$

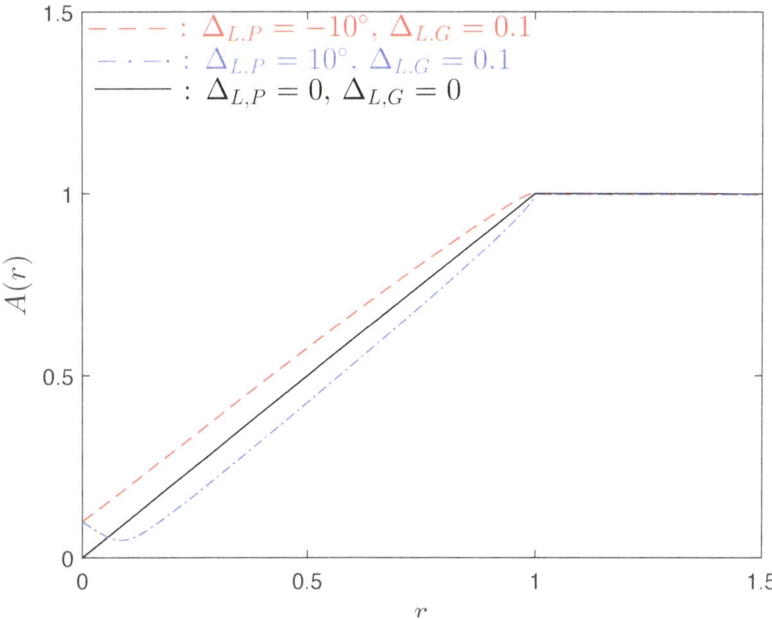

Figure 5.31 AM/AM conversion function for the bandpass nonlinearity of a LINC transmitter

i.e., the bandpass nonlinearity turns into an ideal envelope clipping function (see (4.1)). However, in practice, the amplifiers may present both gain imbalances (i.e., when $|G_{\mathrm{L},1}| \neq |G_{\mathrm{L},2}|$) and/or phase imbalances (i.e., when $\arg(G_{\mathrm{L},1}) \neq \arg(G_{\mathrm{L},2})$). In the following, these gain and phase imbalances are denoted as $\Delta_{L,G}$ and $\Delta_{L,P}$, respectively. Fig. 5.31 shows the AM/AM conversion function of a LINC transmitter for different values of $\Delta_{L,G}$ and $\Delta_{L,P}$. From the results depicted in the figure it can be clearly noted that when the amplifiers are balanced, the nonlinearity associated with a LINC transmitter structure turns into an envelope clipping. However, when there are gain or phase imbalances, the AM/AM curve deviates from the clipping function. These "deviations" accentuate the magnitude of the nonlinear distortion effects, which implies that the analytical characterization of the transmitted signals may require a very large number of IMPs if the truncated IMP approach is employed to obtain the autocorrelation of nonlinearly distorted signal (and the corresponding PSD). This is illustrated in Fig. 5.32, which shows the PSD for a nonlinearly distorted OFDM signal that passes through an unbalanced LINC transmitter. The gain imbalance is $\Delta_{L,G} = 0.1$ and the phase imbalance is $\Delta_{L,P} = 10°$. The normalized clipping level is $s_M/\sigma = 1.0$ and each OFDM signal has $N_u = 256$ and $O = 4$. In this figure, it can be seen that a very large number of IMPs is required to obtain a good match between the theoretical and the simulated PSDs. When $n_\gamma = 10$, the error is

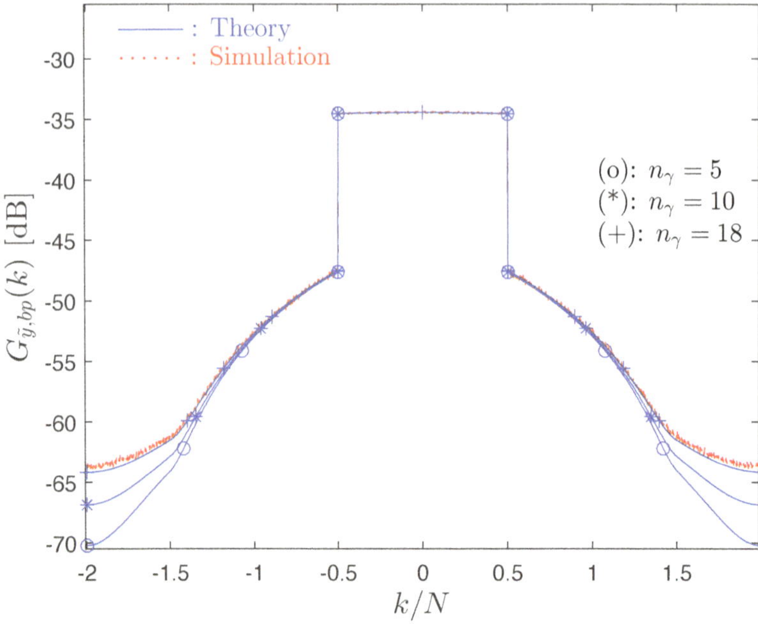

Figure 5.32 PSD of a nonlinearly distorted OFDM that passes through an imbalanced LINC transmitter considering the truncated IMP approach.

approximately 3 dB and can even reach a value around 6 dB when $n_\gamma = 5$. In fact, the error only becomes negligible when $n_\gamma = 18$, i.e., when a very large number of IMPs is taken into account. However, when equivalent nonlinearities are used to substitute the nonlinear functions associated with the LINC process, accurate PSDs can be obtained even with $\gamma_{\max} = 8$, as can be seen in Fig. 5.33. These results were generated with the same transmission scenario of the previous figure, considering not only the truncated IMP approach, but also the equivalent nonlinearity approach with same number of IMP, i.e., $n_\gamma = \gamma_{\max} = 8$. Clearly, the use of the equivalent nonlinearity allows to obtain considerable accuracy gains regarding the spectral characterization of the nonlinearly distorted signals. In fact, to obtain the same level of accuracy using the truncated IMP approach, one may need to consider the contribution of more than 10 IMPs, i.e., a value of $n_\gamma = 18$ should be considered (see Fig. 5.32).

The expression used to obtain the optimum asymptotic performance of (5.19) may present accuracy problems for very strong nonlinearities such as the ones of the LINC process. For this reason, we also consider the use of equivalent nonlinearities to obtain the optimum asymptotic performance. Fig. 5.34 shows the average asymptotic gain for the optimum detection of OFDM schemes that have LINC transmitters obtained both by simulation and theoretically. The two amplifying branches have $\Delta_{L,G} = 0$ and different values of $\Delta_{L,P}$. From the results shown in this

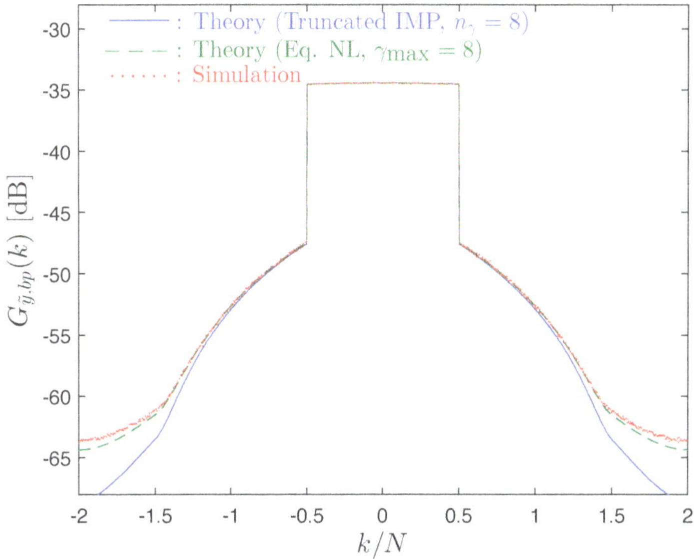

Figure 5.33 PSD of a nonlinearly distorted OFDM that passes through an imbalanced LINC transmitter structure considering different approaches.

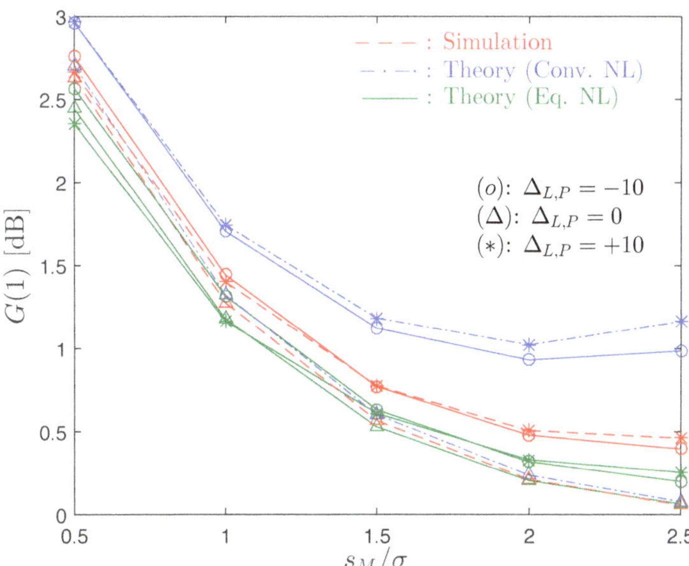

Figure 5.34 Average asymptotic gain associated with the optimum detection considering $\Delta_{L,G} = 0$ and different values of $\Delta_{L,P}$.

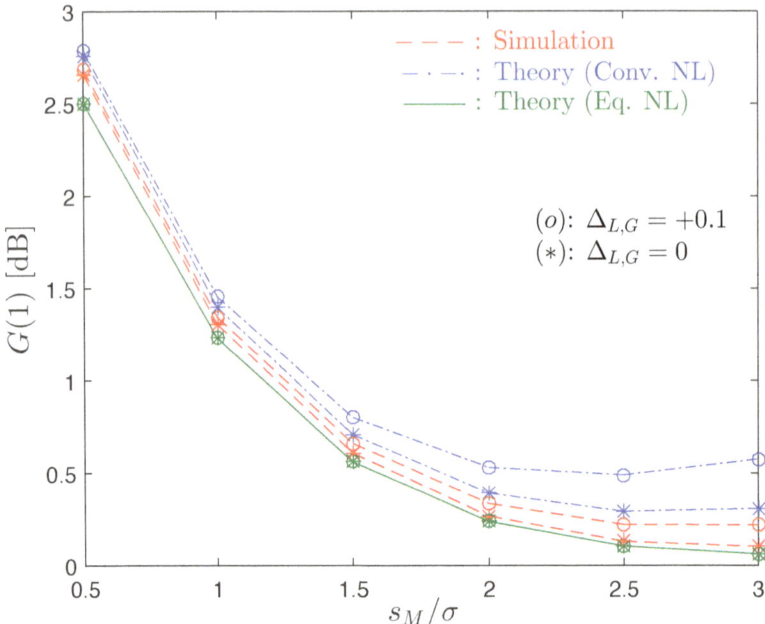

Figure 5.35 Average asymptotic gain associated with the optimum detection considering $\Delta_{L,P} = -5^\circ$ and different values of $\Delta_{L,G}$.

figure, the existence of potential asymptotic gains can be confirmed. Although when $\Delta_{L,G} = \Delta_{L,P} = 0$, the use of (5.19) with the conventional LINC nonlinearity leads to accurate results, this is not the case when amplifiers are unbalanced, since the theoretical asymptotic gain differs considerably from the simulated one, especially for low clipping levels. However, when equivalent nonlinearities are considered, the theory matches the simulation and there is a maximum error of only approximately 0.2 dB in the worst case, i.e, when $\Delta_{L,P} = \pm 10^\circ$. Fig. 5.35 shows the average asymptotic gain associated with the optimum detection for $\Delta_{L,P} = -5^\circ$ and different values of $\Delta_{L,G}$. Once again, it can be noted that the use of equivalent nonlinearities allows to obtain better estimates of the potential asymptotic gains than when the conventional LINC nonlinearities are considered. It should also be pointed out that when we have gain imbalances, the corresponding nonlinear distortion effects are stronger, which means that the optimum performance may present higher potential asymptotic gains.

5.3 CONSTANT-ENVELOPE OFDM

In addition to the LINC and the CEPB-OFDM techniques, the CE-OFDM was recently proposed as a highly efficient technique for extremely high frequency (EHF) communications [111, 112]. As the name indicates, CE-OFDM signals have

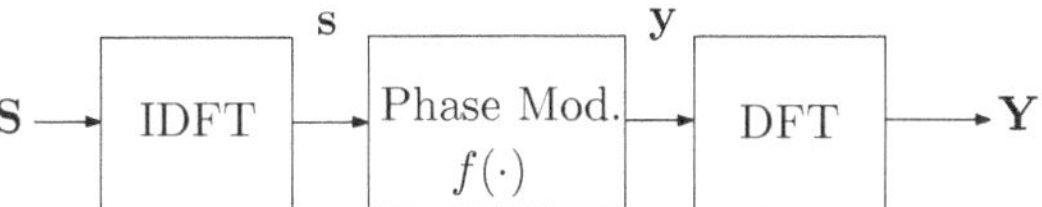

Figure 5.36 Equivalent, subcarrier-level scenario of a CE-OFDM transmitter.

constant envelope, since they are generated by submitting a real-valued OFDM signal to a phase modulator. This means that they can be amplified by highly efficient, nonlinear power amplifiers such as the ones of class D, E or F. However, since the phase modulator is a nonlinear device, it can lead to substantial spectral widening due to the out-of-band radiation levels associated with IMPs. In addition, it can also lead to significant BER degradation due to in-band nonlinear distortion, which includes an unwanted DC component. However, as aforementioned, the nonlinear distortion terms (naturally, excluding the DC component) have information inherent to the transmitted symbols that can be employed to improve the performance, provided that optimum or even sub-optimum receivers are employed.

In this section, we present the analytical spectral characterization as well as the performance evaluation of CE-OFDM. For this purpose, we take advantage of the Gaussian-like nature of OFDM signals and employ the results of Subsection 3.2.1. To employ those results, we assume that the phase modulation process is a nonlinear transformation of a conventional OFDM signal. This analytical characterization is then employed to compute the PSD of CE-OFDM signals. Additionally, using the results of Subsection 4.4.2, we present analytical results for the asymptotic optimum performance of CE-OFDM [113–115].

The equivalent, subcarrier-level scenario of a CE-OFDM transmitter is depicted in Fig. 5.36. We considered that each useful subcarrier carries a QPSK symbol. It should be mentioned that as the corresponding time-domain samples are submitted to a phase modulation process, they must be real-valued. Therefore, before the addition of zeros that is inherent to the oversampling operation, the data symbols are constrained to have Hermitian symmetry, i.e., the frequency-domain samples obey to the following relation

$$S_k = \begin{cases} 0, & k = 0, N_u/2 \\ S_{N_u-k} = S_k^*, & \text{otherwise.} \end{cases} \tag{5.29}$$

Under these conditions, after the IDFT, we have a real-valued OFDM signal whose the samples are represented by $\mathbf{s} = [s_0\, s_1\, \ldots\, s_{N-1}]^T \in \mathbb{C}^N$. Note that between the N subcarriers, only $N_u - 2$ are effectively used to transmit data. Therefore, by considering the same approach of (5.3), it can be shown that the variance of the real and imaginary parts of each sample s_n is $\sigma^2 = (N_u - 2)/N^2$. These time-domain samples are submitted to a phase modulator, that is represented by the following nonlinear function

$$f(s_n) = \exp(j2\pi h s_n/\sigma), \tag{5.30}$$

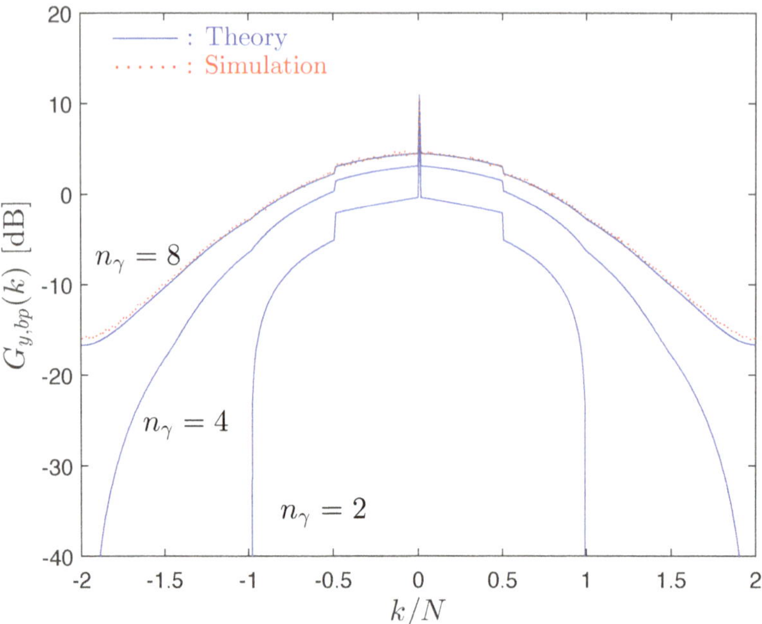

Figure 5.37 PSD of a CE-OFDM signal considering the truncated IMP approach and different values of n_γ.

where h denotes the modulation index. Naturally, regardless of the modulation index, the resultant CE-OFDM signal has a PAPR of 0 dB. Nevertheless, the phase modulation process leads to nonlinear distortion effects on the transmitted signals. Therefore, the impact of this nonlinear distortion on the system's performance should be evaluated. In the following, we consider the truncated IMP approach for obtaining the analytical characterization of CE-OFDM signals. Fig. 5.37 shows the simulated and theoretical PSD of a CE-OFDM signal with $N_u = 128$, $O = 4$ and modulation index $2\pi h = 2.0$. In this figure it can be seen that when $n_\gamma = 8$, the theoretical PSD matches the simulated PSD with an error near 0 dB. As expected, the accuracy is as higher as higher is the number of considered IMPs, i.e., it increases with n_γ. The existence of a DC component should also be noted. This DC component, introduced by the phase modulation process, can substantially degrade the system's performance. As can be seen in Fig. 5.38, the power of the DC component is higher for lower modulation indexes. In fact, under these conditions, the signal at the output of the nonlinearity is essentially a complex exponential with constant and almost zero phase. Fig. 5.39 shows the simulated and theoretical PSD associated with the distortion term of a CE-OFDM signal considering $N_u = 64$, different oversampling factors and $n_\gamma = 8$. Regardless of the oversampling factor O, it can be confirmed that the accuracy of the truncated IMP approach is very high. This high accuracy can also be observed in

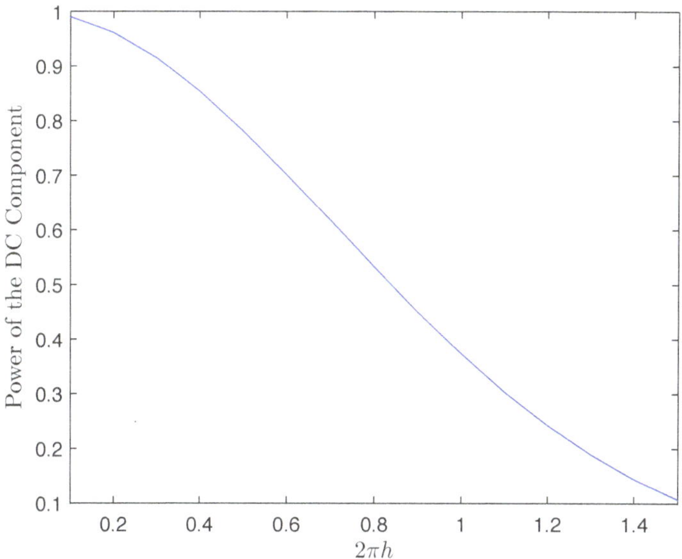

Figure 5.38 Evolution of the power spent in the DC component in a CE-OFDM transmission.

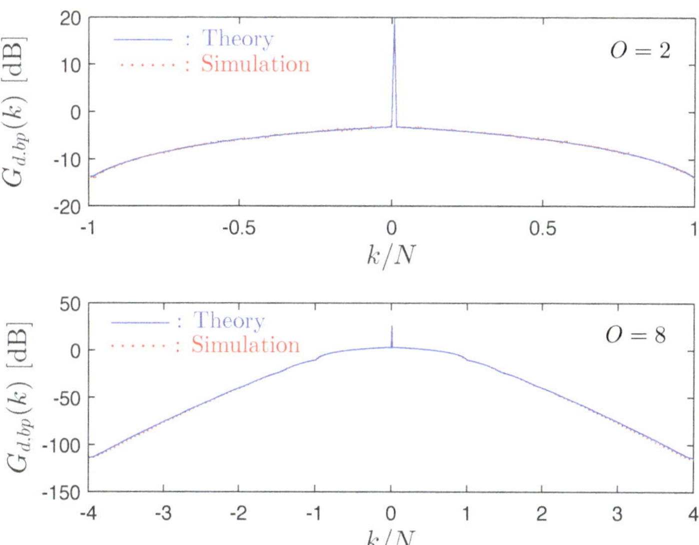

Figure 5.39 PSD of the distortion term of a CE-OFDM signal considering different oversampling factors.

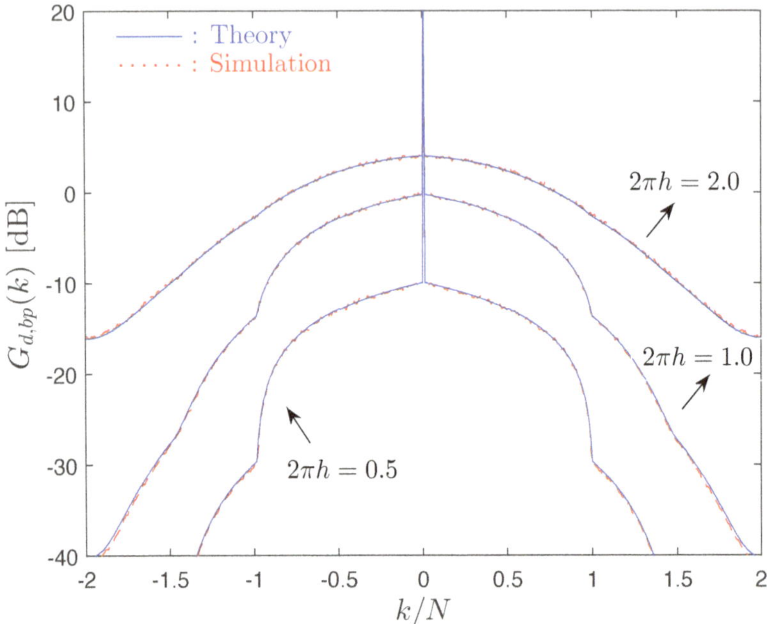

Figure 5.40 PSD of the distortion term of a CE-OFDM signal considering different modulation indexes.

Fig. 5.40, which shows the PSD of the distortion term under the same conditions of the previous figure, but considering different modulation indexes.

Let us now focus on the detection procedures of CE-OFDM schemes. Conventionally, the detection of CE-OFDM signals involves a phase detector, i.e., involves a phase demodulation process [111]. However, receivers based on a phase demodulation process are only suitable when the phase does not have excursions over $\pm\pi$, which can not always be guaranteed, especially, when large modulation indexes are considered. This is demonstrated in Fig. 5.41 where one can clearly see that the probability of having phase excursions outside the interval $[-\pi, \pi]$ increases with the modulation index. On the other hand, it is worth to mention that when lower modulation indexes are employed, the power of the DC component can be very large, which leads to low energy inefficiency and large performance penalties.

As aforementioned, the phase modulation process can be seen as a nonlinear transformation of a conventional OFDM signal. For this reason, following the results that point out to performance gains associated with the optimum detection of nonlinearly distorted OFDM signals, it is important to evaluate what is the optimum performance of CE-OFDM. In the following, the asymptotic gains for optimum detection of CE-OFDM signals are studied considering both AWGN and frequency-selective channels.

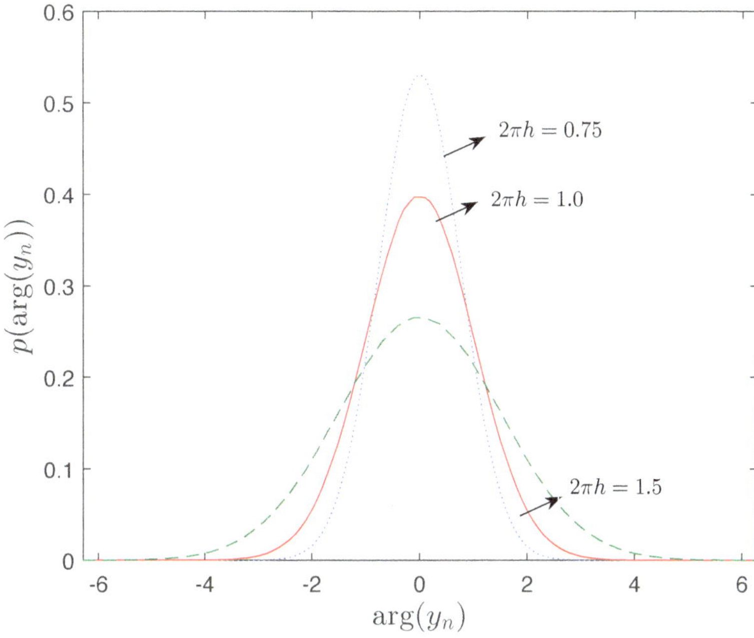

Figure 5.41 Distribution of the phase of CE-OFDM signals considering different modulation indexes.

As demonstrated in Subsection 4.4, the average asymptotic gain for the optimum detection of nonlinearly distorted, real-valued OFDM signals is given by

$$G(\mu) = \frac{\mu (d_{\text{adj}} \sigma)^2}{4} \frac{\int_{-\infty}^{+\infty} |f'(s)|^2 p(s) ds}{\int_{-\infty}^{+\infty} |f(s)|^2 p(s) ds}. \tag{5.31}$$

Considering the specific case of the phase modulation process and replacing $f(s)$ by (5.30) (and $f'(s)$ by its corresponding derivative) in (5.31), we have

$$\begin{aligned} G(\mu) &= \frac{\mu (d_{\text{adj}} \sigma)^2}{4} \left(\frac{2\pi h}{\sigma}\right)^2 \int_{-\infty}^{+\infty} p(s) ds \\ &= \mu (\pi h d_{\text{adj}})^2. \end{aligned} \tag{5.32}$$

Fig. 5.42 shows the simulated and theoretical asymptotic gain (obtained with (5.32)) for CE-OFDM signals with $N_u = 512$, $O = 4$, different values of μ and different

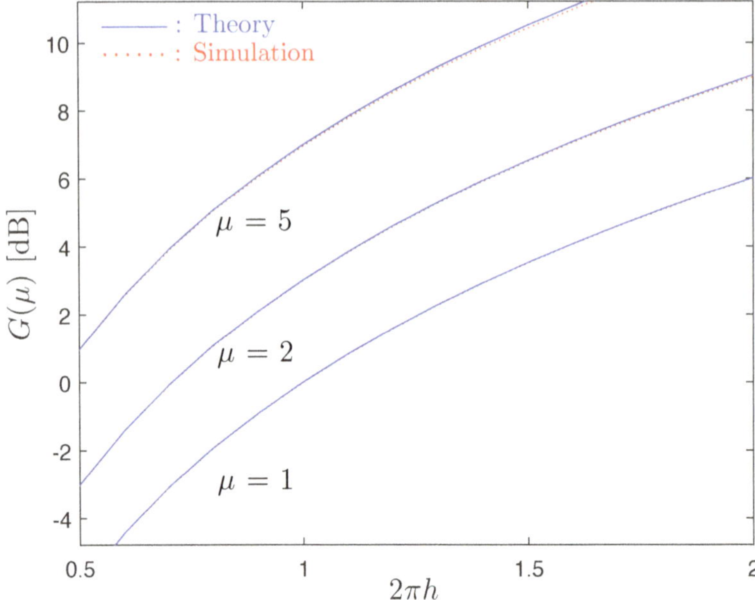

Figure 5.42 Asymptotic gain for the optimum detection of CE-OFDM signals obtained both by simulation and theoretically considering different values of μ.

modulation indexes. From the results depicted in the figure one can see that, regardless of the number of μ bit differences between the signals, (5.32) presents high accuracy. Additionally, it can be noted that the average asymptotic gain for the optimum performance of CE-OFDM signals is strongly related to the modulation index h. Actually, for low values of $2\pi h$, we do not have a gain but a degradation instead. When $2\pi h$ increases, however, the magnitude of the nonlinear distortion effects increases and the power of the DC component decreases, leading to gains relatively to the performance of conventional, linear OFDM schemes.

Let us now compare the optimum asymptotic performance with the performance of the conventional phase demodulator. The nth time-domain sample of the phase modulated signal is

$$y_n = f(s_n) = \exp(j2\pi h s_n/\sigma). \tag{5.33}$$

For ideal AWGN channels, the received signal is

$$\begin{aligned} z_n &= y_n + \nu_n \\ &= \exp(j2\pi h s_n/\sigma) + \nu_n, \end{aligned} \tag{5.34}$$

where ν_n is the nth noise sample and $\mathbb{E}[|\nu_n|^2] = N_0$. In the phase demodulation process, the phase detector extracts the phase of (5.34). After that, a DFT is employed

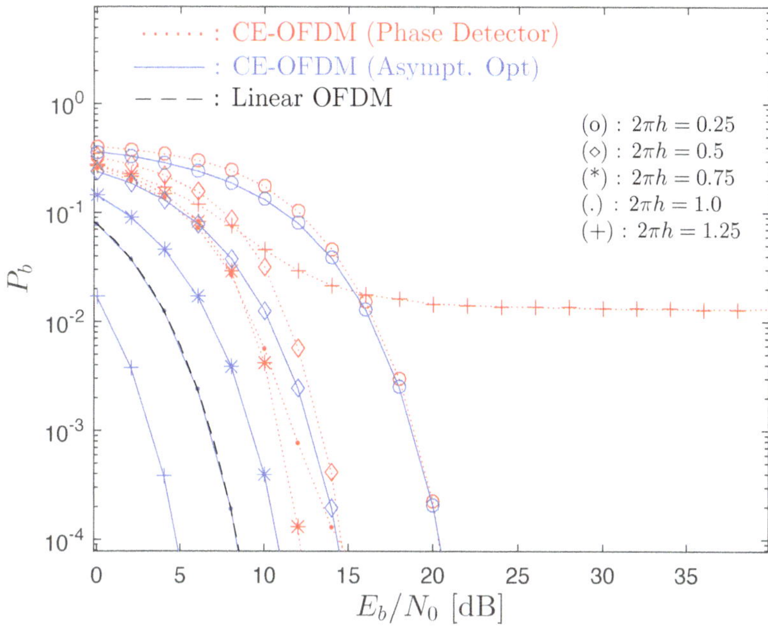

Figure 5.43 Simulated BER of the phase detector and optimum asymptotic BER considering different modulation indexes.

prior to the detection, that works on a subcarrier-by-subcarrier basis. Note additionally that, by considering that $\exp(jx) \approx 1 + jx$ for low values of x, we can approximate (5.34) as

$$z_n \approx 1 + j2\pi h s_n/\sigma + v_n, \tag{5.35}$$

which is a tight approximation for low modulation indexes. After some lengthy but straightforward manipulations, it can be shown that the asymptotic BER of the phase demodulation process is approximately given by [111]

$$P_b \approx Q\left(\sqrt{(2\pi h)^2 \frac{2E_b}{N_0}}\right). \tag{5.36}$$

From the previous equation, it seems that a receiver based on a phase demodulator is able to obtain the optimum performance, since, when $\mu = 1$ and QPSK constellations are considered, (5.32) turns into $G(1) = (2\pi h)^2$. Thus, considering (4.53), it can be noted that the optimum asymptotic performance is also given by (5.36). In fact, as demonstrated in the following, this is only true for low modulation indexes, since the use of high modulation indexes leads to phase excursions over $\pm\pi$ and substantial performance degradations. For this reason, the phase detection is only optimum for a limited range of modulation indexes. This effect is illustrated in Fig. 5.43, which shows the simulated BER of the phase detector, as well as the optimum asymptotic

BER, both obtained in ideal AWGN channels. The CE-OFDM signals have $N_u = 512$, $O = 4$ and different modulation indexes are considered.

From the figure it can be noted when the modulation index is small, the phase detector has approximately the optimum performance (i.e., its performance is given by (5.36)). However, when $2\pi h > 0.5$, its performance starts to degrade substantially when compared to the optimum performance. Additionally, for small modulation indexes, it can be observed that the performance improves with the modulation index. However, this is not true for higher values of h, since the phase detector's performance starts to degrade. In fact, the required E_b/N_0 to achieve a target BER of $P_b = 10^{-3}$ is approximately 10.8 dB when $2\pi h = 0.75$, but increases to a value around 11.6 dB when $2\pi h = 1.0$. Moreover, when $2\pi h = 1.25$, we can even observe an accentuated error floor of $P_b = 1.3 \times 10^{-2}$, which is due to the phase excursions beyond the interval $[-\pi, \pi]$. Naturally, due to the Gaussian nature of the samples s_n, phase excursions outside this interval can always occur, even for small values of h (see Fig. 5.41). However, they become more common for large modulation indexes. This means that, in practice, the phase detector has an applicability zone restricted to values where $2\pi h \leq 0.5$. Therefore, the phase detector can only be employed when the power efficiency is low since, for small modulation indexes, the power of the DC component introduced by the phase modulation process is high. This justifies the demand for other receivers such as the ones based on the optimum detection. Note that, regarding the optimum detection, we only have asymptotic gains relatively to linear OFDM when $2\pi h > 1.0$, which can also be observed in the evolution of the asymptotic gain shown in Fig. 5.42. Fig. 5.44 shows the required E_b/N_0 for obtaining different target BERs, considering the optimum asymptotic performance and different modulation indexes. As expected, the higher the modulation index, the higher the performance gain. In fact, since we only have gains when $2\pi h > 1.0$, the figure clearly identifies the gains and degradations region.

Let us now consider the optimum asymptotic performance in frequency-selective channels. Fig. 5.45 shows the distribution of the asymptotic gain associated with the optimum detection of CE-OFDM signals obtained both by simulation and theoretically with (4.129), for different modulation indexes. Additionally, in order to promote a fairly comparison between the linear and nonlinear cases, we also include the distribution of $|H_k|^2$ and denote it as "equivalent fading factor". The CE-OFDM signals have $N_u = 512$ and $O = 4$. The frequency-selective channel has $L = 32$ uncorrelated multipath components. From the figure, one can note that (4.129) constitutes an accurate expression for obtaining the distribution of the optimum asymptotic gains. As in the case of ideal AWGN channels, the asymptotic gains tend to increase with the modulation index, i.e., when the magnitude of the nonlinear distortion effects increase. In fact, although with lower modulation indexes the gain can be smaller than 1, it is substantially better than the equivalent fading factor that is the "gain" associated with linear OFDM transmissions. Fig. 5.46 shows the average asymptotic BER for the optimum detection of CE-OFDM signals obtained with the distributions of the asymptotic gain (see (4.60)) and considering $N_u = 512$, $O = 4$ and different modulation indexes. From the depicted results, it can be noted that the

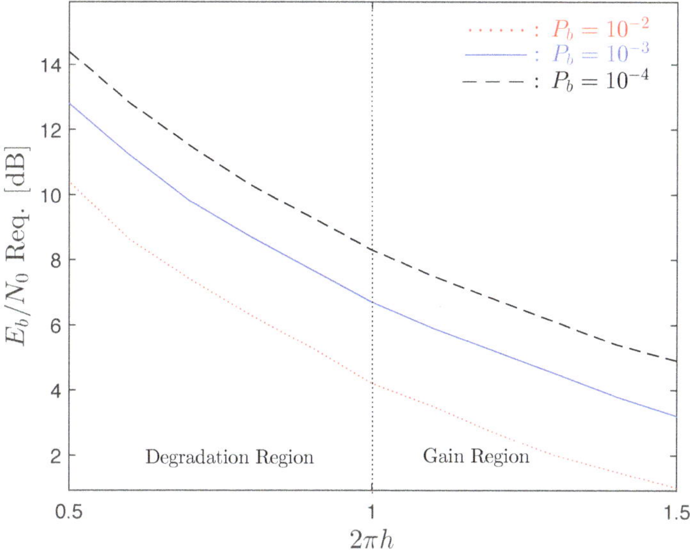

Figure 5.44 Required E_b/N_0 for obtaining a target BER considering the optimum asymptotic performance of CE-OFDM and different modulation indexes.

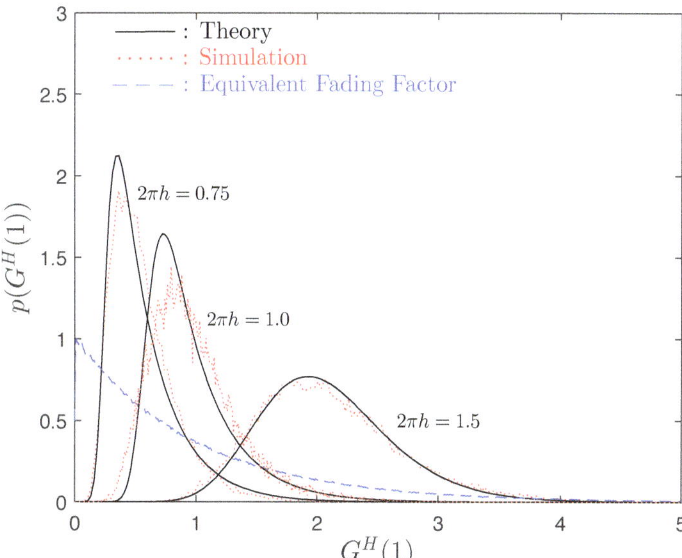

Figure 5.45 Distribution of the asymptotic gains for CE-OFDM schemes in frequency-selective channels considering different modulation indexes.

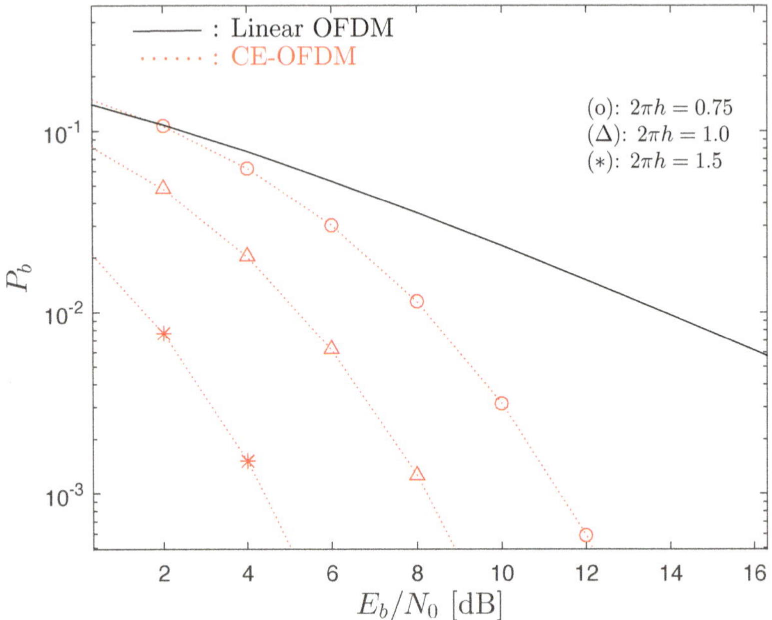

Figure 5.46 Average asymptotic BER of the optimum detection of CE-OFDM signals in frequency-selective channels for different modulation indexes.

optimum performance can be even better than the performance associated with linear OFDM transmissions, where the BER is largely conditioned on the subcarriers that are in deep fade. As mentioned before, the performance improvements are larger for frequency-selective channels, since the optimum receiver analyzes the entire OFDM block, being able to take advantage of the correlation between the subcarriers.

5.4 AMPLIFY-AND-FORWARD RELAY OFDM SYSTEMS

Relay-based techniques have two main important advantages: (i) they allow significant robustness against fading and/or (ii) increase the coverage of wireless communication systems. For this reason, they received considerable attention over the last ten years [116, 117]. In the literature, two main relay-based techniques can be identified: the AF [116] techniques and the decode and forward (DF) [117] techniques. Although DF techniques allow better performance in general, they require a more complex relay, not to mention the existence of higher delays at the relay than AF techniques. In addition, the simpler AF techniques do not require channel knowledge at the relay. For this reason, the relay in such techniques can be simply regarded as passive repeater that not only increases the signal power, but also provides additional multipath diversity.

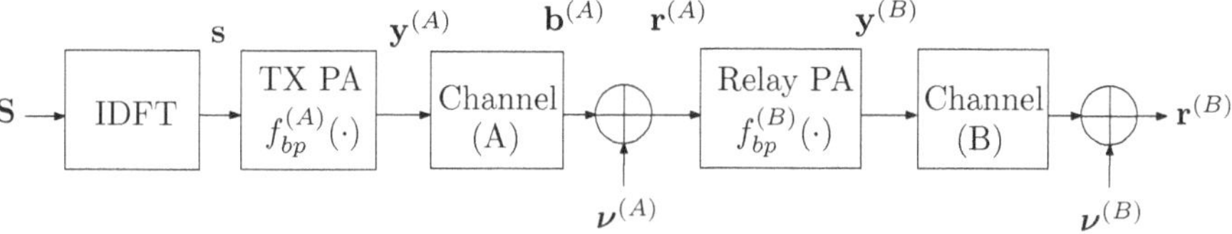

Figure 5.47 AF relay system model.

AF relay techniques combined with OFDM modulations are being considered for broadband wireless systems since they allow simple relay systems suitable for severely time-dispersive channels. However, due to the large envelope fluctuations of OFDM signals, these techniques are prone to nonlinear distortion effects that can take place not only at the transmitter but also at the relay. In this section, we consider OFDM-based AF relay systems with strong nonlinear distortion effects at both the transmitter and the relay nodes and we present an analytical method to characterize the corresponding nonlinearly distorted signals [118].

The model considered for the AF relay OFDM-based system is depicted in Fig. 5.47. Contrarily to other scenarios previously considered in this work, this nonlinear OFDM system has two bandpass nonlinearities[3], which are denoted as $f_{bp}^{(A)}(\cdot)$ and $f_{bp}^{(B)}(\cdot)$. These two nonlinearities model the transmitter and the relay amplifiers, respectively, and are assumed to be ideal envelope clippings (see (4.1)).

Let us focus on the system's description regarding the characterization of the signals along the AF relay system. Each OFDM signal has $N_u = 256$ useful subcarriers with QPSK constellations ($\mathbb{E}\left[|S_k|^2\right] = 2$) and oversampling factor $O = 4$. Regarding the time-domain, the variance of its real and imaginary parts is σ^2. At the transmitter's amplifier output, the time-domain samples are represented by $\mathbf{y}^{(A)} = [y_0^{(A)}\ y_1^{(A)}\ y_2^{(A)}\ \ldots\ y_{N-1}^{(A)}]^T \in \mathbb{C}^N$. Fig. 5.48 shows the simulated and theoretical PSD of the nonlinearly distorted OFDM signal after being amplified in the transmitter, considering a normalized clipping level of $s_M/\sigma = 1.0$. Clearly, the PSD obtained theoretically is almost equal to the PSD obtained by simulation. After the amplification process in the transmitter, the nonlinearly distorted OFDM signal is sent to a time-dispersive channel characterized by the channel matrix $\mathbf{H}^{(A)} = \text{diag}\left([H_0^{(A)}\ H_1^{(A)}\ H_2^{(A)}\ \ldots\ H_{N-1}^{(A)}] \in \mathbb{C}^N\right)$ (see (2.75)). This channel models the link between the transmitter and the relay and is characterized by $I^{(A)}$ uncorrelated multipath components. In addition, its frequency response are defined in such a way that $\mathbb{E}[|H_k^{(A)}|^2] = 1$. After pass through the frequency-selective channel, the signal of the kth subcarrier can be written as

$$B_k^{(A)} = \alpha_{bp}^{(A)} H_k^{(A)} S_k + H_k^{(A)} D_k^{(A)}, \tag{5.37}$$

[3]In this section the superscript (A) and (B) are associated with the link between the transmitter and the relay and the link between the relay and the receiver, respectively.

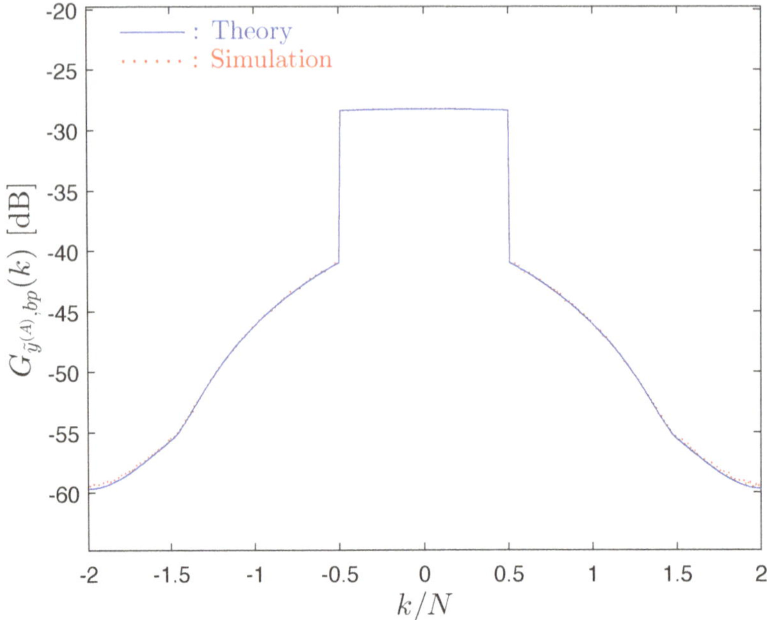

Figure 5.48 PSD of the nonlinearly distorted signal at the transmitter.

where $\alpha_{bp}^{(A)}$ is the scale factor of $f_{bp}^{(A)}(\cdot)$. The AWGN that is added to (5.37) is characterized in the frequency-domain by the block $\mathbf{N}^{(A)} = [N_0^{(A)}\ N_1^{(A)}\ N_2^{(A)}\ ...\ N_{N-1}^{(A)}]^T \in \mathbb{C}^N$, where $\mathbb{E}[|N_k^{(A)}|^2] = N_0$ is chosen according to a specific channel SNR. After addition of AWGN, the time-domain samples of the resultant signal are represented by $\mathbf{r}^{(A)} = [r_0^{(A)}\ r_1^{(A)}\ r_2^{(A)}\ ...\ r_{N-1}^{(A)}]^T \in \mathbb{C}^N$. This signal is then amplified in the relay node and sent to the receiver. The link between the relay and the receiver is modeled as a time-dispersive channel with frequency responses $\mathbf{H}^{(B)} = \mathrm{diag}\left([H_0^{(B)}\ H_1^{(B)}\ H_2^{(B)}\ ...\ H_{N-1}^{(B)}] \in \mathbb{C}^N\right)$, characterized by $I^{(B)}$ multipath rays and $\mathbb{E}[|H_k^{(B)}|^2] = 1$. The frequency responses of the second block of AWGN are represented by $\mathbf{N}^{(B)} = [N_0^{(B)}\ N_1^{(B)}\ N_2^{(B)}\ ...\ N_{N-1}^{(B)}]^T \in \mathbb{C}^N$ and $\mathbb{E}\left[\left|N_k^{(B)}\right|^2\right] = N_0$.

It should be noted that, although the samples of the nonlinearly distorted signal at the transmitter's output $\mathbf{y}^{(A)} = [y_0^{(A)}\ y_1^{(A)}\ y_2^{(A)}\ ...\ y_{N-1}^{(A)}]^T \in \mathbb{C}^N$ do not have a Gaussian distribution, after passing through the frequency-selective channel between the transmitter and the relay, $\mathbf{H}^{(A)}$, they become approximately Gaussian. In fact, the variance of the real and imaginary parts of the resultant samples $b_n^{(A)}$ is given by the power of the signal at the nonlinearity's output, i.e.,

$$\sigma_b^2 = \sum_{\gamma=0}^{+\infty} P_{2\gamma+1}^{bp,(A)}, \tag{5.38}$$

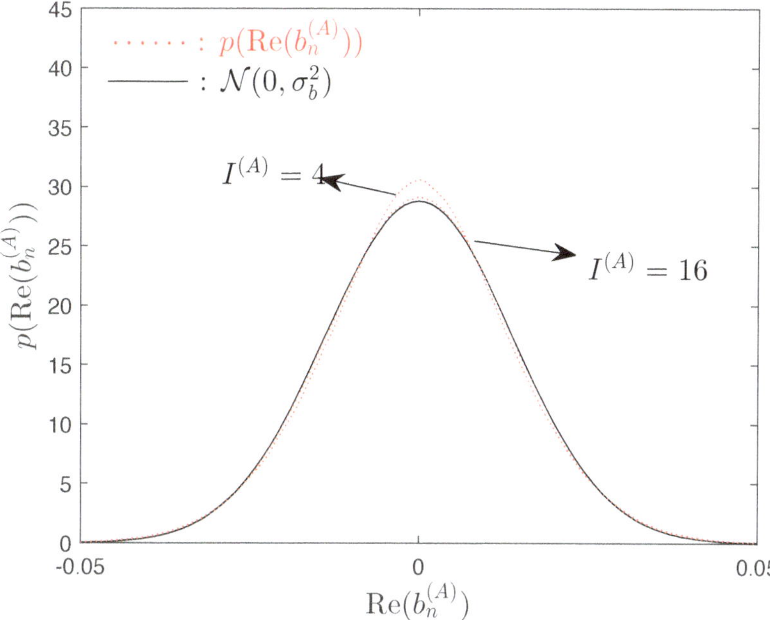

Figure 5.49 Distribution of the samples of the nonlinearly distorted signal after passing through a frequency-selective channel.

where $P_{2\gamma+1}^{bp,(A)}$ represents the power of the IMP of order γ at the output of the bandpass nonlinearity $f_{bp}^{(A)}(\cdot)$. Fig. 5.49 demonstrates the validity of this Gaussian approximation by showing the PDF associated with the real part of the samples $b_n^{(A)}$ considering different values of $I^{(A)}$. Clearly, even when a low-to-moderate number of multipath components is considered, the PDF is almost Gaussian. To characterize the distortion level between the transmitter and the relay, we define the equivalent signal-to-noise plus self-interference ratio (ESNR) associated with the kth subcarrier as

$$\mathrm{ESNR}_k^{(A)} = \frac{|\alpha_{bp}^{(A)} H_k^{(A)}|^2 \mathbb{E}\left[|S_k|^2\right]}{\left|H_k^{(A)}\right|^2 \mathbb{E}\left[\left|D_k^{(A)}\right|^2\right] + \mathbb{E}\left[\left|N_k^{(A)}\right|^2\right]}$$
$$= \frac{\left|\alpha_{bp}^{(A)} H_k^{(A)}\right|^2}{\left|H_k^{(A)}\right|^2 \frac{\mathbb{E}\left[\left|D_k^{(A)}\right|^2\right]}{\mathbb{E}[|S_k|^2]} + \frac{N_0}{\mathbb{E}[|S_k|^2]}}. \tag{5.39}$$

Fig. 5.50 shows the simulated and the theoretical $\mathrm{ESNR}_k^{(A)}$ considering $O = 4$, $N_u = 512$, $I^{(A)} = 128$, SNR $= 10$ dB, $s_M/\sigma = 1.0$ and a given channel realization.

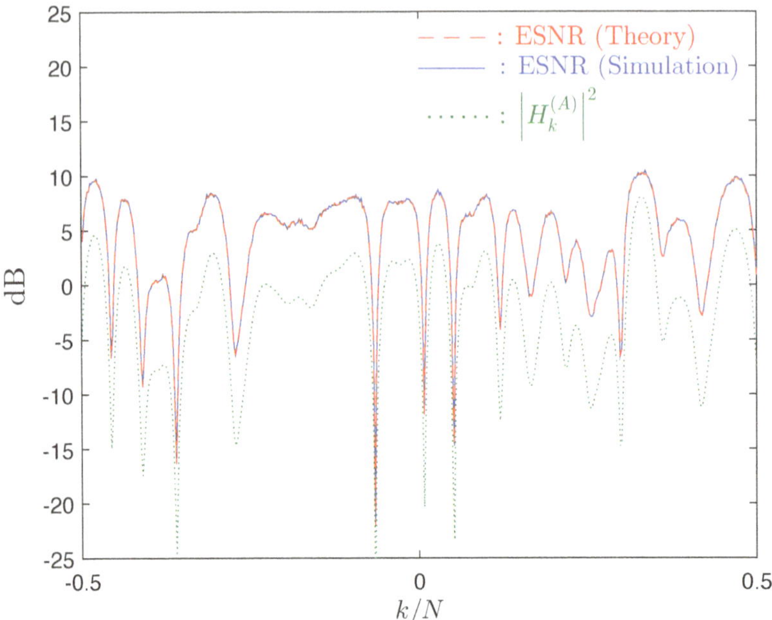

Figure 5.50 Evolution of $\text{ESNR}_k^{(A)}$ and $\left|H_k^{(A)}\right|^2$ when the channel SNR is 10 dB.

The theoretical evolution of the spectral distribution of the nonlinear distortion terms $\mathbb{E}[|D_k^{(A)}|^2]$ was obtained through the truncated IMP approach. From the results depicted in the figure, one can that the simulated and theoretical results match. Additionally, it can also be seen that evolution of $\text{ESNR}_k^{(A)}$ follows closely the evolution of the channel frequency responses. Considering the Gaussian approximation validated in (5.49), it can be noted that the samples of the input to the relay's nonlinearity $\mathbf{r}^{(A)} = [r_0^{(A)}\ r_1^{(A)}\ r_2^{(A)}\ \ldots\ r_{N-1}^{(A)}]^T \in \mathbb{C}^N$ are approximately Gaussian distributed with variance $\sigma_r^2 = \sigma_b^2 + \sigma_n^2$, where $\sigma_n^2 = N_0/(2N)$. This means that the signal at the output of the relay's amplifier can be statistically characterized considering the truncated IMP or the equivalent nonlinearity approach. At the output of the relay's amplifier, the signal for the kth subcarrier is given by

$$Y_k^{(B)} = \alpha_{bp}^{(B)} \alpha_{bp}^{(A)} H_k^{(A)} S_k + \alpha_{bp}^{(B)} H_k^{(A)} D_k^{(A)} + \alpha_{bp}^{(B)} N_k^{(A)} + D_k^{(B)}, \tag{5.40}$$

where $\alpha_{bp}^{(B)}$ and $D_k^{(B)}$ denote the scale factor and the nonlinear distortion term associated with the relay's nonlinearity. Fig. 5.51 depicts the evolution of $\left|H_k^{(A)}\right|^2$, as well as the PSD-associated distortion component produced by the relay's amplifier,

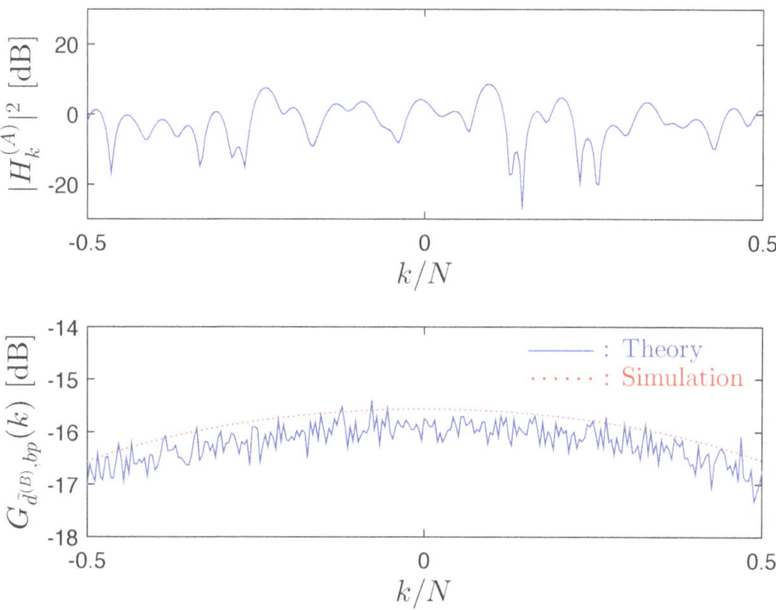

Figure 5.51 Evolution of $|H_k^{(A)}|^2$ and $G_{\tilde{d}^{(B)},bp}(k)$ when the channel SNR is 10 dB.

$G_{\tilde{d}^{(B)},bp}(k)$, considering $O = 4$, $N_u = 256$, $I^{(A)} = 64$ and $s_M/\sigma = s_M/\sigma_r = 1.5$. From the figure, one can note that due to the existence of intermodulation products, the PSD of the distortion term $\mathbb{E}[|D_k^{(B)}|^2]$ is approximately constant, which means that it is approximately independent of $|H_k^{(A)}|^2$. Once again, one can also observe that there is a good match between the simulated and the theoretical PSDs. Fig. 5.52 presents the PSD of the nonlinearly distorted signal after the relay's amplifier, i.e., the PSD of (5.40), obtained both theoretically and by simulation. It is considered that $O = 4$, $N_u = 128$, SNR $= 10$ dB and $s_M/\sigma = s_M/\sigma_r = 1.0$. At the receiver, the signal at the kth subcarrier is

$$\begin{aligned} Z_k^{(B)} &= H_k^{(B)} Y_k^{(B)} + N_k^{(B)} \\ &= \underbrace{\alpha_{bp}^{(B)} \alpha_{bp}^{(A)} H_k^{(A)} H_k^{(B)} S_k}_{\text{Useful term}} + \underbrace{\alpha_{bp}^{(B)} H_k^{(A)} H_k^{(B)} D_k^{(A)}}_{\text{Distortion term}} \\ &\quad + \underbrace{\alpha_{bp}^{(B)} H_k^{(B)} N_k^{(A)} + H_k^{(B)} D_k^{(B)} + N_k^{(B)}}_{\text{Distortion term}}. \end{aligned} \tag{5.41}$$

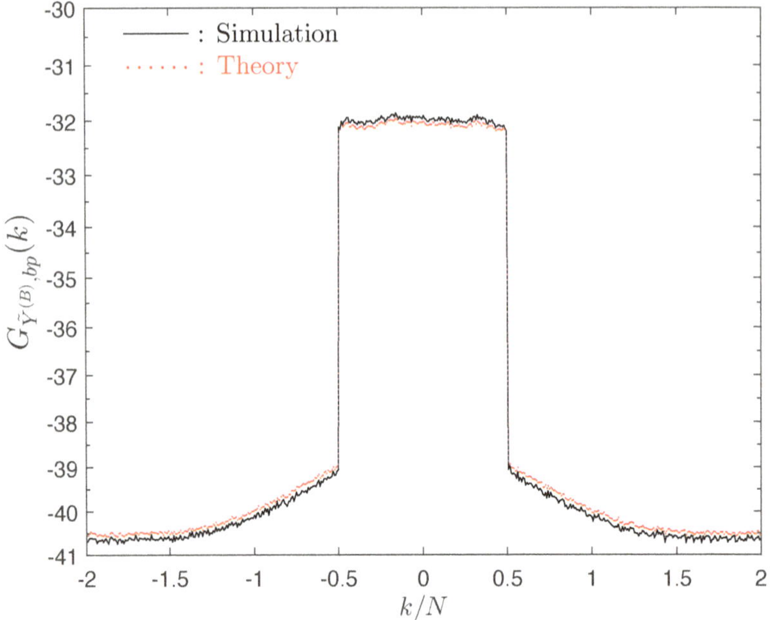

Figure 5.52 PSD of the nonlinearly distorted signal after the relay's node amplifier when the SNR = 10 dB.

Therefore, the ESNR of the kth subcarrier is given by

$$\mathrm{ESNR}_k^{(B)} = \frac{\left|\alpha_{bp}^{(B)}\alpha_{bp}^{(A)}H_k^{(A)}H_k^{(B)}\right|^2 \mathbb{E}\left[|S_k|^2\right]}{\left|\alpha_{bp}^{(B)}H_k^{(A)}H_k^{(B)}\right|^2 \mathbb{E}\left[\left|D_k^{(A)}\right|^2\right] + \left|H_k^{(B)}\right|^2 \Gamma_k + N_0}, \tag{5.42}$$

with $\Gamma_k = \left(|\alpha_{bp}^{(B)}|^2 N_0 + \mathbb{E}\left[\left|D_k^{(B)}\right|^2\right]\right)$. Fig. 5.53 shows the $\mathrm{ESNR}_k^{(B)}$ considering $N_u = 256$, $O = 4$, $I^{(A)} = I^{(B)} = 64$, $s_M/\sigma = s_M/\sigma_r = 1.5$ and SNR = 10 dB. The values of $\mathbb{E}\left[\left|D_k^{(B)}\right|^2\right]$ and $\mathbb{E}\left[\left|D_k^{(A)}\right|^2\right]$ were obtained both by simulation and theoretically.

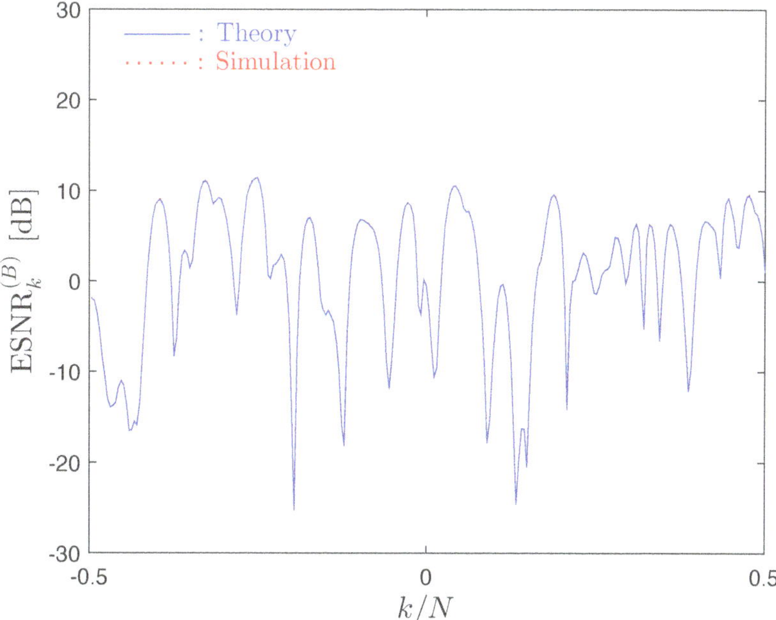

Figure 5.53 $\mathrm{ESNR}_k^{(B)}$ obtained theoretically and by simulation when both channels have an SNR of 10 dB.

From the figure it can be noted that the distortion levels at the subcarrier level can be computed analytically with very high accuracy, this can be explained by the high accuracy associated with the analytical characterization of the nonlinear distortion terms of both nonlinearities.

6 MIMO-OFDM and Optical OFDM Systems

The larger and larger demands for higher throughputs and stronger link reliabilities have become key goals for present and future wireless broadband communications systems. By exploiting the spatial domain, multiple antenna systems such as MIMO systems were proposed to achieve diversity, higher data-rates, and/or increased capacity, and are being widely used in the modern wireless transceivers.

In a MIMO system, multiple antennas can be placed at the transmitter and at receiver. These antennas should be spatially separated to promote and enrich the spatial diversity between the different streams. Essentially, MIMO techniques can be used to obtain: (i) a diversity gain to mitigate the fading or (ii) an increased capacity. Depending on the intended goal, different MIMO implementations can be adopted. To maximize the spatial diversity and the power efficiency, space-time techniques such as space-time block codes (STBC) [119, 120] can be considered. On the other hand, to fully maximize the capacity, techniques such as V-Bell Laboratories layered space time (BLAST) [121, 122] may be implemented.

Initially, MIMO techniques were proposed for flat-fading channels, although they were later extended to frequency-selective channels [123, 124]. As seen in Chapter 2, OFDM can convert frequency-selective channels into a set of parallel flat-fading channels. Therefore, the combination of MIMO with OFDM arises naturally, allowing to explore the advantages of MIMO and OFDM in a unified way. This combination makes possible the attainment of high data-rates in wireless multipath channels, where there is strong frequency-selective fading. For this reason, MIMO-OFDM schemes are employed in different wireless communication standards [23, 25, 125].

Recently, to take further advantage of the gains inherent to multiple antenna schemes, the so-called massive MIMO systems [126] have been proposed. In these systems, a very large number of antenna elements (a number that can range from several tens to even hundreds) are placed at the transmitter and receiver. However, the use of a very large number of antennas can lead to implementation difficulties and complex transmission chains, not to mention the complexity associated with the signal processing schemes, that grows rapidly with the number of antennas. Nevertheless, despite of these implementation challenges, massive MIMO is being proposed for 5th generation (5G) cellular networks [127] and a large attention is being given to low-complexity, massive MIMO systems [128, 129].

In this chapter, the analytical characterization of Chapters 2 and 3 is extended to MIMO and massive MIMO systems impaired by strong nonlinear distortion effects. Section 6.1 focuses on nonlinearly distorted MIMO-OFDM systems. In that section, the nonlinear distortion effects of both conventional MIMO-OFDM as well as to massive MIMO-OFDM systems are considered. Section 6.2 presents results for

DOI: 10.1201/9781032708744-6

nonlinear, baseband multicarrier schemes such as DMT schemes that have clipping and/or quantization operations on their transmission chain. Finally, Section 6.3 focuses on optical OFDM schemes. This section is subdivided into two subsections: in Subsection 6.3.1, fiber optical OFDM systems impaired by nonlinear phase noise are analyzed and, in Subsection 6.3.2, wireless optical OFDM schemes such as DCO-OFDM and asymmetrical clipping optical (ACO)-OFDM systems that have asymmetrical clipping operations on their transmission chain are characterized.

6.1 MIMO AND MASSIVE MIMO-OFDM

Let us start by introducing a MIMO system composed of T transmit antennas and R receive antennas. This MIMO system employs an SVD technique [130]. This combination leads to capacity gains that arise from the exploitation of the spatial multiplexing capability that is inherent to multiple antenna systems. Indeed, if the channel is known at the transmitter and receiver, then the SVD allows to convert the frequency-selective channel associated with a given MIMO-OFDM system into a set of narrow-band, flat-fading channels at the "subcarrier-level", which is achieved by using unitary pre-processing and decoding matrices at the transmitter and receiver, respectively. Under these conditions, the MIMO-OFDM system can be regarded as a set of independent SISO-OFDM[1] systems that transmit individual data streams in parallel, which means that large capacity gains can be obtained, specially when massive MIMO-OFDM systems are considered. Note that with the SVD, a value up to $P = \min(T,R)$ independent OFDM data streams can be multiplexed onto the MIMO channel. In general, our scenario admits $T > R$, which means that $P = R$. This MIMO system is represented in Fig. 6.1.

We consider a wideband, frequency-selective channel with uncorrelated Rayleigh fading and L resolvable taps that is represented by

$$\mathbf{h}(t) = \sum_{l=1}^{L} \mathbf{h}^{(l)} \delta(t - \tau_l), \tag{6.1}$$

where τ_l and $\mathbf{h}^{(l)}$ represent the delay and the channel matrix of the lth tap, respectively. For the lth path, the $R \times T$ channel matrix is represented as

$$\mathbf{h}^{(l)} = \begin{bmatrix} h_{1,1}^{(l)} & h_{1,2}^{(l)} & \cdots & h_{1,T}^{(l)} \\ h_{2,1}^{(l)} & \cdots & \cdots & h_{2,t}^{(l)} \\ \vdots & \cdots & \ddots & \vdots \\ h_{R,1}^{(l)} & \cdots & \cdots & h_{R,T}^{(l)} \end{bmatrix}, \tag{6.2}$$

where $h_{r,t}^{(l)}$ is the Rayleigh fading coefficient between the rth receive antenna and the tth transmit antenna for the lth tap.

---

[1] It should be noted that the analysis and conclusions presented previously regarding the analytical characterization of nonlinearly distorted SISO-OFDM systems hold for each individual "stream" of the MIMO-OFDM system.

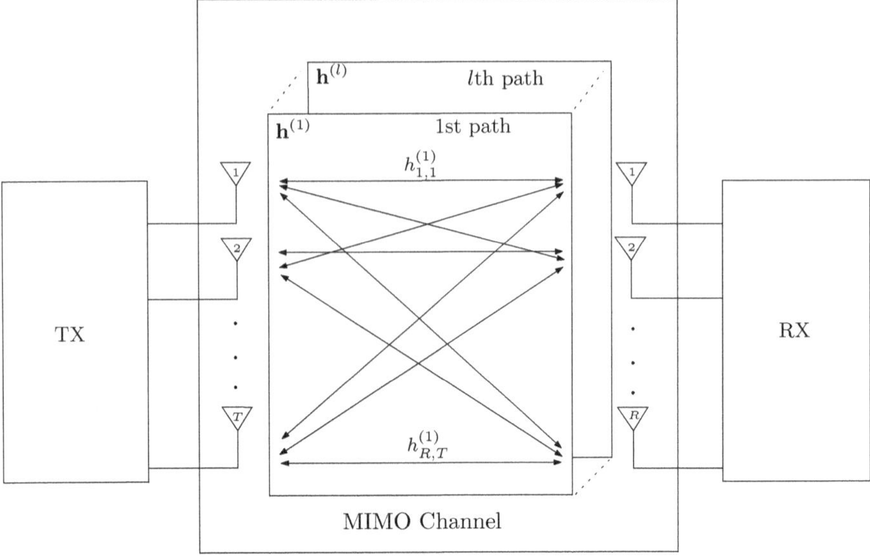

Figure 6.1 MIMO-OFDM scenario.

Let us now analyze the transmitter and receiver blocks of Fig. 6.1 in detail. Firstly, we consider a linear MIMO-OFDM transmitter and then we extend our results to the nonlinear transmission case. Each one of the P OFDM streams is composed of N_u useful data symbols plus $N_g = N_u(O-1)$ subcarriers for oversampling purposes. The total number of subcarriers of each stream $N = N_g + N_u$. The "global" OFDM block to be transmitted is represented by the $P \times N$ matrix

$$\mathbf{S} = \begin{bmatrix} S_{1,1} \; S_{1,2} \cdots & S_{1,N} \\ S_{2,1} \;\; \cdots \;\; \cdots & S_{2,N} \\ \vdots \;\; \cdots \;\; \ddots & \vdots \\ S_{P,1} \;\; \cdots \;\; \cdots & S_{P,N} \end{bmatrix}. \tag{6.3}$$

Each data symbol is selected from a QPSK constellation. Without loss of generality, we consider normalized QPSK symbols of the form $\pm 1 \pm j$. Therefore, the symbol to be transmitted on the kth subcarrier of the pth stream is defined as

$$S_{p,k} = \begin{cases} \pm 1 \pm j, & \frac{N-N_u}{2} \leq k \leq \frac{N-N_u}{2} + N_u \\ 0, & \text{otherwise.} \end{cases} \tag{6.4}$$

To represent the set of data symbols transmitted on the kth subcarrier, we use the block $\mathbf{S}(k) = [S(k)_1 \; S(k)_2 \; \cdots \; S(k)_P]^T \in \mathbb{C}^N$, where $S(k)_p = S_{p,k}$. On the other hand, we define the set of data symbols of the pth stream as $\mathbf{S}^{(p)} = [S_1^{(p)} \; S_2^{(p)} \; \ldots \; S_N^{(p)}]^T \in$

$\mathbb{C}^N$, where $S_k^{(p)} = S_{p,k}$. The time-domain version of the pth OFDM stream is given by the IDFT of $\mathbf{S}^{(p)}$, i.e., $\mathbf{s}^{(p)} = \mathbf{F}^{-1}\mathbf{S}^{(p)} = [s_1^{(p)}\ s_2^{(p)}\ \cdots\ s_N^{(p)}]^T \in \mathbb{C}^N$. As in the case of a single OFDM stream (see (2.26)), the real and imaginary part of the time-domain samples $s_n^{(p)}$ have a zero-mean Gaussian distribution with variance σ^2.

As mentioned in the previous section, the use of OFDM together with a proper CP larger than the channel impulsive response L, allows to convert a given frequency-selective channel into a set of flat-fading channels. Therefore, the MIMO frequency-selective channel represented in (6.1) can be decomposed into N flat-fading channels at the subcarrier level. Under these conditions, the corresponding channel of the kth subcarrier is characterized by the $R \times T$ matrix

$$\mathbf{H}(k) = \begin{bmatrix} H(k)_{1,1} & H(k)_{1,2} & \cdots & H(k)_{1,T} \\ H(k)_{2,1} & \cdots & \cdots & H(k)_{2,T} \\ \vdots & \cdots & \ddots & \vdots \\ H(k)_{R,1} & \cdots & \cdots & H(k)_{R,T} \end{bmatrix}. \tag{6.5}$$

Without loss of generality, we assume a unitary channel gain in each subcarrier, i.e., $\mathbb{E}[|H_{r,t}^{(k)}|^2] = 1$. The main purpose of employing the SVD technique is to decouple the MIMO channel into several parallel channels for spatial multiplexing. In order to do that, the channel matrix of the kth subcarrier $\mathbf{H}(k)$ is decomposed as the multiplication of three matrices, i.e.,

$$\mathbf{H}(k) = \mathbf{U}(k)\Lambda(k)\mathbf{V}^H(k), \tag{6.6}$$

where $\mathbf{U}(k)$ and $\mathbf{V}^H(k)$ are two unitary matrices used for decoding and precoding operations, respectively, and $\Lambda(k) = \mathrm{diag}([\Lambda(k)_1\ \cdots\ \Lambda(k)_P])$ is a diagonal matrix composed by P non-zero singular values of $\mathbf{H}(k)$ in decreasing order. $\Lambda(k)_p$ represents the "channel gain" for the pth stream of kth subcarrier. To obtain the block of data symbols to be transmitted, the set of data symbols of the kth subcarrier $\mathbf{S}(k)$ is precoded by the $T \times R$ matrix $\mathbf{V}(k)$, resulting

$$\mathbf{X}(k) = \mathbf{V}(k)\mathbf{S}(k), \tag{6.7}$$

where $\mathbf{X}(k) = [X(k)_1\ X(k)_2\ \cdots\ X(k)_T]^T \in \mathbb{C}^T$ denotes the set of precoded data symbols of the kth subcarrier. On the other hand, the tth OFDM stream is denoted as $\mathbf{X}^{(t)} = [X_0^{(t)}\ X_2^{(t)}\ \cdots\ X_{N-1}^{(t)}]^T$ where $X_k^{(t)} = X(k)_t$ and, regarding the time-domain, the samples of that stream are represented by the block $\mathbf{x}^{(t)} = \mathbf{F}^{-1}\mathbf{X}^{(t)} = [x_0^{(t)}\ x_2^{(t)}\ \cdots\ x_{N-1}^{(t)}]^T \in \mathbb{C}^N$. After the precoding process, it should be noted that the real and imaginary parts of $x_n^{(t)}$ are still Gaussian distributed. However, their variance is given by σ_x^2, where

$$\sigma_x^2 = \frac{R}{T}\sigma^2. \tag{6.8}$$

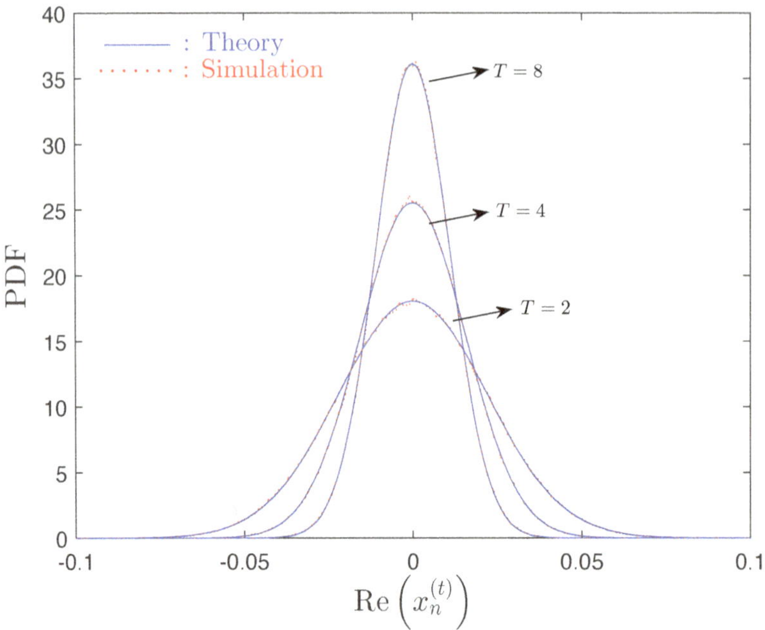

Figure 6.2 PDF of the real part of the time-domain samples of a given OFDM stream considering $R = 2$ and different values of T.

Fig. 6.2 shows the distribution of the real part of the time-domain samples of a given OFDM stream after the precoding operation, considering $R = 2$ and different values of T. Clearly, the Gaussian approximation for the distribution of the samples of precoded OFDM signal is very tight. The variance of this distribution is dependent on the ratio R/T and equal to the variance of OFDM signal before the precoding operation if $T = R$. After the precoding operation, the T OFDM streams are amplified and sent to a frequency-selective channel (see (6.1)). At the subcarrier level, we may represent the received signal as $\mathbf{W}(k) = [W(k)_1\ W(k)_2\ \cdots\ W(k)_R]^T \in \mathbb{C}^R$, where

$$\begin{aligned}\mathbf{W}(k) &= \mathbf{H}(k)\mathbf{X}(k) + \mathbf{N}(k)\\ &= \mathbf{H}(k)\mathbf{V}(k)\mathbf{S}(k) + \mathbf{N}(k)\\ &= \mathbf{U}(k)\Lambda(k)\mathbf{V}^H(k)\mathbf{V}(k)\mathbf{S}(k) + \mathbf{N}(k)\\ &= \mathbf{U}(k)\Lambda(k)\mathbf{S}(k) + \mathbf{N}(k),\end{aligned} \tag{6.9}$$

with $\mathbf{N}(k) = [N(k)_1\ N(k)_2\ \cdots\ N(k)_R]^T \in \mathbb{C}^R$ denoting the block of frequency-domain AWGN noise samples associated with the kth subcarrier, where $\mathbb{E}[|N(k)_r|^2] = 2\sigma_N^2$. At the reception, the post-processing operation consists of a multiplication of the

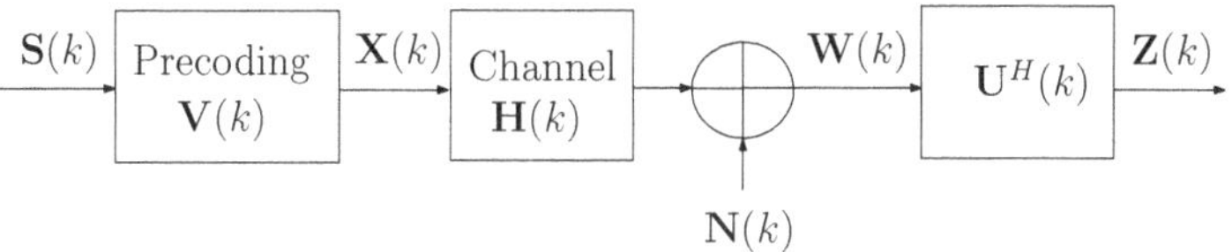

Figure 6.3 Equivalent, subcarrier-level model for the linear MIMO-OFDM system considering the SVD technique.

received signal $\mathbf{W}(k)$ by the $R \times R$ decoding matrix $\mathbf{U}^H(k)$, resulting

$$\begin{aligned}\mathbf{Z}^{(k)} &= \mathbf{U}^H(k)\mathbf{W}(k) \\ &= \mathbf{U}^H(k)\mathbf{U}(k)\Lambda(k)\mathbf{S}(k) + \mathbf{U}^H(k)\mathbf{N}(k) \\ &= \Lambda(k)\mathbf{S}(k) + \underbrace{\mathbf{U}^H(k)\mathbf{N}(k)}_{\mathbf{N}'(k)} \\ &= \Lambda(k)\mathbf{S}(k) + \mathbf{N}'(k). \end{aligned} \tag{6.10}$$

As the matrix $\mathbf{U}^H(k)$ is unitary, i.e., $\mathbf{U}(k)\mathbf{U}^H(k) = \mathbf{I}_R$, the decoding operation does not change the power of the noise samples. The equivalent, subcarrier-level model for the considered linear MIMO-OFDM system employing an SVD technique is depicted in Fig. 6.3. The received signal associated with the kth subcarrier of the pth stream is given by

$$Z(k)_p = \Lambda(k)_p S(k)_p + N'(k)_p. \tag{6.11}$$

In fact, this means that each stream can be regarded as an individual SISO-OFDM stream with flat-fading, where the flat-fading coefficient of the kth subcarrier of the pth stream is $|\Lambda(k)_p|^2$. Indeed, there is an equivalent, subcarrier-level model similar to the one represented in Fig. 2.11 for each one of the streams. After some lengthy but straightforward manipulations, it can be shown that the BER for a given channel realization can be written as

$$P_b(H) \approx \frac{1}{N_u P} \sum_{k=1}^{N_u} \sum_{p=1}^{P} Q\left(\sqrt{|\Lambda(k)_p|^2 \frac{2E_b}{N_0}}\right). \tag{6.12}$$

Fig. 6.4 shows the average BER of a MIMO-OFDM system with $R = 2$ receive antennas, a variable number of transmit antennas T, and several channel realizations with $L = 64$ multipath components. The OFDM signals have $N_u = 64$ and $O = 1$. With these combinations of T and R, we can have up to $P = R$ streams spatially multiplexed in the same MIMO channel. From the results depicted in the figure, it can be seen that the BER considerably decreases as the number of transmit antennas increase for a fixed number of streams, since the reliability of each link increases. When $T = R = 2$, the BER is identical to the one associated with a SISO-OFDM system and, to obtain a target BER of $P_b = 10^{-3}$, we need an E_b/N_0 of approximately

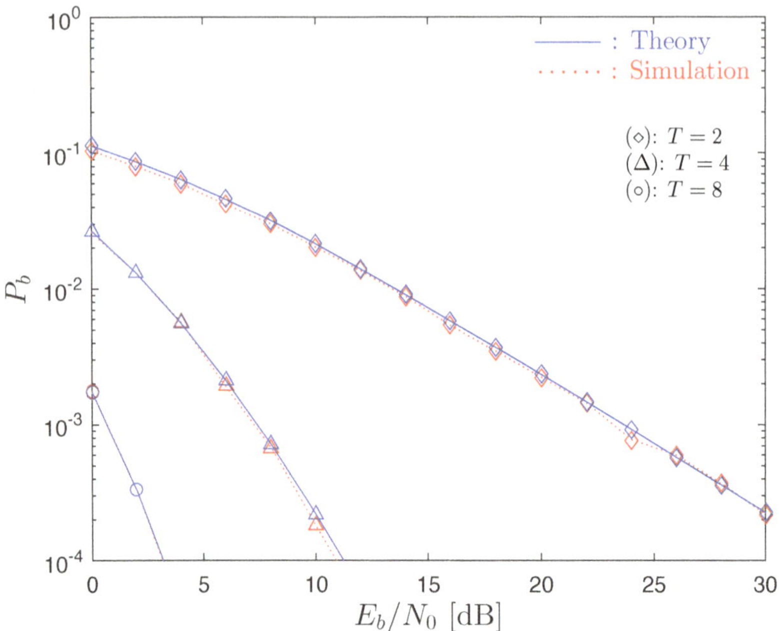

Figure 6.4 Average BER of a MIMO-OFDM system with SVD considering $R = 2$ and different values of T.

23.6 dB. However, the required E_b/N_0 decreases to approximately 7.4 dB and to a value around 0.67 dB to achieve the same BER when $T = 4$ and $T = 8$, respectively.

Let us now consider a nonlinear MIMO-OFDM system where there is a nonlinear power amplifier (otherwise stated, modeled by an envelope clipping operation with clipping level s_M/σ_x) in each transmitter branch and evaluate the impact of the corresponding nonlinear distortion effects on the system's performance. In order to do that, we present SIR expressions for the different OFDM streams, which allows to estimate the performance penalty associated with each stream [131]. To obtain the SIR of each stream analytically, we take advantage of the Gaussian nature of the samples of precoded signal.

After the amplification process, the nonlinearly distorted OFDM signal of the tth branch can be represented by

$$\mathbf{y}^{(t)} = \alpha_{bp}^{(t)}\mathbf{x}^{(t)} + \mathbf{d}^{(t)}, \tag{6.13}$$

where $\mathbf{d}^{(t)} = [d_0^{(t)}\ d_1^{(t)}\ \cdots\ d_{N-1}^{(t)}]^T \in \mathbb{C}^N$ is the set of nonlinear distortion terms of the tth branch and $\alpha_{bp}^{(t)}$ is the scale factor for the tth branch. This scale factor can be obtained as

$$\alpha_{bp}^{(t)} = \frac{\mathbb{E}\left[x_n^{(t)} y_n^{*(t)}\right]}{\mathbb{E}\left[|x_n^{(t)}|^2\right]} = \frac{\mathbb{E}\left[x_n^{(t)} f_{bp}^{*(t)}\left(x_n^{(t)}\right)\right]}{2\sigma_x^2}. \tag{6.14}$$

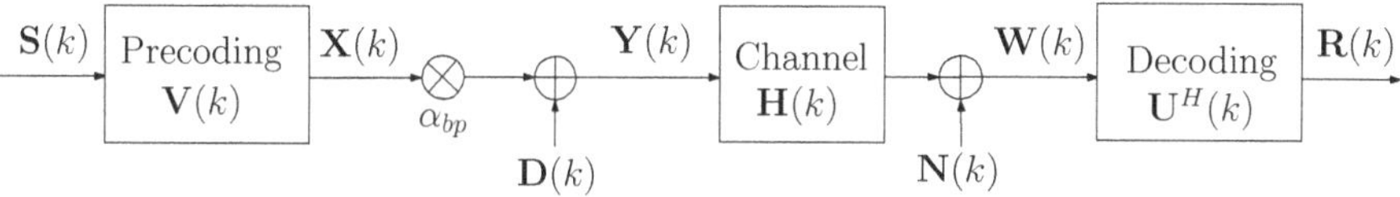

Figure 6.5 Equivalent, subcarrier-level model for a nonlinear MIMO-OFDM system with the SVD technique.

In fact, as it is assumed that the nonlinear characteristics of the power amplifiers are equal and the power of precoded signals is the same for all branches, the scale factor of (6.15) is independent of t, i.e., $\alpha_{bp}^{(t)} = \alpha_{bp}\ \forall\ t$. Note also that the DFT of (6.13) yields the frequency-domain version of the signal to be transmitted $\mathbf{Y}^{(t)} = \mathbf{F}\mathbf{y}^{(t)}$. Once again, from the Bussgang's theorem, we can separate $\mathbf{Y}^{(t)}$ into two uncorrelated components, leading to

$$\mathbf{Y}^{(t)} = \alpha_{bp}\mathbf{X}^{(t)} + \mathbf{D}^{(t)}, \tag{6.15}$$

where $\mathbf{D}^{(t)} = \mathbf{F}\mathbf{d}^{(t)}$ represents the nonlinear distortion terms along the tth branch. Under these conditions, the nonlinearly distorted signal associated with the kth subcarrier is

$$\mathbf{Y}(k) = \alpha_{bp}\mathbf{X}(k) + \mathbf{D}(k), \tag{6.16}$$

where $\mathbf{D}(k) = [D(k)_1\ D(k)_2\ \cdots\ D(k)_T]^T \in \mathbb{C}^T$ is the set of nonlinearly distorted terms of the kth subcarrier. The nonlinearly amplified signals are transmitted through the frequency-selective channel represented in (6.5). The equivalent, subcarrier-level model for the nonlinear MIMO-OFDM system is depicted in Fig. 6.5. In the case of a nonlinear transmission in each transmitter branch, the received signal for the kth subcarrier expressed in (6.9) becomes

$$\begin{aligned}\mathbf{W}(k) &= \mathbf{H}(k)\mathbf{Y}(k) + \mathbf{N}(k)\\ &= \mathbf{H}(k)(\alpha_{bp}\mathbf{X}(k) + \mathbf{D}(k)) + \mathbf{N}(k),\end{aligned} \tag{6.17}$$

and the decoded signal can be expressed as

$$\begin{aligned}\mathbf{R}(k) &= \mathbf{U}^H(k)\mathbf{W}(k) &(6.18)\\ &= \mathbf{U}^H(k)(\mathbf{H}(k)\mathbf{Y}(k) + \mathbf{N}(k))\\ &= \mathbf{U}^H(\mathbf{U}(k)\Lambda(k)\mathbf{V}^H(k)\mathbf{Y}(k) + \mathbf{N}(k))\\ &= \mathbf{U}^H(\mathbf{U}(k)\Lambda(k)\mathbf{V}^H(k)\left(\alpha_{bp}\mathbf{X}(k) + \mathbf{D}(k)\right) + \mathbf{N}(k))\\ &= \alpha_{bp}\Lambda(k)\mathbf{S}(k) + \Lambda(k)\underbrace{\mathbf{V}^H(k)\mathbf{D}(k)}_{\mathbf{D}'(k)} + \mathbf{N}'(k)\\ &= \underbrace{\alpha_{bp}\Lambda(k)\mathbf{S}(k)}_{\text{Useful part}} + \underbrace{\Lambda(k)\mathbf{D}'(k)}_{\text{Distortion part}} + \underbrace{\mathbf{N}'(k)}_{\text{Noise part}}, &(6.19)\end{aligned}$$

where $\mathbf{D}'(k) = [D'(k)_0\ D'(k)_1\ \cdots\ D'(k)_P]^T \in \mathbb{C}^C$ and $\mathbf{N}'(k) = [N'(k)_0\ N'(k)_1\ \cdots\ N'(k)_P]^T \in \mathbb{C}^C$ represent the set of equivalent nonlinear distortion and noise terms, respectively. As the detection is made in frequency-domain, we are specially interested on the spectral characterization of nonlinearly distorted OFDM signals after the SVD decomposition. In the following, results of the SIR after the decoding process are shown. The SIR is obtained as the ratio between the useful and the distortion part of the decoded signal (see (6.18)) and, for a given subcarrier k and stream c, is given by [131]

$$\mathrm{SIR}_{k,p}^{\mathrm{MIMO}} = \frac{|\alpha_{bp}|^2 |\Lambda(k)_p|^2 \mathbb{E}\left[|S(k)_p|^2\right]}{|\Lambda(k)_p|^2 \mathbb{E}\left[|D'(k)_p|^2\right]} = \frac{|\alpha_{bp}|^2 \mathbb{E}\left[|S(k)_p|^2\right]}{\mathbb{E}\left[\left|\sum_{t=1}^{T} V^H(k)_{p,t} D(k)_t\right|^2\right]}. \tag{6.20}$$

For performance evaluation purposes, it would be interesting to compare the SIR of a given stream of a nonlinear MIMO-OFDM system with the SIR of a nonlinearly distorted SISO-OFDM signal with the same power and that passes through the same nonlinearity. In fact, it can be shown that the SIR for a given stream of a nonlinear MIMO-OFDM system decreases with the number of transmit antennas, provided that the number of streams is fixed [131]. Indeed,

$$\mathbb{E}\left[\left|\sum_{t=1}^{T} V^H(k)_{p,t} D(k)_t\right|^2\right] \approx \frac{T}{R} \mathbb{E}\left[|D(k)|^2\right], \tag{6.21}$$

i.e., the distortion levels decrease with the ratio T/R. Therefore, by denoting the spectral distribution of the nonlinear distortion terms in a SISO-OFDM system as $\mathbb{E}\left[|D(k)|^2\right]$, we have that

$$\mathrm{SIR}_{k}^{\mathrm{SISO}} = \frac{|\alpha_{bp}|^2 \mathbb{E}\left[|S(k)|^2\right]}{\mathbb{E}\left[|D(k)|^2\right]}, \tag{6.22}$$

which means that (6.20) can be also written as

$$\mathrm{SIR}_{k,p}^{\mathrm{MIMO}} \approx \frac{|\alpha_{bp}|^2 \mathbb{E}\left[|S(k)_p|^2\right]}{\frac{T}{R} \mathbb{E}\left[|D(k)|^2\right]} = \frac{T}{R} \mathrm{SIR}_{k}^{\mathrm{SISO}}. \tag{6.23}$$

Fig. 6.6 shows the spectral distribution of the nonlinear distortion term concerning the first stream[2] ($p = 1$), i.e., $\mathbb{E}\left[|D'(k)_1|^2\right]$, for $R = 8$ and different values of T, as

[2] This analysis is independent of the stream, since the spectral distribution of the nonlinear distortion term is equal for all the P streams.

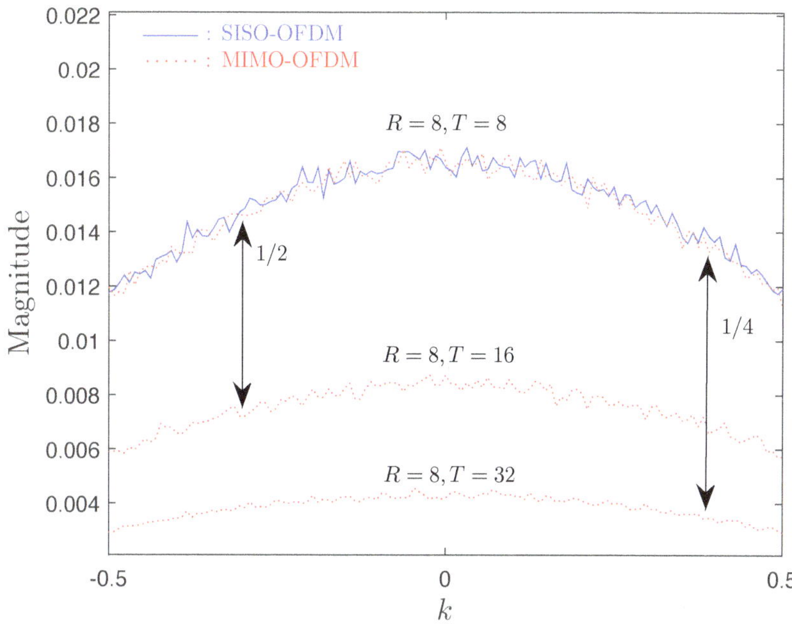

Figure 6.6 Spectral distribution of the nonlinear distortion term considering a SISO-OFDM and a MIMO-OFDM system with $R=8$ and different values of T.

well as the spectral distribution of the nonlinear distortion term in a SISO-OFDM system. The OFDM signals have $N_u = 256$ and $O = 4$ and the normalized clipping level is $s_M/\sigma_x = 0.5$. From the results shown in the figure, it can be noted that for $R = T = 8$, we have $T/R = 1$ and $\mathbb{E}\left[|D'(k)_1|^2\right] \approx \mathbb{E}\left[|D(k)|^2\right]$. On the other hand, it can also be observed that the spectral distribution of the nonlinear distortion term decreases by a factor of $T/R = 1/2$, when $T = 16$, and by a factor of $T/R = 1/4$, when $T = 32$. This confirms that the distortion level can be greatly reduced when $T \gg R$. Fig. 6.7 shows the simulated and the theoretical evolution of the SIR for the first stream (i.e., for $p = 1$) considering OFDM signals with $N_u = 256$, $O = 4$, $s_M/\sigma_x = 0.5$, $R = 1$ and several values of T. The frequency-selective channel has $L = 64$ multipath components. The figure shows that, regardless of the number of transmit antennas, very accurate estimates of the SIR can be obtained theoretically. In addition, it should be pointed out that the higher the number of transmit antennas, the better the SIR. In fact, there is a gain relatively to the case when $T = 1$ and $R = 1$ (i.e., relatively to a SISO-OFDM system) in the SIR, that is approximately given by $10\log_{10}(T)$ dB, which means that the SIR increases approximately 3 dB when T is doubled. This effect is also illustrated in Fig. 6.8, where it is shown the theoretical average value of SIR_k, $\mathbb{E}[\text{SIR}_k^{\text{MIMO}}]$ for different normalized clipping levels and different number of transmit antennas. As expected, the results shown in the

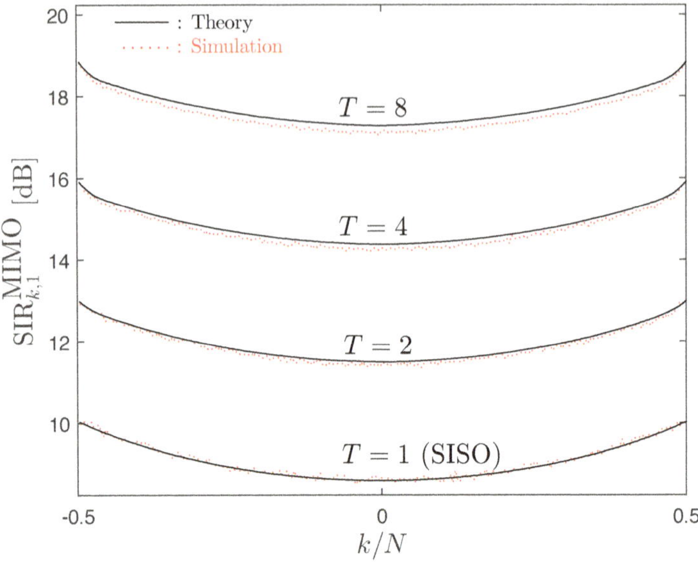

Figure 6.7 Evolution of SIR for the first stream ($p = 1$) considering $R = 1$ and different values of T.

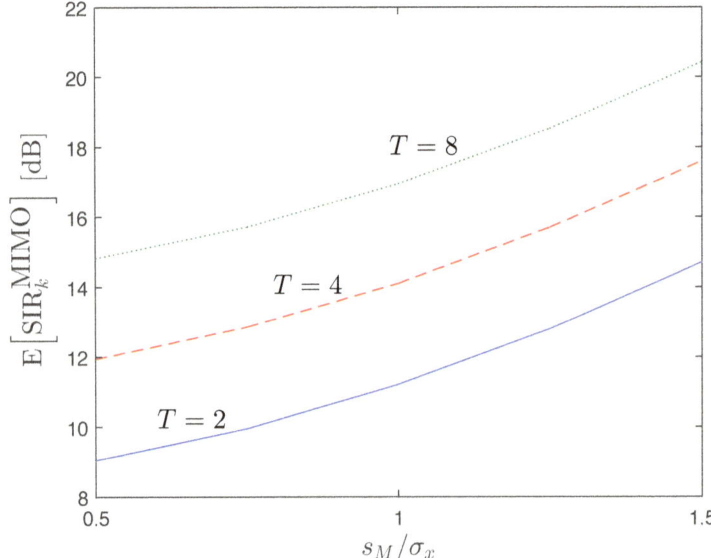

Figure 6.8 Theoretical average value of $\mathrm{SIR}_k^{\mathrm{MIMO}}$ considering different values of T and different normalized clipping levels s_M/σ_x.

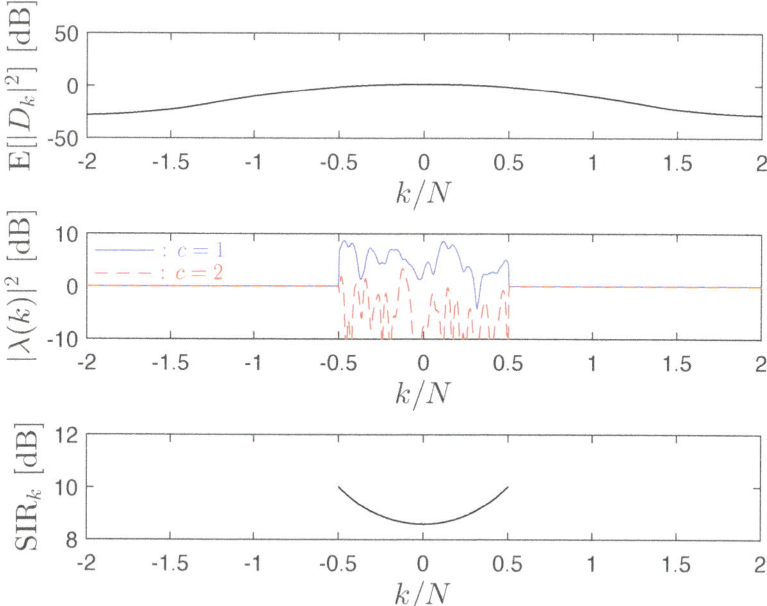

Figure 6.9 Evolution of the SIR, power of the singular values and distortion at the detection level considering a MIMO-OFDM system with $T = R = 2$ and a normalized clipping level $s_M/\sigma_x = 0.5$.

figure show that the average value of SIR grows when the normalized clipping level increases, since the magnitude of nonlinear distortion is lower. Additionally, it can be seen that, regardless of the clipping level, the SIR augments when T increases, for a fixed number of R streams. Fig. 6.9 shows the PSD of nonlinear distortion term, the evolution of singular values $|\blacksquare(k)|^2$ and the SIR_k for a given channel realization and considering a MIMO system with $T = R = 2$. The OFDM signals have $N_u = 256$, $O = 4$ and the normalized clipping level is $s_M/\sigma_x = 0.5$. From the results shown in the figure, it can be seen that although $|\blacksquare(k)|^2$ varies with frequency, and besides the existence of singular values with low amplitudes (especially the ones associated with the second stream $p = 2$), the values of SIR_k and the distortion at the detection level have relatively small fluctuations in the frequency (less than 2 dB from the center of the band to the edge of the band). In addition, SIR_k is equal for both streams ($p = 1$ and $p = 2$), which means that even when there are weaker singular values in a given stream, the corresponding SIR_k does not decrease.

Taking advantage of the conclusions presented above, that suggest that the nonlinear distortion effects can be reduced with the number of transmit antennas, provided that the number of streams is fixed we propose, in the following, a very low-complexity, massive MIMO-OFDM transmitter with T transmit antennas

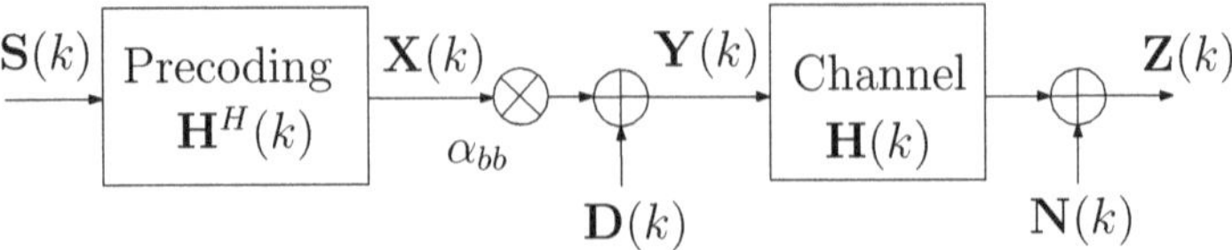

Figure 6.10 Equivalent, subcarrier-level scenario of a nonlinear, massive MIMO-OFDM scheme considering the MRT technique.

communicating with R receive antennas [132]. Such low complexity is achieved with 1-bit DACs at each transmitting branch and a low-complexity MRT technique for user separation, which has much lower signal processing requirements than SVD or zero forcing (ZF) techniques. Although the use of 1-bit DACs leads to severe nonlinear distortion effects in the transmitted signals, we considered a scenario where $T \gg R$, showing that, in those conditions, the impact of the resultant severe nonlinear distortion effects can be mitigated. Fig. 6.10 shows the equivalent, subcarrier-level model of the system. Note that the quantization operation is performed at the baseband regarding separately the real and imaginary parts of the precoded signal. For this reason, the quantization is modeled as a baseband memoryless nonlinearity. At the quantizer output, we have

$$\mathbf{Y}(k) = \alpha_{bb}\mathbf{X}(k) + \mathbf{D}(k), \tag{6.24}$$

where α_{bb} is a real-valued scale factor. Note that, differently from the case of the SVD decomposition, the precoded signal associated with the kth subcarrier is obtained as

$$\mathbf{X}(k) = \mathbf{H}^H(k)\mathbf{S}(k), \tag{6.25}$$

i.e., the precoding matrix is $\mathbf{H}^H(k)$. At reception, the received signal for the kth subcarrier is

$$\begin{aligned}\mathbf{Z}(k) &= \mathbf{H}(k)\mathbf{Y}(k) + \mathbf{N}(k) \\ &= \mathbf{H}(k)(\alpha_{bb}\mathbf{X}(k) + \mathbf{D}(k)) + \mathbf{N}(k) \\ &= \alpha_{bb}\underbrace{\mathbf{H}(k)\mathbf{H}^H(k)}_{\mathbf{P}(k)}\mathbf{S}(k) + \underbrace{\mathbf{H}(k)\mathbf{D}(k)}_{\mathbf{B}(k)} + \mathbf{N}(k).\end{aligned} \tag{6.26}$$

Under these conditions, for the kth subcarrier of the rth user, we have

$$\begin{aligned}Z(k)_r &= \alpha_{bb}\sum_{j=1}^{R} P(k)_{r,j}S(k)_j + B(k)_r + N(k)_r \\ &= \alpha_{bb}P(k)_{r,k}S(k)_j + \underbrace{\alpha_{bb}\sum_{j=1,j\neq k}^{R} P(k)_{r,j}S(k)_j}_{\text{Inter-user intereference}} + \underbrace{B(k)_r + N(k)_r}_{\text{Nonlinear distortion and noise}},\end{aligned} \tag{6.27}$$

where

$$B(k)_r = \sum_{j=1}^{T} H(k)_{r,j} D(k)_j, \tag{6.28}$$

represents the nonlinear distortion term scaled by the channel frequency responses. As we are specifically interested in study the impact of the nonlinear distortion on the transmitted signals, we consider a very large SNR and neglect $\mathbb{E}[|N(k)_r|^2]$. Under these conditions, the SIR for the kth subcarrier is

$$\mathrm{SIR}_{k,r}^{\mathrm{MIMO}} = \frac{|\alpha_{bb}|^2 \mathbb{E}[|P(k)_{r,k} S(k)_j|^2]}{\mathbb{E}\left[\left|\sum_{j=1, j\neq k}^{R} P(k)_{r,j} S(k)_j\right|^2\right] + \mathbb{E}[|B(k)_r|^2]}. \tag{6.29}$$

According to (6.28), the average value of the nonlinear distortion term can be computed as

$$\begin{aligned} \mathbb{E}[|B(k)_r|^2] &= \mathbb{E}[B(k)_r B(k)_r^*] \\ &= \mathbb{E}\left[\left(\sum_{j=1}^{T} H(k)_{r,j} D(k)_j\right)\left(\sum_{j'=1}^{T} H(k)_{r,j} D(k)_j\right)^*\right] \\ &= \mathbb{E}\left[\sum_{j=1}^{T}\sum_{j'=1}^{T} H(k)_{r,j} H(k)_{r,j'}^* D(k)_j D(k)_{j'}^*\right]. \end{aligned} \tag{6.30}$$

As the nonlinear distortion elements between different transmit branches are uncorrelated, i.e., $\mathbb{E}\left[D(k)_j D(k)_{j'}^*\right] = 0$ for $j \neq j'$, we may write

$$\begin{aligned} \mathbb{E}[|B(k)_r|^2] &= \mathbb{E}\left[\sum_{j=1}^{T} |H(k)_{r,j}|^2 |D(k)_j|^2\right] \\ &= T\mathbb{E}\left[|D(k)_r|^2\right]. \end{aligned} \tag{6.31}$$

Therefore, the SIR for the kth subcarrier becomes

$$\mathrm{SIR}_{k,r}^{\mathrm{MIMO}} = \frac{|\alpha|^2 \mathbb{E}[|P(k)_{r,k} S(k)_r|^2]}{\mathbb{E}[|\sum_{j=1, j\neq k}^{R} P(k)_{r,j} S(k)_j|^2] + T\mathbb{E}\left[|D(k)_r|^2\right]}. \tag{6.32}$$

Clearly, to avoid interference amongst users, it is required that $P(k)_{r,j} = 0$ for $j \neq k$. In other words, $\mathbf{P}(k)$ should be diagonal, i.e.,

$$\mathbf{P}(k) = \beta \mathbf{I}, \tag{6.33}$$

where β is a constant. However, in general, the matrix $\mathbf{P}(k) = \mathbf{H}(k)\mathbf{H}^H(k)$ is not diagonal. Nevertheless, when a large number of transmit antennas is considered, i.e.,

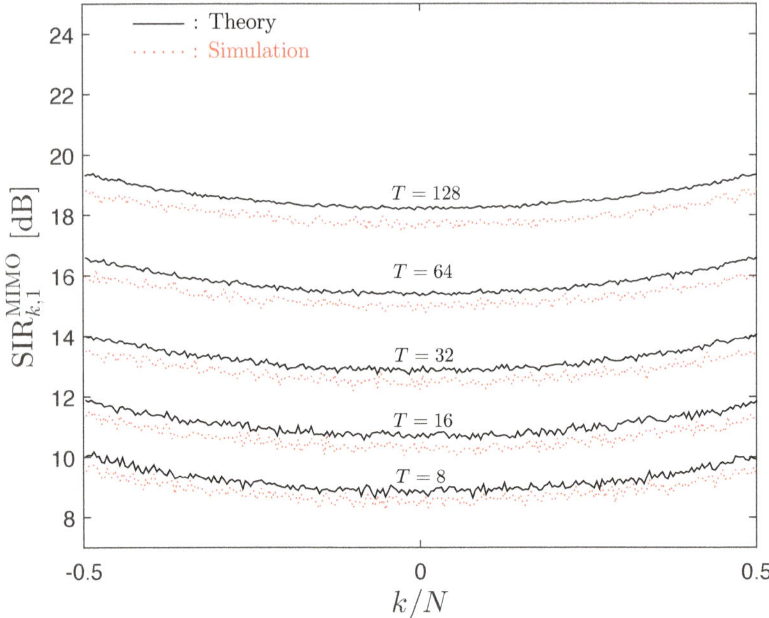

Figure 6.11 Evolution of the SIR for a given user considering $R = 8$, different values of T and a 1-bit quantization process.

$T \gg R$, and the links between the transmit and receive antennas are uncorrelated, the elements outside the diagonal are much smaller than the elements inside the diagonal [126]. In fact, when $T \gg R$, the following approximation can be made

$$\mathbf{P}(k) \approx T\mathbf{I}, \tag{6.34}$$

meaning that in such scenarios the interference amongst different users can be almost neglected. Under these conditions, (6.32) can be rewritten as

$$\mathrm{SIR}_{k,r}^{\mathrm{MIMO}} \approx \frac{|\alpha_{bb}|^2 T^2 \mathbb{E}[|S(k)_r|^2]}{T\mathbb{E}[|D(k)_r|^2]} = \frac{|\alpha_{bb}|^2 T \mathbb{E}[|S(k)_r|^2]}{\mathbb{E}[|D(k)_r|^2]}. \tag{6.35}$$

Comparing (6.35) with (6.22), one can note that there is a relation similar to (6.23), i.e.,

$$\mathrm{SIR}_{k,r}^{\mathrm{MIMO}} \approx \frac{T}{R}\mathrm{SIR}_{k}^{\mathrm{SISO}}, \tag{6.36}$$

provided that $T \gg R$. Fig. 6.11 shows the simulated and theoretical SIR for a given user (the same results are observed for other users, i.e., for other streams). The theoretical SIR was obtained by considering the equivalent nonlinearities for obtaining the PSD of the distortion component. The simulation runs with OFDM signals with

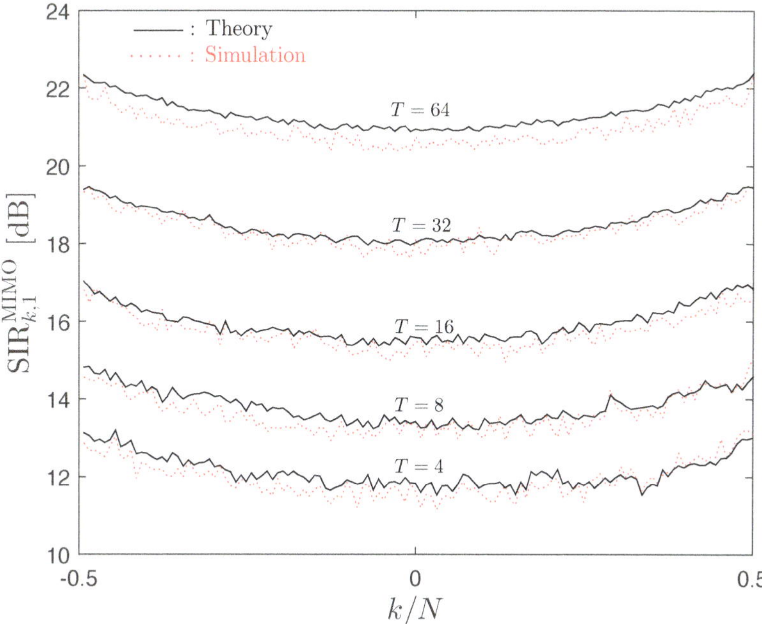

Figure 6.12 Evolution of the SIR considering $R = 4$, different values of T and an envelope clipping operation.

$N_u = 256$ and $O = 4$. We present the SIR for massive MIMO-OFDM systems with $R = 8$ users and different values of T.

Clearly, one can observe that the SIR increases when T increases. In fact, although the SIR does not increase by a factor of T/R for all values of T, this is clearly seen for larger values of T (for instance, when T increases from $T = 32$ to $T = 64$ or when T increases from $T = 64$ to $T = 128$), where the interference among users can be almost neglected (note that, under these conditions (6.34) holds). This means that even when 1-bit highly nonlinear quantizers are employed, the corresponding nonlinear distortion effects can have a low magnitude, provided that $T \gg R$. Fig. 6.12 shows the simulated and theoretical SIR for the first user $\mathrm{SIR}_{k,1}^{\mathrm{MIMO}}$, considering the impact of an envelope clipping for PAPR reduction [133], instead of considering a low-resolution quantization process. The simulation runs with OFDM signals with $N_u = 256$, $O = 4$ and a normalized clipping level of $s_M/\sigma_x = 0.5$. The frequency-selective channel has $L = 64$ multipath components. The massive MIMO-OFDM system has $R = 4$ and different values of T. The results show that accurate estimates of the SIR can be obtained. Additionally, although it can be seen that the SIR does not double when T is doubled for all values of T, we can observe the validity of (6.36) from a given value of T, since (6.34) becomes tight for $T \gg R$. As an example, it can be noted that when T increases from $T = 4$ to $T = 8$, the difference between the two corresponding values of $\mathrm{SIR}_{k,r}^{\mathrm{MIMO}}$ is approximately 2.5 dB. However, when T increases from $T = 16$ to $T = 32$, the SIR increases almost 3 dB, which also is verified when T is doubled to $T = 64$. Therefore, it can be concluded that the magnitude of the nonlinear distortion effects associated with the clipping operation decreases as T increases, which means that when massive MIMO-OFDM systems and MRT techniques are considered, we can have acceptable nonlinear distortion levels even when 1-bit quantization processes or severe clipping operations are employed. This is an important result that reveals that the transmitting branches of massive MIMO-OFDM systems may be composed of non-ideal, low-complexity devices that nonlinearly distorts the signals, while at the same time one can guarantee that the performance penalty inherent to the use of such devices is low.

6.2 DMT

DMT schemes are multicarrier schemes employed for baseband data transmission [21]. As other multicarrier signaling techniques, they are very sensitive to nonlinear distortion effects due to the large envelope fluctuations of their signals. For PAPR reduction in DMT schemes, clipping techniques are widely used, not only due to their effectiveness, but also due to their simplicity. However, as the clipping acts separately regarding the real and imaginary parts of the baseband DMT signal, we do not have envelope clipping functions (modeled as bandpass memoryless nonlinearities), but instead Cartesian clipping techniques, modeled as baseband memoryless nonlinearities (see Section 3.1.1). Additionally, other memoryless nonlinearities such as the ones associated with low resolution quantizers can also take place in DMT transceivers. Therefore, it is very likely that DMT signals face nonlinear distortion effects. Thus, for performance evaluation purposes, the impact of this distortion should be accurately characterized. In this section, we consider the analytical characterization of DMT signals submitted to baseband memoryless nonlinearities and we present an accurate spectral characterization of nonlinearly distorted DMT signals. In addition, results regarding the asymptotic optimum performance of nonlinearly distorted DMT signals are also shown.

Let us start by considering the analytical spectral characterization of quantized DMT signals. Due to the severeness of the quantization operation, especially when the number of bits of resolution is low, the equivalent nonlinearity approach is considered (see Subsection 3.3.1) [90, 134]. We consider DMT signals with N_u useful subcarriers and oversampling factor O. The real and imaginary parts of the time-domain samples $\mathbf{s} = [s_0\, s_1\, \ldots\, s_{N-1}]^T \in \mathbb{C}^N$ are Gaussian distributed with zero mean and variance σ^2. The quantizer is characterized by n_b bits of resolution and normalized clipping level s_M/σ. Fig. 6.13 shows quantization characteristics with normalized clipping level $s_M/\sigma = 1.0$ and a different number of bits resolution. In fact, the results shown in the figure reveal that when n_b is low, the corresponding nonlinear function associated with the quantization process has accentuated discontinuities, which might lead to nonlinear distortion effects with a high magnitude. For this reason, the truncated IMP approach may present convergence and complexity problems

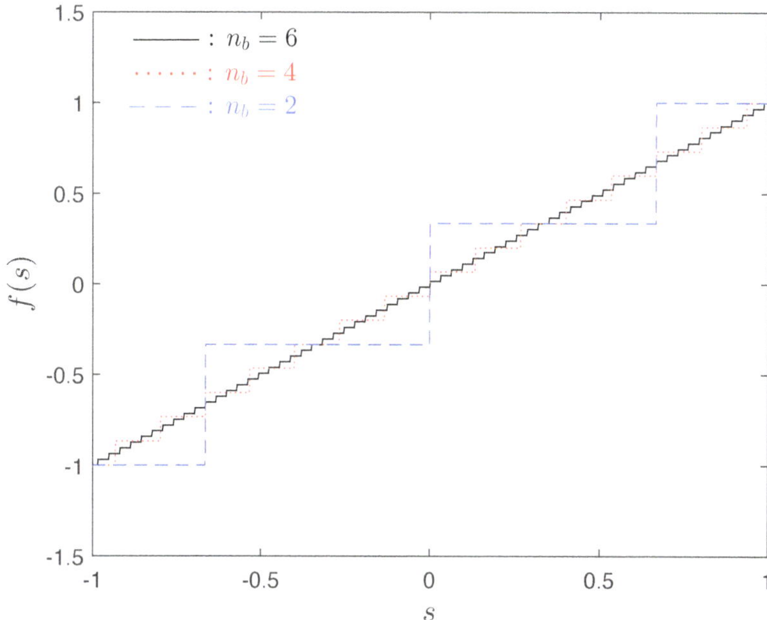

Figure 6.13 Quantization characteristics with the same normalized clipping level and a different number of bits of resolution.

when employed for obtaining the spectral characterization of DMT signals quantized with a low number of bits of resolution. Fig. 6.14 shows the theoretical and simulated PSDs of a quantized DMT signal with $N_u = 512$ and $O = 4$. The quantizer has $n_b = 4$ bits of resolution and clipping level $s_M/\sigma = 1.0$. Clearly, there are accuracy gains when the equivalent nonlinearity approach is considered for the theoretical computation of the PSD. In fact, it can be observed that the maximum error between the simulated and the theoretical PSDs is approximately 6 dB when the truncated IMP approach is considered. However, this error decreases to approximately 1 dB when an equivalent nonlinearity is considered to substitute the conventional quantization characteristic. Fig. 6.15 shows the maximum error between the theoretical and simulated PSDs considering a quantizer with a normalized clipping level of $s_M/\sigma = 2.0$ and different resolutions. Clearly, when the number of resolution's bits decreases, the nonlinearity becomes more severe, which means that the distortion is higher and a larger number of IMPs should be considered for the characterization of the nonlinearly distorted signals. Additionally, this figure confirms that the equivalent nonlinearity approach presents much more accuracy than the truncated IMP approach. In fact, when the IMP approach is considered, we can observe the existence of convergence problems that are traduced by the existence of "error floors" in the figure. For instance, when $n_b = 5$ bits of resolution are considered, the error floor

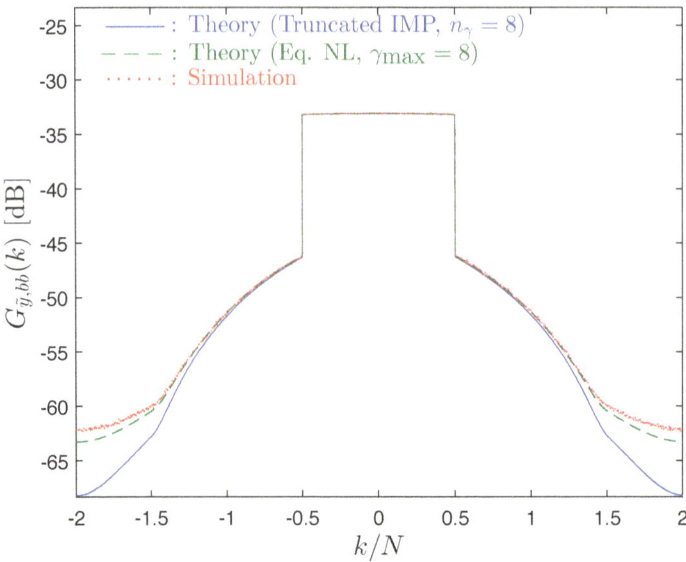

Figure 6.14 Theoretical and simulated PSD of a quantized DMT signal.

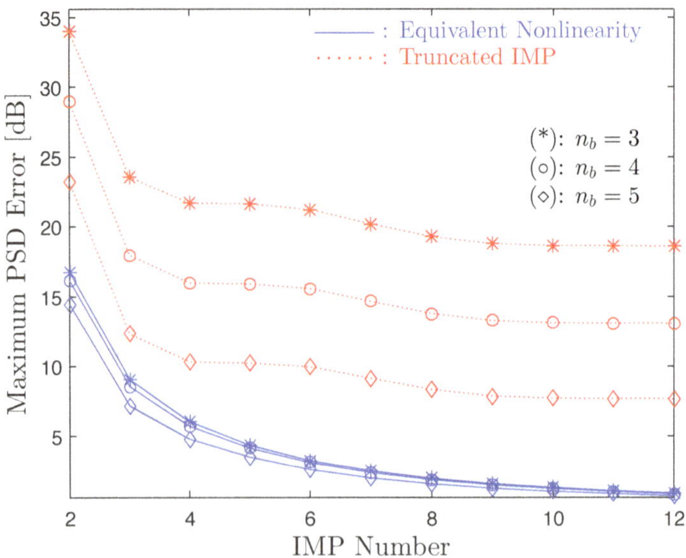

Figure 6.15 Maximum error between the simulated and theoretical PSDs of a quantized DMT signal considering a quantization process with different bits of resolution.

is around 7.6 dB. However, this error can reach a value of approximately 13 dB and 18.5 dB when $n_b = 4$ and $n_b = 3$, respectively. This means that even increasing the number of IMPs, we might be not able to improve the accuracy of the corresponding PSD. On the other hand, when the equivalent nonlinearity approach is considered, it can be noted that this maximum error decays at a much more higher "rate", not to mention that it tends do be 0, which means that the simulated and theoretical PSDs tend to match and we do not have error floors as when the truncated IMP approach is employed.

In the following, results concerning the optimum detection of nonlinearly distorted DMT signals are presented. The memoryless nonlinearity considered here is a Cartesian clipping with normalized clipping level s_M/σ, whose nonlinear function is represented by

$$f(s) = \begin{cases} s_M/\sigma, & s > s_M/\sigma \\ s, & -s_M/\sigma \leq s \leq s_M/\sigma \\ -s_M/\sigma, s < -s_M/\sigma. \end{cases} \tag{6.37}$$

As demonstrated in Subsection 4.4, the average asymptotic gain associated with the optimum detection of multicarrier signals impaired by complex-valued, baseband nonlinearities in AWGN channels can be obtained as

$$G(1) = \frac{(d_{\text{adj}}\sigma)^2}{4} \frac{\int_{-\infty}^{+\infty} f'^2(s)\, p(s)ds}{\int_{-\infty}^{+\infty} f^2(s)p(s)ds}. \tag{6.38}$$

Fig. 6.16 shows the asymptotic gain of a clipped DMT signal, considering different normalized clipping levels s_M/σ, N_u useful subcarriers and oversampling factor $O = 4$. This figure confirms the high accuracy of (6.38). This accuracy increases with the number of subcarriers, which is justified by the higher accuracy of the Taylor approximations employed to obtain (6.38). Fig. 6.17 shows the average value of the asymptotic gain considering the same conditions of the previous figure, but including also scenarios where the DMT signals differ in more than 1 bit (see the expression for $G(\mu)$ in (4.104)). From the results shown in the figure, one can note that (6.38) is also accurate when $\mu > 1$, which is the typical scenario of DMT schemes employing channel coding mechanisms. This reveals that even in those scenarios the optimum performance can be a good alternative to the conventional detection. Fig. 6.18 shows the potential asymptotic gain associated with the optimum detection of clipped DMT signals in frequency-selective channels. The gain distribution was obtained with (4.129). From the figure, it can be noted that the optimum performance in frequency-selective channels can be substantially better than the performance associated with the conventional, linear DMT schemes. This can be observed by comparing the distribution of the asymptotic gain with the distribution of the equivalent fading factor (i.e., the distribution of $|H_k|^2$). In fact, although the asymptotic gain can be lower than 1, this is also the case of the equivalent fading factor, that can assume a value lower than 1 with higher probability.

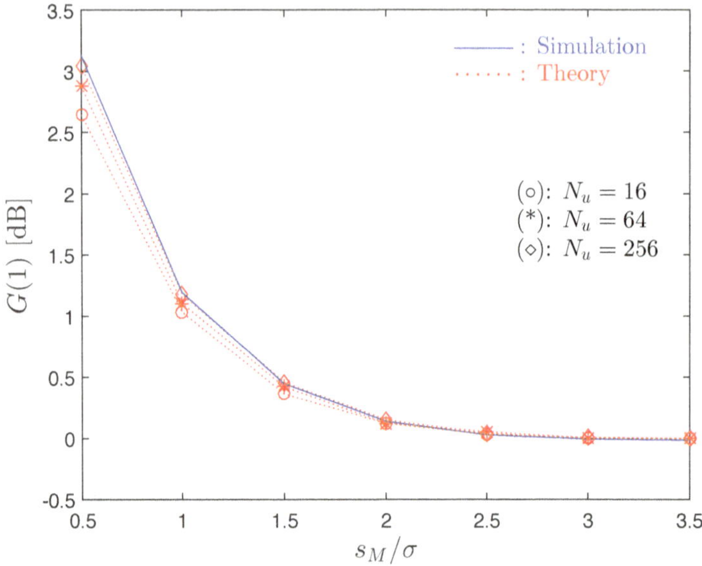

Figure 6.16 Average asymptotic gain for the optimum detection of clipped DMT signals considering different values of N_u.

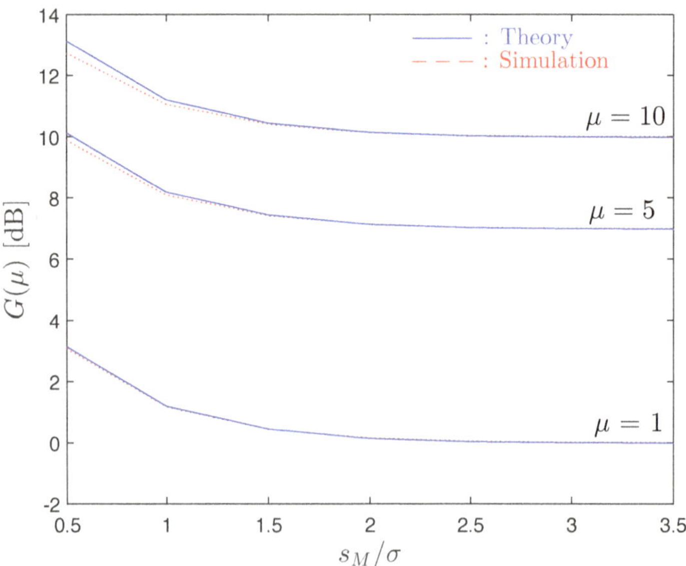

Figure 6.17 Average asymptotic performance gain associated with the optimum detection of clipped DMT signals considering different values of μ.

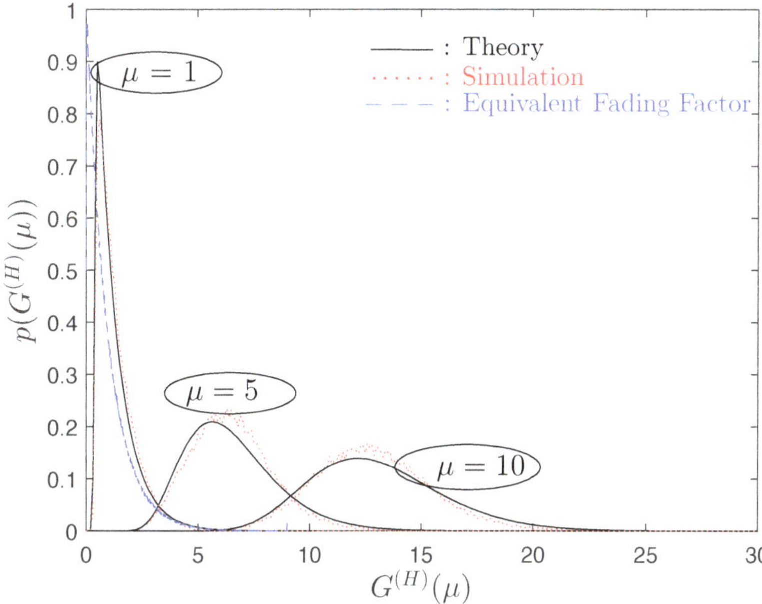

Figure 6.18 Distribution of the potential asymptotic gain considering frequency-selective channels and different values of μ.

6.3 OPTICAL OFDM

6.3.1 COHERENT OPTICAL OFDM

The coherent optical (CO)-OFDM schemes [135] are widely used in long-haul wired communications, specially due to the high data rates that can be achievable without the need of complex receivers. However, these schemes are very sensitive to one of the key limitations of the fiber optic communication channels: the presence of NLPN, caused by the Kerr's effect [136, 137]. Most of the proposed techniques to mitigate this problem involves the use of pre-distortion and/or post-compensation of the nonlinear distortion [9]. As an alternative, it can simply accept the existence of nonlinear effects in the transmission. In this section, we consider CO-OFDM transmissions with strong NLPN distortion effects. We present an accurate spectral characterization of CO-OFDM signals as well as a study regarding the optimum of nonlinearly distorted CO-OFDM signals. It is shown that, as with other nonlinear OFDM schemes, the nonlinear phase distortion can lead to performance improvements relatively to the ideal, linear case. This optimum performance can be achieved with relatively simple sub-optimum receivers [138].

The CO-OFDM scenario considered here is similar to the one presented in Fig. 5.1, since the nonlinear phase noise is regarded as a bandpass memoryless

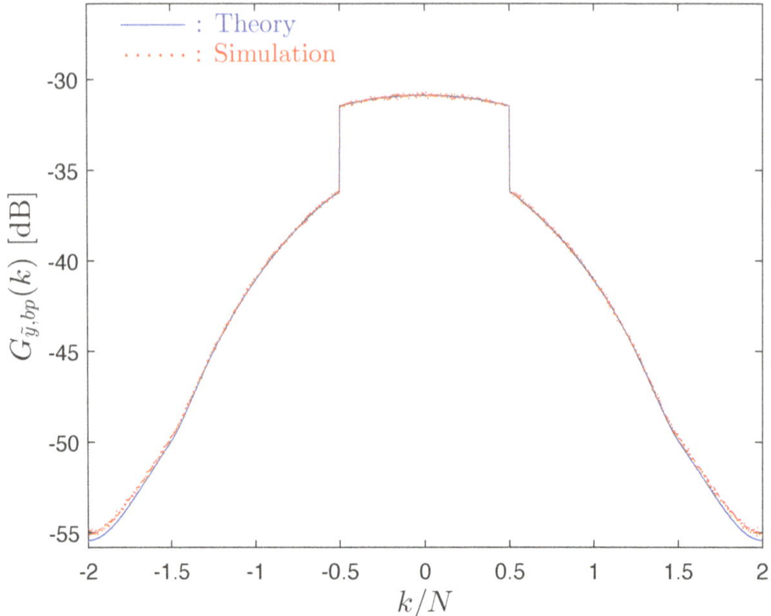

Figure 6.19 Theoretical and simulated PSD of a CO-OFDM signal

nonlinearity. Each CO-OFDM has N_u subcarriers and an oversampling factor of O. The QPSK symbols S_k form the block $\mathbf{S} = [S_0\ S_1\ ...\ S_{N-1}]^T \in \mathbb{C}^N$. The variance of the real and imaginary parts of the corresponding time-domain samples $\mathbf{s} = [s_0\ s_1\ ...\ s_{N-1}]^T \in \mathbb{C}^N$ is σ^2.

The bandpass nonlinearity that models the phase noise associated with the Kerr's effect is given by

$$f_{bp}(r_n) = r_n \exp(j2\pi k_\theta r_n/\sigma), \tag{6.39}$$

where $r_n = |s_n|$ and k_θ is a parameter that controls the magnitude of the NLPN distortion effects. It should be mentioned that this nonlinearity only distorts the phase of the time-domain samples and, consequently, has a null AM/AM conversion function (i.e., $A(r_n) = r_n$). On the other hand, the AM/PM conversion function is given by

$$\Theta(r_n) = 2\pi k_\theta r_n/\sigma. \tag{6.40}$$

Note also that under ideal conditions (i.e., when transmission chain is linear), we have $k_\theta = 0$ and, consequently, $\Theta(r_n) = 0$. In the following, we consider the approach of Subsection 3.2.2 to present results regarding the analytical spectral characterization of CO-OFDM signals [138, 139]. Fig. 6.19 shows the simulated and theoretical PSD associated with a CO-OFDM signal with $N_u = 512$ and $O = 4$. The NLPN parameter is $k_\theta = 0.2$. The theoretical PSD was obtained with the truncated IMP approach and $n_\gamma = 5$.

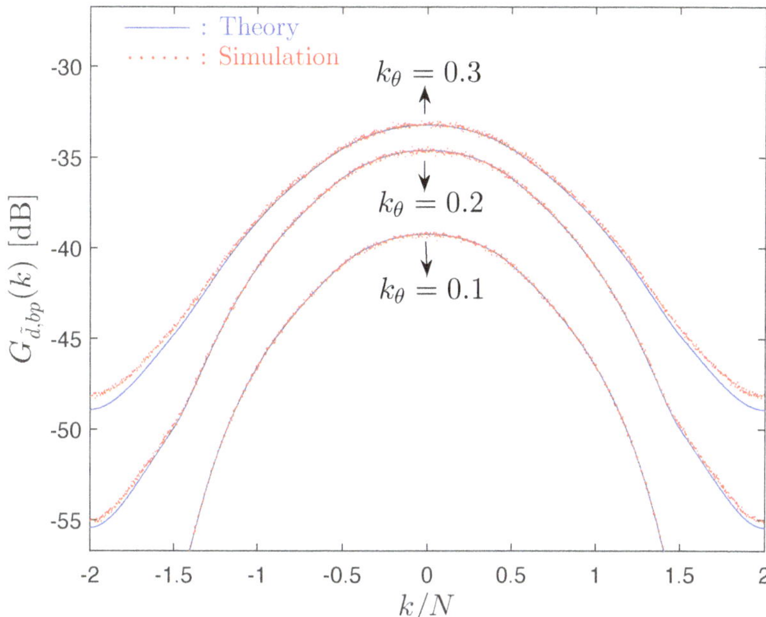

Figure 6.20 Theoretical and simulated PSD associated with the nonlinear distortion term of a CO-OFDM signal.

From the figure, it can be seen that the truncated IMP approach presents a considerable accuracy when $n_\gamma = 5$. This high accuracy can also be observed in Fig. 6.20, where are shown the simulated and the theoretical PSDs of a CO-OFDM signal with $N_u = 512$, $O = 4$ and different values of k_θ. Besides the accuracy of the theoretical PSD, it can also be seen that higher values of k_θ lead to stronger nonlinear distortion effects, since the magnitude of the nonlinear distortion term increases.

In the following, results regarding the optimum detection of CO-OFDM signals are presented. Fig. 6.21 shows the average asymptotic gain obtained by simulation and theoretically with (4.92), considering $O = 4$ as well as different values of μ and N_u. From the results shown in the figure, one can see that (4.92) is accurate for the NLPN distortion. In fact, although the combination of $\mu = 5$ and a value of $N_u = 16$ subcarriers leads to some discrepancies between the simulated and the theoretical asymptotic gain, when $N_u = 256$, very accurate results can be obtained, regardless the number of different bits between the signals. Fig. 6.22 shows the asymptotic BER associated with the optimum detection of CO-OFDM signals in AWGN channels for $N_u = 64$, $O = 4$ and different values of k_θ. From the figure, it can be noted that the performance gains increase with k_θ, which is an expected result from the results depicted in Fig. 6.21. More concretely, for a target BER of $P_b = 10^{-3}$, the performance gain of the optimum detection is approximately 0.4 dB. However, this performance

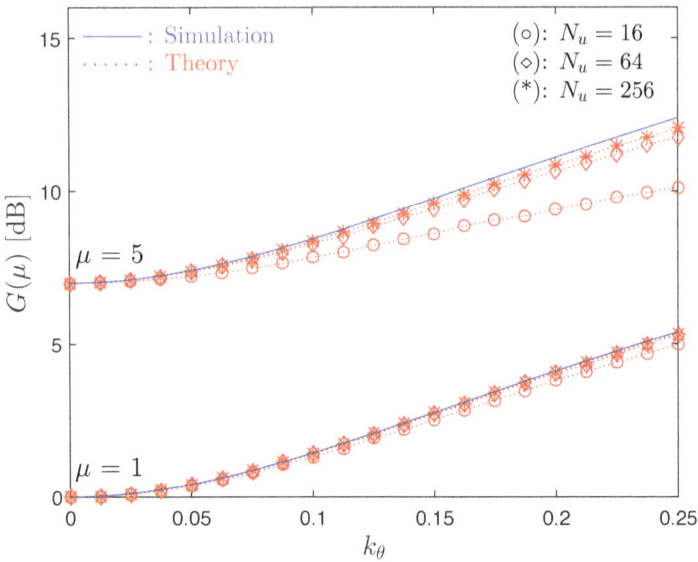

Figure 6.21 Average asymptotic gain associated with the optimum detection of CO-OFDM signals in AWGN channels.

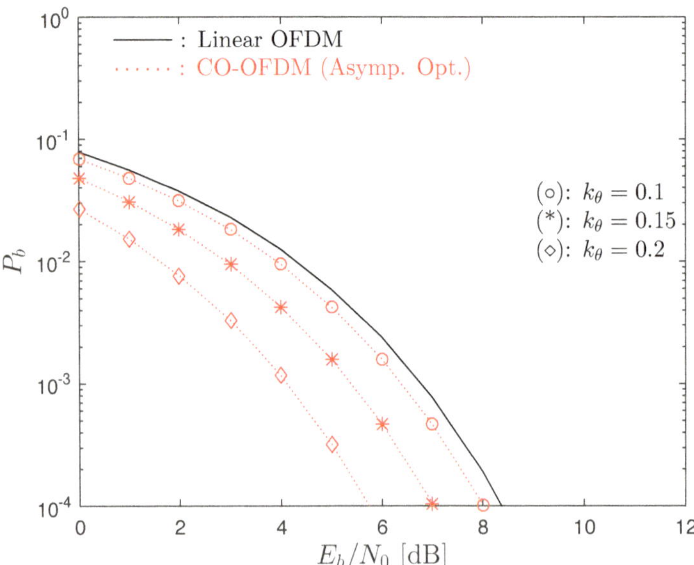

Figure 6.22 Asymptotic BER of the optimum detection of CO-OFDM signals in AWGN channels.

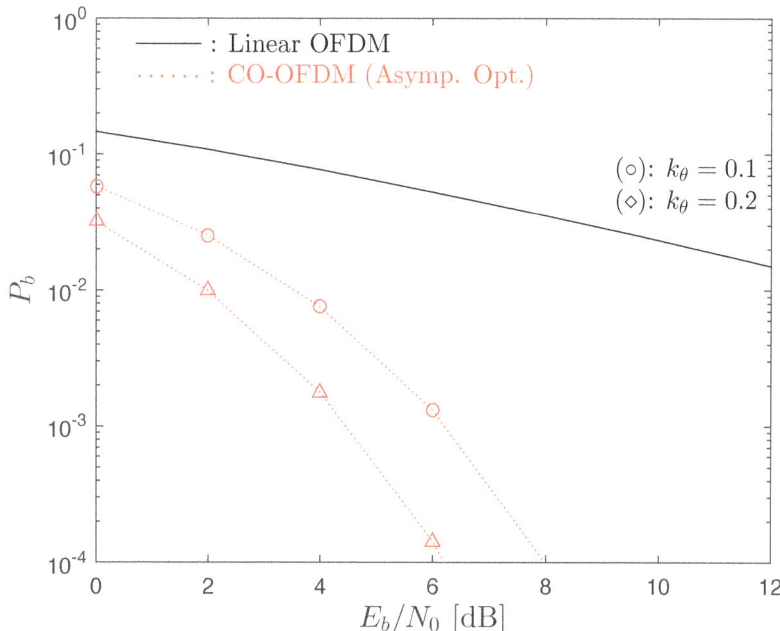

Figure 6.23 Asymptotic BER for the optimum detection of CO-OFDM signals in frequency-selective channels.

gain increases to approximately 1.4 dB and 2.7 dB when $k_\theta = 0.15$ and $k_\theta = 0.2$, respectively. Potential asymptotic gains can also be observed when frequency-selective channels are considered as shown in Fig. 6.23, which depicts the asymptotic BER for the optimum detection of CO-OFDM signals in frequency-selective channels assuming $N_u = 64$, $O = 4$ and $L = 16$ uncorrelated multipath components. Fig. 6.24 shows the simulated BER for the sub-optimum receiver described in Subsection 5.1.3, considering a CO-OFDM transmission with $N_u = 64$, $O = 4$ and $k_\theta = 0.1$. The number of cycles of bit modifications is $V = 2$. For the case of frequency-selective channels, $L = 16$ multipath components are considered. From the results shown in the figure, one can note that even when a sub-optimum receiver is considered and the number of analyzed sequences (and the corresponding complexity) relatively to the full optimum detection is substantially reduced, there are substantial performance improvements. These potential performance gains are observed in both ideal AWGN and frequency-selective channels although, in this latter scenario, the performance gains are higher.

6.3.2 OPTICAL WIRELESS OFDM

Recently, OFDM modulations have been considered to be used in optical wireless communications (OWC) [140]. This is mainly due to the fact that these modulations

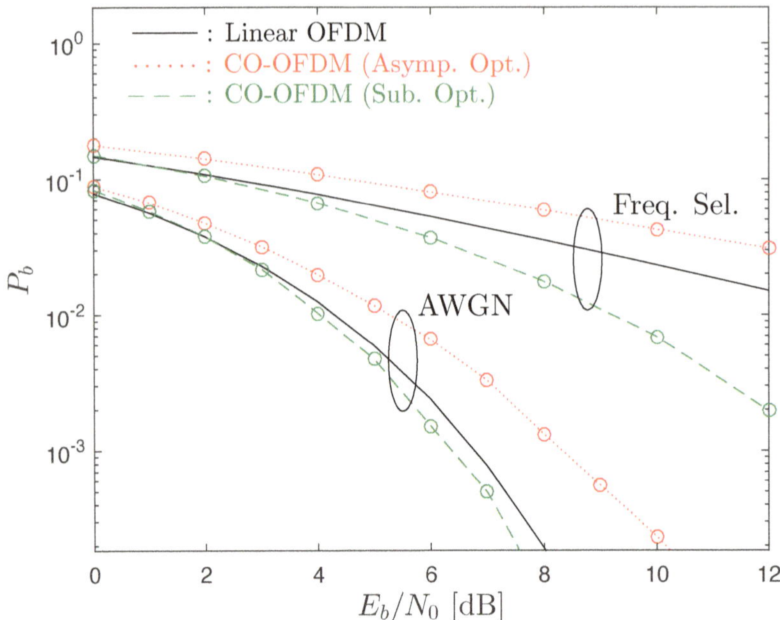

Figure 6.24 Sub-optimum receiver's BER considering a CO-OFDM transmission in both AWGN and frequency-selective channels.

are suitable for severely time-dispersive channels since they can cope with strong ISI levels of the multipath propagation. However, contrary to what is common in radio-frequency wireless communications, in OWC it is better to employ intensity modulation techniques where the OFDM signal modulates the light's intensity. For this reason, the OFDM signal must be constrained to have only non-negative, real values. There are two main approaches to transform a conventional OFDM signal (that has a complex envelope spanning all over the complex plane and the corresponding real-valued signals associated with its in-phase and quadrature components are equally likely to be positive or negative) into a real-valued, non-negative signal [141]. One approach is DCO-OFDM [142, 143], where a DC component is added to a clipped version of the original OFDM signal (actually it is an asymmetric clipping, since usually we only clip the signal below the symmetric of the DC component). Another approach is the ACO-OFDM [144, 145], where the original OFDM signal is deliberately clipped at zero (actually, this is also an asymmetric clipping, but with a nature slightly different of the one used in DCO-OFDM).

In this section, are considered both the theoretical spectral characterization and the optimum detection of real-valued OFDM signals submitted to asymmetrical clipping operations such as the ones of ACO-OFDM and the DCO-OFDM

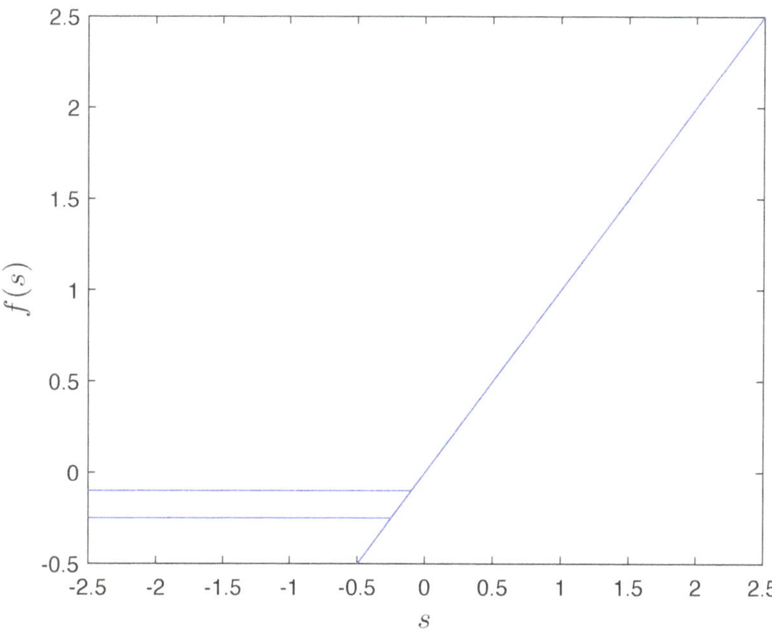

Figure 6.25 Nonlinear function of the asymmetrical clipping operation.

systems [146, 147]. To obtain real-valued OFDM signals, we considered OFDM frequency-domain blocks $\mathbf{S} = [S_0\ S_1\ \ldots\ S_{N-1}]^T \in \mathbb{C}^N$ with Hermitian symmetry. The variance of corresponding time-domain samples $\mathbf{s} = [s_0\ s_1\ \ldots\ s_{N-1}]^T \in \mathbb{C}^N$ is σ^2.

The asymmetric clipping function operating on the real-valued samples of an OFDM signal is defined as

$$f(s_n) = \begin{cases} s_n, & s_n > -s_M/\sigma \\ -s_M/\sigma, & s_n \leq -s_M/\sigma. \end{cases} \tag{6.41}$$

This nonlinear function is shown in Fig. 6.25 for different normalized clipping levels. Although this case is not represented in the figure, it should be noted that when ACO-OFDM systems are considered, the normalized clipping level is $s_M/\sigma = 0$. Additionally, it should be mentioned that due to the nonlinear characteristic of optical devices used in DCO-OFDM, schemes such as light emitting diodes (LEDs), it might also be desirable to have an upper clipping level to minimize nonlinear distortion effects. In this case, both positive and negative parts of the signal can be clipped. However, the clipping operation might still be asymmetric since the upper and the lower clipping levels do not have to be equal. Although we only considered a lower clipping level, our approach is still valid when both lower and upper clipping levels are taken into account. It should be also noticed that in DCO-OFDM schemes, to

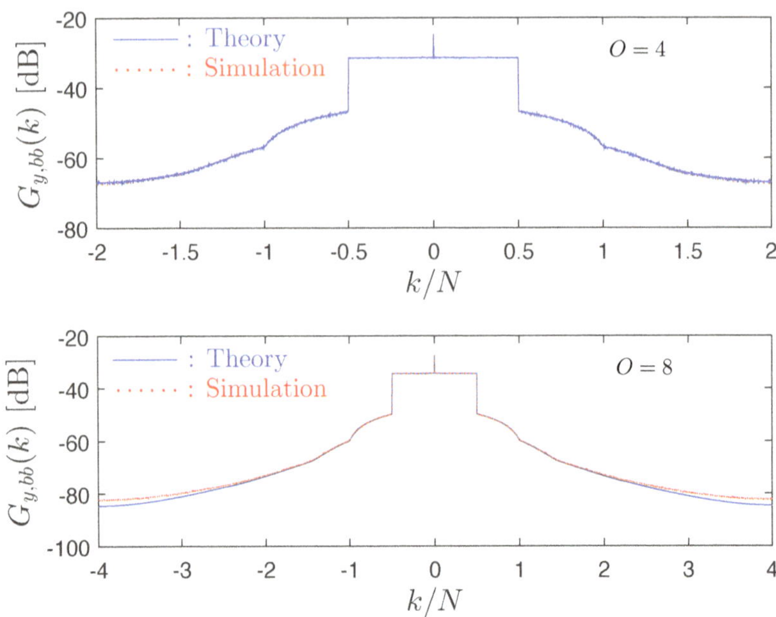

Figure 6.26 Simulated and theoretical PSD of a DCO-OFDM signal considering different oversampling factors.

obtain a real-valued and positive OFDM signal, a DC component equal to the symmetrical of the clipping level should be added after the clipping operation. Fig. 6.26 shows PSD of a DCO-OFDM signal obtained theoretically with the truncated IMP approach and by simulation. It is considered that $N_u = 512$, $s_M/\sigma = 1.0$ and different oversampling factors. Regardless of the oversampling factor, the results clearly show the high accuracy of the truncated IMP approach when the nonlinearity is an asymmetrical clipping. This allows us to obtain accurate estimates of the SIR, at subcarrier level, as shown in Fig. 6.27. Fig. 6.28 represents the simulated and theoretical evolution of the asymptotic gain associated with the optimum detection of a DCO-OFDM signal with $N_u = 512$, $O = 4$ and different clipping levels. In this figure it can be seen that regardless of μ, the average potential asymptotic gains can be obtained with accuracy considering (4.115). This unveils that even when channel coding schemes with an appropriate interleaving are considered, the optimum detection might present considerable performance gains over the conventional, linear COFDM.

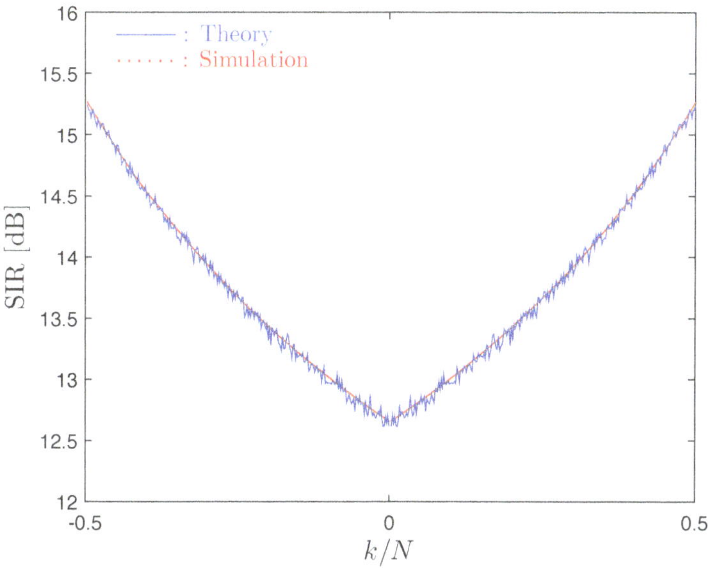

Figure 6.27 SIR at the subcarrier level regarding a DCO-OFDM signal with $s_M/\sigma = 1.0$.

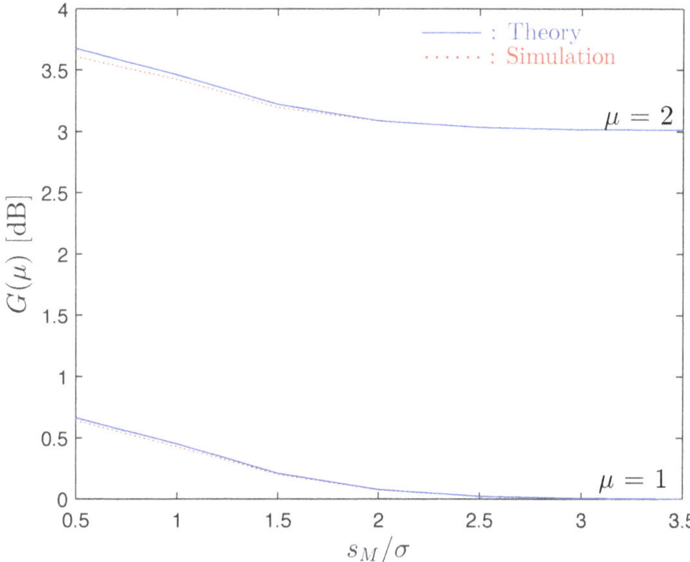

Figure 6.28 Average asymptotic gain for the optimum detection of DCO-OFDM signals in AWGN channels

7 Conclusions

The analytical characterization and the optimum detection of nonlinearly distorted multicarrier signals were the two main goals of this book. Regarding the former, we presented the analytical characterization of several nonlinear multicarrier schemes that can be encountered in actual communication systems. This analytical characterization was based on IMP tools, directly by employing the truncated IMP approach or through the novel concept of equivalent nonlinearities that is suitable even for severe nonlinearities, where the truncated IMP approach is not adequate due to accuracy and convergence problems. In relation to the optimum detection, the optimum asymptotic performance was obtained for several nonlinear multicarrier schemes, showing that it can be a good alternative, not only for the conventional detection of nonlinear multicarrier schemes, but also for the conventional detection of linear, multicarrier schemes.

Chapter 2 was dedicated to the characterization of multicarrier schemes as well as to its most important particularities and drawbacks. It included an introduction to multicarrier schemes in general and a statistical characterization of the signals at the receiver and at the transmitter for the particular case of OFDM. The main drawback of multicarrier schemes - the large envelope fluctuations and the corresponding amplification problems - was also analyzed in this chapter. It was pointed out that the amplification process involves trade-offs between spectral and energy efficiency and that usually, PAPR-reducing techniques are employed prior to that process. It was identified that the most simple and effective PAPR-reducing techniques are the ones that involve a clipping operation, although they are distortion techniques that introduce nonlinear distortion effects in the transmitted signals.

Chapter 3 presented the theory behind the analytical characterization and the optimum detection of nonlinearly distorted multicarrier signals. Different nonlinearities that can have either a baseband or bandpass nature were identified and characterized. The conventional spectral characterization made through IMP tools, although not original, was presented to introduce and motivate the concept of equivalent nonlinearities. These equivalent nonlinearities are polynomial nonlinearities that can substitute the conventional ones. However, although being smoother and suitable to be employed in an IMP analysis, they give rise to signals with the same spectral characterization than that of the ones distorted by the original nonlinearities. This means that they can be employed to obtain a simple, accurate spectral characterization of nonlinearly distorted multicarrier signals.

Chapter 4 presented a study on the optimum receiver's performance. The study started with the introduction of the optimum receiver's principle. Further, the optimum performance regarding linear OFDM schemes was derived. It was shown that an insight on the optimum performance can be computed through the squared Euclidean distance between the signals. In fact, when the transmission is linear, this distance is equal regardless of the subcarrier that has the symbol error. However,

DOI: 10.1201/9781032708744-7

when the transmission is nonlinear, the squared Euclidean distance depends on the information spread along the signal band. For this reason, and due to the large number of possible transmitted sequences, the distribution of the squared Euclidean distance (and of the corresponding asymptotic gain) was obtained. It was noted that when the number of subcarriers is large, the variance of this distribution decreases, meaning that for a very large number of subcarriers, the squared Euclidean distance between two nonlinearly distorted signals assumes a unique value. The asymptotic gain was then obtained theoretically for different types of nonlinearity. For the case of frequency-selective channels, the squared Euclidean distance between the nonlinearly distorted signals is random even if the number of subcarriers is very large, since it is dependent on the channel's frequency responses, which presents a Rayleigh distribution. For this reason, we included a statistical characterization of the channel's frequency responses. This statistical characterization was then used to obtain the distribution of the asymptotic gain associated with the optimum detection.

Chapter 5 was concerned with applications for which the analytical characterization and the asymptotic optimum performance presented in Chapters 3 and 4 can be employed. It started with the conventional OFDM impaired by either an envelope clipping or nonlinear power amplifiers. For that system, it was shown that the equivalent nonlinearities may be used to obtain an accurate spectral characterization of the corresponding nonlinearly distorted signals and that the optimum performance present considerable gains when compared to the traditional detection of nonlinear and even of linear, OFDM schemes. In addition, results regarding a sub-optimum receiver were presented considering receive diversity, clipping and filtering techniques and M-QAM constellations, showing that even when a sub-optimum detection is considered, the potential performance gains associated with the optimum detection might be obtained. After that, the LINC techniques were analyzed. These techniques allow an efficient amplification through power amplifiers of class D or E, but their use may lead to nonlinear distortion effects in the transmitted signals. For this reason, we presented both the analytical characterization as well as the optimum detection of the corresponding nonlinearly distorted signals. Another alternative for the conventional OFDM that allows the use of highly efficient amplifiers is the CE-OFDM. It was noted that the severe nonlinear distortion effects created by the phase modulation process of CE-OFDM schemes can be accurately characterized using IMP tools and that the optimum detection outperforms the conventional detection based on a phase demodulator, especially in scenarios where the power efficiency is crucial. We also presented the analytical spectral characterization for the signals along the chain of an OFDM-based AF relay system, i.e., in a scenario where there are nonlinearities both at the transmitter and at the relay. Nonlinear, multiple antenna OFDM systems were also considered in this chapter.

Chapter 6 presented the analytical characterization of nonlinear, MIMO-OFDM systems with an envelope clipping operation employing an SVD decomposition and massive MIMO-OFDM systems with MRT approach, employing both Cartesian, 1-bit quantizers and nonlinear amplifiers. For both systems, it was demonstrated that the nonlinear distortion at the subcarrier level decreases when the number of transmit antennas increase, provided that the number of streams is fixed. This means that when we have much more transmit antennas than receive antennas, the nonlinear distortion can be substantially reduced. Therefore, massive MIMO-OFDM may have nonlinear amplification processes and/or employ low-complex, low-resolution quantizers that can meet their need to have simple and low-cost transmitting branches, since the corresponding performance penalty associated with the use of such nonlinear devices can be tolerated. We also considered baseband, nonlinear DMT schemes that have a quantization operation on the transmission chain. Due to the severity of the nonlinearity of the quantization operation, especially when the resolution is low, we consider the use of equivalent nonlinearities to provide an accurate spectral characterization of quantized, multicarrier signals. Additionally, the optimum asymptotic performance was derived and its performance gains over the linear DMT schemes were quantified through a set of performance results. Chapter 6 ends with the analytical characterization and a study on the optimum detection of optical OFDM signals and includes the characterization of nonlinear distortion effects on fiber optical systems, i.e., associated with the NLPN, as well as the characterization of the distortion effects related to optical wireless OFDM systems, such as DCO-OFDM systems, where asymmetrical clipping operations take place.

Besides these particular applications, the theory behind the analytical characterization of nonlinearly distorted multicarrier signals presented in Chapter 3 can employed more generally in sampled, Gaussian signals [89, 148], regardless of their multicarrier or single carrier nature. In fact, the accuracy associated with the use of equivalent nonlinearities was already demonstrated in a nonlinear, single carrier system [149].

A Baseband Representation of Bandpass Nonlinearities

In fact, when $s(t)$ is a narrow-band, bandpass signal with $f_c \gg B_s$, its complex envelope $\tilde{s}(t)$ changes at a much lower rate than the carrier frequency and $s(t)$ is approximately periodic. Therefore, if $s(t)$ is submitted to a memoryless nonlinearity characterized by the function $f(\cdot)$, its nonlinearly distorted version is (omitting the dependence with t)

$$z = f(r\cos(\psi)). \tag{A.1}$$

Due to the periodicity of the signal $r\cos(\psi)$, the nonlinearity output can be expanded in a Fourier series, resulting

$$z = \sum_{n=0}^{\infty} \left(a_n \cos(n\psi) + b_n \sin(n\psi)\right), \tag{A.2}$$

where a_n and b_n are the Fourier coefficients. Note that as $f_c \gg B_s$, the contribution of the harmonics created at nf_c on the spectrum located at $(n-1)f_c$ is almost negligible. In fact, as we are only interested in the output centered at the carrier frequency ($n =$ 0) (also known as first-zone), we can consider that a zonal filter is applied around f_c, which means that all other spectral components can be removed. Under these conditions, the output of the filtered signal can be written as

$$y = a_1(r)\cos(\psi) + b_1(r)\sin(\psi), \tag{A.3}$$

where

$$a_1(r) = \frac{1}{\pi}\int_0^{2\pi} f(r\cos(\psi))\cos(\psi)d\psi, \tag{A.4}$$

and

$$b_1(r) = \frac{1}{\pi}\int_0^{2\pi} f(r\cos(\psi))\sin(\psi)d\psi. \tag{A.5}$$

It should be mentioned that if all harmonics generated by the nonlinearity were taken into account, we do not have a complex baseband representation of the bandpass output signal. However, if only the output of the first zone is considered, we do have a complex envelope representation, and the complex envelope of the output can be expressed as

$$\begin{aligned} \tilde{y} &= (a_1(r) - jb_1(r))\exp(j\theta) \\ &= f_{bp}(r)\exp(j\theta). \end{aligned} \tag{A.6}$$

DOI: 10.1201/9781032708744-A

In fact, (A.6) is equal to (3.23), i.e., from this approach one can also note that the complex envelope of the corresponding bandpass nonlinearly distorted signal only depends on the nonlinear bandpass function $f_{bp}(r)$. Note also that the bandpass, nonlinearly distorted output signal can be written as

$$\begin{aligned} y &= \mathrm{Re}(\tilde{y}\exp(j2\pi f_c t)) \\ &= \mathrm{Re}\left(f_{bp}(r)\exp(j\psi)\right). \end{aligned} \tag{A.7}$$

B Distribution of the Optimum Asymptotic Gain in Frequency-Selective Channels

In [151], a general expression for the distribution of the sum of n independent random variables with Gamma distribution, $Y = X_1 + X_2 + ... + X_n$, with different scale and shape parameters and in the form $X_i \sim \Gamma(\alpha_i = \omega_i, \beta_i = \frac{1}{\upsilon_i})$, is given. In our specific case, we have $n = 2$, since $G^{(H)}(\mu)$ is constituted by the sum of two gamma variables. In these conditions, the general expression in [151] can be particularized to yield

$$p(y) = Cy^{\alpha_1+\alpha_2-1} \int_0^1 \exp(-yC_{\beta_1,\beta_2}(t))B_{\alpha_1,\alpha_2}(t)dt, \tag{B.1}$$

with C given by

$$C = \frac{\beta_1^{\alpha_1}\beta_2^{\alpha_2}}{\Gamma(\alpha_1+\alpha_2)}, \tag{B.2}$$

$$C_{\beta_1,\beta_2}(t) = \beta_1 t + \beta_2(1-t), \tag{B.3}$$

$$B_{\alpha_1,\alpha_2}(t) = \frac{1}{B_2(\alpha_1,\alpha_2)} t^{\alpha_1-1}(1-t)^{\alpha_2-1}, \tag{B.4}$$

and

$$B_2(\alpha_1,\alpha_2) = \frac{\Gamma(\alpha_1)\Gamma(\alpha_2)}{\Gamma(\alpha_1+\alpha_2)}. \tag{B.5}$$

By replacing the last equations in (B.1), we may write

$$\begin{aligned} p(y) &= \frac{\beta_1^{\alpha_1}\beta_2^{\alpha_2}}{\Gamma(\alpha_1+\alpha_2)} y^{\alpha_1+\alpha_2-1} \int_0^1 \exp(-y(\beta_1 t + \beta_2(1-t))) \frac{1}{B_2(\alpha_1,\alpha_2)} t^{\alpha_1-1}(1-t)^{\alpha_2-1} dt \\ &= \frac{\Gamma(\alpha_1+\alpha_2)}{\Gamma(\alpha_1)\Gamma(\alpha_2)} \frac{\beta_1^{\alpha_1}\beta_2^{\alpha_2}}{\Gamma(\alpha_1+\alpha_2)} y^{\alpha_1+\alpha_2-1} \exp(-y\beta_2) \int_0^1 \exp(yt(\beta_2-\beta_1)) t^{\alpha_1-1}(1-t)^{\alpha_2-1} dt. \end{aligned} \tag{B.6}$$

DOI: 10.1201/9781032708744-B

Further, by introducing the variable $K = y(\beta_2 - \beta_1)$, we have

$$p(y) = \frac{\beta_1^{\alpha_1}\beta_2^{\alpha_2}}{\Gamma(\alpha_1+\alpha_2)} y^{\alpha_1+\alpha_2-1}\exp(-y\beta_2)\frac{\Gamma(\alpha_1+\alpha_2)}{\Gamma(\alpha_1)\Gamma(\alpha_2)}\int_0^1 \exp(Kt)t^{\alpha_1-1}(1-t)^{\alpha_2-1}dt. \tag{B.7}$$

Additionally, by considering the definition of the Kummer's function of the first kind, $M(a,b,z)$, that is given by [83]

$$M(a,b,z) = \frac{\Gamma(b)}{\Gamma(a)\Gamma(b-a)}\int_0^1 \exp(zu)u^{a-1}(1-u)^{b-a-1}du, \tag{B.8}$$

and replacing $y = G^{(H)}(\mu)$, $a = \alpha_1$, $b = \alpha_1 + \alpha_2$, $\alpha_1 = \mu$, $\alpha_2 = L$, $\beta_1 = \frac{\mu}{G_d(\mu)}$ and $\beta_2 = \frac{L}{G_c(\mu)}$, we have

$$p(G^{(H)}(\mu)) = \frac{\left(\frac{\mu}{G_d(\mu)}\right)^{\mu}\left(\frac{L}{G_c(\mu)}\right)^{L}}{\Gamma(L+\mu)} G^{(H)}(\mu)^{L+\mu-1}\exp\left(-\frac{G^{(H)}(\mu)L}{G_c(\mu)}\right) \times M\left(\mu, L+\mu, G^{(H)}(\mu)\left(\frac{L}{G_c(\mu)} - \frac{\mu}{G_d(\mu)}\right)\right). \tag{B.9}$$

References

1. G. Kortuem, F. Kawsar, V. Sundramoorthy, and D. Fitton. Smart objects as building blocks for the internet of things. *IEEE Internet Computing*, 14(1):44–51, December 2009.
2. B. Sklar. Rayleigh fading channels in mobile digital communication systems part I: Characterization. *IEEE Communications Magazine*, 35(7):90–100, July 1997.
3. L. Cimini. Analysis and simulation of a digital mobile channel using orthogonal frequency division multiplexing. *IEEE Transactions on Communications*. 33(7):665–675, July 1985.
4. J. Cooley and J. Tukey. An algorithm for the machine calculation of complex Fourier series. *Mathematics of Computation*, 19(90):297–301, April 1965.
5. M. Isaksson, D. Wisell, and D. Ronnow. A comparative analysis of behavioral models for RF power amplifiers. *IEEE Transactions on Microwave Theory and Techniques*, 54(1):348–359, January 2006.
6. Y. Rahmatallah and S. Mohan. Peak-to-average power ratio reduction in OFDM systems: a survey and taxonomy. *IEEE Communications Surveys & Tutorials*, 15(4):1567–1592, February 2013.
7. R. Dinis and A. Gusmão. A new class of signal processing schemes for bandwidth-efficient OFDM transmission with low envelope fluctuation. In *Proc. of IEEE VTC'01* (Spring), pages 1–5, Rhodes, Greece.
8. T. Araújo. *Analytical Evaluation of Nonlinear Distortion Effects in Multicarrier Signals*. PhD thesis, Instituto Superior Técnico, Univ. Técnica de Lisboa, 1 2012.
9. A. Lau and J. Kahn, "Signal design and detection in presence of nonlinear phase noise", *J. Lightwave Technol.*, Vol. 45, n. 10, pp. 3008–3016, Sep. 2007.
10. H. Rowe. Memoryless nonlinearities with gaussian input: elementary results. *Bell Sys. Tech. J.*, 61(7):1519–1526, September 1982.
11. G. Stette. Calculation of intermodulation from a single carrier amplitude characteristic. 22(3):319–323, March 1974.
12. O. Shimbo. Effects of intermodulation, AM-PM conversion, and additive noise in multicarrier TWT systems. *Proc. IEEE*, 59(2):230–238, February 1971.
13. J. Tellado, L. Hoo, and J. Cioffi. Maximum likelihood detection of nonlinearly distorted multicarrier symbols by iterative decoding. 51(2):218–228, February 2003.
14. F. Peng and E. Ryan. MLSD bounds and receiver designs for clipped OFDM channels. 7(9):3568–3578, September 2008.
15. J. Guerreiro, R. Dinis, and P. Montezuma. Optimum and sub-optimum receivers for OFDM signals with strong nonlinear distortion effects. 61(9):3830–3840, September 2013.
16. M. Schwartz. The origins of carrier multiplexing: Major George Owen Squier and AT&T. 46(5):20–24, May 2008.
17. S. Weinstein. The history of orthogonal frequency-division multiplexing. 47(11):26–35, November 2009.
18. R. Chang. Synthesis of band-limited orthogonal signals for multichannel data transmission. *Bell Syst. Tech. J.*, 45(10):1775–1797, December 1966.

19. B. Saltzberg. Performance of an efficient parallel data transmission system. 15(6):805–811, December 1967.
20. S. Weinstein and P. Ebert. Data transmission by frequency-division multiplexing using the discrete Fourier transform. 19(5):628–634, October 1971.
21. J. Bingham. Multicarrier modulation for data transmission: an idea whose time has come. 28(5):5–14, May 1990.
22. A. Batra, J. Balakrishnan, G. Aiello, J. Foerster, and A. Dabak. Design of a multiband OFDM system for realistic UWB channel environments. 52(9):2123–2138, September 2004.
23. E. Perahia. IEEE 802.11n development: History, process, and technology. 46(7):48–55, July 2008.
24. IEEE standard for local and metropolitan area networks – part 16: air interface for fixed broadband wireless access systems, October 2004.
25. 3rd generation partnership project: technical specification group radio access network; physical layers aspects for evolved UTRA, September 2006.
26. Digital video broadcasting (DVB); framing structure, channel coding and modulation for digital terrestrial television, January 2009.
27. Radio broadcasting systems; digital audio broadcasting (DAB) to mobile, portable and fixed receivers, January 2006.
28. D. Falconer. Frequency domain equalization for single-carrier broadband wireless systems. 40(4):58–66, April 2002.
29. S. Wei, D. Goeckel, and P. Kelly. The complex envelope of a bandlimited OFDM signal converges weakly to a Gaussian random process. 56(10):4893–4904, October 2010.
30. R. Dinis. *Técnicas Multiportadora para Rádio Móvel de Alto Débito.* PhD thesis, Instituto Superior Técnico, Univ. Técnica de Lisboa, 7 2001.
31. A. Jayalath and C. Tellambura. Reducing the out-of-band radiation of OFDM using an extended guard interval. In *Proc. of VTC'01 (Fall)*, pages 829–833, Atlantic City, NJ, USA.
32. T. Waterschoot, V. Le Nir, J. Duplicy, and M. Moonen. Analytical expressions for the power spectral density of CP-OFDM and ZP-OFDM signals. 17(4):371–374, January 2010.
33. J. Proakis. *Digital Communications.* McGraw-Hill, 4th ed edition, 2001.
34. A. Peled and A. Ruiz. Frequency domain data transmission using reduced computational complexity algorithms. In *Proc. of ICASSP'80*, pages 964–967, Denver, Colorado, USA.
35. T. Araújo and R. Dinis. Efficient detection of zero-padded OFDM signals with large blocks. In *Proc. of SIP'06*, pages 1–5, Honolulu, Hawaii, USA.
36. P. Sivakumar, S. Vidhya, and M. Rajaram. On the equalization scheme for ZP-OFDM using MMSE criteria over deep fading channels. In *Proc. of IEEE ICSCCN'11*, pages 96–99, Thuckafay, India.
37. S. Haykin. *Digital Communications.* Wiley, New York, 4th ed. 2001.
38. K. Ahmed, S. Majumder, and B. Rahman. Performance analysis of an OFDM wireless communication system with convolutional coding. In *Proc. of ICACT 2009*, pages 829–833, Phoenix Park, Ireland.
39. B. Floch and C. Berrou. Coded orthogonal frequency division multiplex. *Proc. IEEE*, 83(6):982–996, June 1995.
40. G. Forney. The Viterbi algorithm. *Proc. IEEE*, 61(3):267–278, March 1973.

41. C. Tellambura. Computation of the continuous-time PAR of an OFDM signal with BPSK subcarriers. 5(5):185–187, May 2001.
42. H. Ochiai and H. Imai. On the distribution of the peak-to-average power ratio in OFDM signals. 49(2):282–289, February 2001.
43. M. Friese. On the achievable information rate with peak-power-limited orthogonal frequency-division multiplexing. 46(7):2579–2587, November 2000.
44. F. Raab, P. Asbeck, S. Cripps, P. Kenington, Z. Popovic, N. Pothecary, J. Sevic, and N. Sokal. Power amplifiers and transmitters for RF and microwave. 50(3):814–826, March 2005.
45. J. Joung, C. Ho, and S. Sun. Spectral efficiency and energy efficiency of OFDM systems: Impact of power amplifiers and countermeasures. 32(2):208–220, February 2014.
46. S. Han and J. Lee. An overview of peak-to-average power ratio reduction techniques for multicarrier transmission. 12(2):56–65, April 2005.
47. R. Dinis and A. Gusmão. A class of signal processing algorithms for good power bandwidth trade offs with OFDM transmission. In *Proc. of IEEE ISIT'00*, pages 1–5, Sorrento, Italy.
48. S. Cha, M. Park, S. Lee, K. Bang, and D. Hong. A new PAPR reduction technique for OFDM systems using advanced peak windowing method. 54(2):405–410, May 2008.
49. X. Huang, J. Li, J. Zheng, K. Letaief, and J. Gu. Companding transform for the reduction of peak-to-average power ratio of OFDM signals. 3(6):2030–2039, November 2004.
50. P. Borjesson, H. Feichtinger, N. Grip, M. Isaksson, N. Kaiblinger, P. Odling, and L. Persson. A low-complexity PAR-reduction method for DMT-VDSL. In *Proc. of DSPCS'99*, pages 1–5, Perth, Australia.
51. H. Ochiai and H. Imai. On clipping for peak power reduction of OFDM signals. In *Proc. of IEEE GLOBECOM'00*, pages 1–5, San Francisco, CA, USA.
52. R. Dinis and A. Gusmão. On the performance evaluation of OFDM transmission using clipping techniques. In *Proc. of IEEE VTC'99 (Fall)*, pages 1–5, Amsterdam, Netherlands.
53. X. Li and L. Cimini. Effects of clipping and filtering on the performance of OFDM. 2(5):131–133, May 1998.
54. J. Armstrong. Peak-to-average power reduction for OFDM by repeated clipping and frequency-domain filtering. *Electronic Lett.*, 38(5):246–257, February 2002.
55. R. O'Neill and L. Lopes. Envelope variations and spectral splatter in clipped multicarrier signals. In *Proc. of IEEE PIMRC'95*, pages 1–5, Toronto, Canada.
56. R. Dinis and A. Gusmão. A new class of signal processing schemes for bandwidth-efficient OFDM transmission with low envelope fluctuation. In *Proc. of IEEE ICC'01*, pages 1–5, Helsinki, Finland.
57. R. Dinis and A. Gusmão. Performance evaluation of peak cancellation schemes for bandwidth-efficient OFDM transmission with low envelope fluctuation. In *Proc. of IEEE ISCTA'01*, pages 1–5, Ambleside, UK.
58. R. Bauml, R. Fisher, and J. Huber. Reducing the peak-to-average power ratio of multicarrier modulation by selected mapping. *Electronic Lett.*, 32(22):2056–2057, October 1996.
59. P. Eetvelt, G. Wade, and M. Tomlinson. Peak to average power reduction for OFDM schemes by selective scrambling. *Electronic Lett.*, 32(21):1963–1964, October 1996.
60. L. Cimini and N. Sollenberger. Peak-to-average power ratio reduction of an OFDM signal using partial transmit sequences. 4(3):86–88, March 2000.

61. T. Wattanasuwakull and W. Benjapolakul. PAPR reduction for OFDM transmission by using a method of tone reservation and tone injection. In *Proc. of ICICSP'05*, pages 1–5, Bangkok, Thailand.
62. B. Krongold and D. Jones. PAR reduction in OFDM via active constellation extension. 49(3):258–268, September 2003.
63. A. Jones, T. Wilkinson, and S. Barton. Block coding scheme for reduction of peak to mean envelope power ratio of multicarrier transmission scheme. *Electronic Lett.*, 30(22):2098–2099, December 1994.
64. B. Popovic. Synthesis of power efficient multitone signals with flat amplitude spectrum. 39(7):1031–1033, July 1991.
65. M. Lin, K. Chen, and S. Li. Turbo coded OFDM system with peak power reduction. In *Proc. of IEEE VTC'03 (Fall)*, pages 1–5.
66. Michel Jeruchim. *Simulation of Communication Systems: Modeling, Methodology and Techniques*. Springer, 2nd ed. 2000.
67. A. Brajal and A. Chouly. Compensation of nonlinear distortions for orthogonal multicarrier schemes using predistortion. In *Proc. GLOBECOM'94*, volume 3, pages 1909–1914, San Francisco, CA, USA, 1994.
68. Won Jeon, Kyung Chang, and Yong Cho. An adaptive data predistorter for compensation of nonlinear distortion in OFDM systems. 45(10):1167–1171, 1997.
69. J. Pedro and S. Maas. A comparative overview of microwave and wireless power-amplifier behavioral modeling approaches. 53(4):1150–1163, 04 2005.
70. Michiel Hazewinkel. *Taylor formula: Encyclopedia of Mathematics*. Springer, 1st edition, 2001.
71. N. Wiener. Response of a nonlinear device to noise. Technical Report R-129, Radiation Lab, MIT, MA, USA, April 1942.
72. V. Volterra. *Theory of Functionals and of Integral and Integro-Differential Equations*. Dover, 1959.
73. R. Raich and G. Zhou. On the modeling of memory nonlinear effects of power amplifiers for communication applications. In *Proc. of IEEE DSP Workshop'02*, pages 7–10, 2002.
74. G. Zhou, H. Qian, L. Ding, and R. Raich. On the baseband representation of a bandpass nonlinearity. 53(8):2953–2957, August 2005.
75. S. Benedetto, E. Biglieri, and R. Daffara. Modeling and performance evaluation of nonlinear satellite links – a Volterra series approach. 15(4):494–507, July 1979.
76. W. Bosch and G. Gatti. Measurement and simulation of memory effects in predistortion linearizers. 37(12):1885–1890, December 1989.
77. A. Saleh. Frequency-independent and frequency-dependent nonlinear models of TWT amplifiers. 29(11):1715–1720, November 1981.
78. A. Ghorbani and M. Sheikhan. The effect of solid state power amplifiers (SSPAs) nonlinearities on MPSK and M-QAM signal transmission. In *Proc. IET DPSC'91*, Loughborough, England.
79. C. Rapp. Effects of HPA-nonlinearity on a 4-DPSK/OFDM-signal for a digital sound broadcasting signal. In *Proc. 2nd Conference on Satellite Communications*, Liege, Belgium.
80. J. Bussgang. Cross-correlation function of amplitude-distorted Gaussian signals. Technical Report R-43, Electronics Lab, MIT, MA, USA, March 1952.
81. R. Price. A useful theorem for nonlinear devices having Gaussian inputs. *IRE Trans. Inf. Theory*, 4(2):69–72, June 1958.

82. G. Szego. *The art of computer programming Volume 23: Orthogonal polynomials.* Colloquium Publications, 1975.
83. M. Abramowitz and I. Stegun. *Handbook of Mathematical Functions.* Dover Publications, 1972.
84. K. Iniewski. *Wireless Technologies: Circuits, Systems, and Devices.* CRC Press, 1st ed. 2007.
85. A. van den Bos. The multivariate complex normal distribution – a generalization. 41(2):537–539, March 1995.
86. N. Blachman. The uncorrelated output components of a nonlinearity. 14(2):250–255, March 1968.
87. A. Prudnikov, Y. Brychkov, and O. Marichev. *Integrals and Series, Volume 2-Special Functions.* Gordon and Breach Science Publishers, 1986.
88. E. Costa, M. Midrio, and S. Pupolin. Impact of amplifier nonlinearities on OFDM transmission system performance. *IEEE Commun. Lett.*, 3:37–39, February 1999.
89. J. Guerreiro, R. Dinis, and P. Montezuma. An efficient method for modelling and evaluating quantization effects on Gaussian signals. In *Proc. of MIC'15*, pages 1–5, Innsbruck, Austria, February.
90. J. Guerreiro, R. Dinis, and P. Montezuma. Use of equivalent nonlinearities for studying quantization effects on sampled multicarrier signals. *Elect. Lett.*, 52(2):151–153, January 2015.
91. J. Guerreiro, R. Dinis, and P. Montezuma. A simplified method for evaluating clipping effects on sampled OFDM signals. In *Proc. of VTC Fall'15*, pages 1–5, Boston, USA, September.
92. J. Guerreiro, R. Dinis, and P. Montezuma. Equivalent nonlinearities for studying nonlinear effects on sampled OFDM signals. 19(4):529–532, April 2015.
93. J. Guerreiro, R. Dinis, and P. Montezuma. Optimum asymptotic performance for nonlinearly amplified OFDM signals. In *Proc. of VTC Spring'13*, pages 1–5, Dresden, Germany.
94. J. Guerreiro, R. Dinis, and P. Montezuma. On the optimum performance of nonlinearly distorted OFDM signals. In *Proc. of VTC Spring'14*, pages 1–5, Seoul, Korea, May.
95. J. Guerreiro, R. Dinis, and P. Montezuma. On the optimum multicarrier performance with memoryless nonlinearities. 63(2):498–509, January 2015.
96. A. Gusmão, R. Dinis, and P. Torres. Low-PMEPR OFDM transmission with an iterative receiver technique for cancellation of nonlinear distortion. In *Proc. of VTC'05 (Fall)*, pages 1–5, Dallas, TX, USA.
97. Z. Kollar, M. Grossmann, and R. Thoma. Convergence analysis of BNC turbo detection for clipped OFDM signaling. In *Proc. of InOWo'08*, pages 1–5, Hamburg, Germany.
98. R. Dinis and P. Silva. Analytical evaluation of nonlinear effects in MC-CDMA signals. 5(8):2277–2284, August 2006.
99. B. Xu, C. Yang, and S. Mao. A carrierless UWB/OFDM system. In *Proc. of IEEE ICSP'04*, pages 1–5, Beijing, China.
100. S. Coleri, M. Ergen, A. Puri, and A. Bahai. Channel estimation techniques based on pilot arrangement in OFDM systems. 48(3):223–229, September 2002.
101. J. Guerreiro, R. Dinis, and P. Montezuma. On the optimum performance of coded OFDM with strongly nonlinear transmitters. In *Proc. of VTC Fall'13*, pages 1–5, Las Vegas, USA.

102. J. Guerreiro, R. Dinis, and P. Montezuma. Approaching the maximum likelihood performance with nonlinearly distorted OFDM signals. In *Proc. of IEEE Vehicular Technology Conference (Spring) (VTC'12)*, pages 1–5, Yokohama, Japan.
103. J. Guerreiro, R. Dinis, and P. Montezuma. Optimum and sub-optimum receivers for OFDM signals with iterative clipping and filtering. In *Proc. of VTC Fall'12*, pages 1–5, Quebec City, Canada.
104. J. Guerreiro, R. Dinis, and P. Montezuma. ML-based receivers for underwater networks using OFDM signals with strong nonlinear distortion effects. In *Proc. of MILCOM'12*, pages 1–7, Orlando, USA.
105. J. Guerreiro, R. Dinis, and P. Montezuma. On quasi-optimum detection of nonlinearly distorted OFDM signals. In *Proc. of DoCEIS'14*, pages 1–9, Costa de Caparica, Portugal, April.
106. D. Cox. Linear amplification with nonlinear components. 22(12):1942–1945, December 1974.
107. R. Dinis and A. Gusmão. Performance evaluation of OFDM transmission with conventional and two-branch combining power amplification schemes. In *Proc. of IEEE GLOBECOM'96*, pages 1–5, London, England.
108. R. Dinis and A. Gusmão. Nonlinear signal processing schemes for OFDM modulations within conventional or LINC transmitter structures. *European Trans. on Telecomm.*, 19(3):257–271, April 2008.
109. R. Dinis and A. Gusmão. CEPB-OFDM: a new technique for multicarrier transmission with saturated power amplifiers schemes. In *Proc. of IEEE ICCS'96*, pages 1–5, Singapore City, Singapore.
110. J. Guerreiro, R. Dinis, and P. Montezuma. A simple method for the analytical characterization of OFDM schemes with LINC transmitter structures. In *Proc. of ISWCS'15*, pages 1–5, Brussels, Belgium, August.
111. S. Thompson. *Constant OFDM envelope phase modulation*. PhD thesis, California Univ, 7 2005.
112. E. Ayanoglu, P. Heydari, and F. Capolino. Millimeter-wave massive MIMO: the next wireless revolution? 52(9):56–62, September 2014.
113. J. Guerreiro, R. Dinis, and P. Montezuma. Optimum performance and spectral characterization of CE-OFDM signals. In *Proc. of VTC Fall'15*, pages 1–5, Boston, USA, September.
114. J. Guerreiro, R. Dinis, and P. Montezuma. On the optimum performance of CE-OFDM schemes in frequency-selective channels. In *Proc. of TEMU'16*, pages 1–5, Crete, Greece, July.
115. J. Guerreiro, R. Dinis, and P. Montezuma. On the detection of CE-OFDM signals. 20(5):1–4, August 2016.
116. M. Torabi, D. Haccoun, and J. Frigon. Relay selection in AF cooperative systems: An overview. 7(4):104–113, December 2012.
117. M. Bhatnagar. On the capacity of decode-and-forward relaying over Rician fading channels. 17(6):1100–1103, May 2013.
118. J. Guerreiro, R. Dinis, and P. Montezuma. Analytical evaluation of nonlinear amplify-and-forward relay systems for OFDM signals. In *Proc. of VTC Fall'14*, pages 1–5, Vancouver, Canada, September.
119. S. Alamouti. A simple transmit diversity technique for wireless communications. 16(8):1451–1458, October 1998.

120. V. Tarokh, H. Jafarkhani, and A. Calderbank. Space-time block codes from orthogonal designs. 45(5):1456–1467, July 1999.
121. G. Foschini and M. Gans. On limits of wireless communications in a fading environment when using multiple antennas. *Wireless Personal Commun.*, 6(3):311–335, March 1998.
122. P. Wolniasky, G. Foschini, G. Golden, and R. Valenzuela. V-BLAST: an architecture for realizing very high data rates over rich-scattering wireless channel. In *Proc. ISSSE'98*, pages 295–300, Pisa, Italy.
123. G. Stuber, J. Barry, S. McLaughlin, Y. Li, M. Ingram, and T. Pratt. Broadband MIMO-OFDM wireless communications. *Proc. IEEE*, 92(2):271–294, February 2004.
124. H. Yang. A road to future broadband wireless access: MIMO-OFDM based air interface. 43(1):53–60, January 2005.
125. M. Juntti, M. Vehkapera, J. Leinonen, V. Zexian, D. Tujkovic, S. Tsumura, and S. Hara. MIMO MC-CDMA communications for future cellular systems. 43(2):118–124, February 2005.
126. Fredrik Rusek, Daniel Persson, Buon Kiong Lau, Erik G. Larsson, Thomas L. Marzetta, Ove Edfors, and Fredrik Tufvesson. Scaling up MIMO: opportunities and challenges with very large arrays. 30(1):40–60, October 2013.
127. V. Jungnickel, K. Manolakis, W. Zirwas, B. Panzner, V. Braun, M. Lossow, M. Sternad, R. Apelfröjd, and T. Svensson. The role of small cells, coordinated multipoint, and massive MIMO in 5G. 52(5):44–51, May 2014.
128. E. Larsson, O. Edfors, F. Tufvesson, and T. Marzetta. Massive MIMO for next generation wireless systems. 52(2):186–195, February 2014.
129. R. Dinis and A. Gusmão. Power efficient low complexity precoding for massive MIMO systems. In *Proc. of IEEE GlobalSIP'96*, pages 647–651, Atlanta, GA, USA.
130. G. Lebrun, J. Gao, and M. Faulkner. MIMO transmission over a time-varying channel using SVD. 4(2):757–764, March 2005.
131. J. Guerreiro, R. Dinis, and P. Montezuma. On the assessment of nonlinear distortion effects in MIMO-OFDM systems. In *Proc. of VTC Spring'16*, pages 1–5, Nanjing, China, May.
132. J. Guerreiro, R. Dinis, and P. Montezuma. On the use of 1-bit DACs in massive MIMO systems. *Elect. Lett.*, 52(9):778–779, April 2016.
133. J. Guerreiro, R. Dinis, and P. Montezuma. On the evaluation of clipping effects in massive MIMO-OFDM systems. In *Proc. of VTC Fall'16*, pages 1–5, Montreal, Canada, September.
134. J. Guerreiro, R. Dinis, and P. Montezuma. On the performance of quantized DMT signals. In *Proc. of ICTC'15*, pages 1–5, Jeju Island, Korea, October.
135. W. Shieh and C. Athaudage. Coherent optical orthogonal frequency division multiplexing. *Electron. Lett*, 42(10):587–589, May 2006.
136. J. Gordon and L. Mollenauer. Phase noise in photonic communications systems using linear amplifiers. *Optics Letters*, 15(23):1351–1353, September 1990.
137. A. Mecozzi. Limits to long-haul coherent transmission set by the Kerr nonlinearity and noise of the in-line amplifiers. 12(11):1993–2000, November 1994.
138. J. Guerreiro, R. Dinis, and P. Montezuma. On the optimum performance of multicarrier optical signals with nonlinear phase distortion. In *Proc. of GLOBECOM'13*, pages 2442–2447.

139. J. Guerreiro, R. Dinis, and P. Montezuma. Analytical characterization and optimum performance of nonlinearly distorted coherent optical OFDM signals. *International Journal of Microwave and Optical Technology*, 10(2):132–142, March 2015.
140. J. Armstrong. OFDM for optical communications. 27(3):189–204, February 2009.
141. D. Dissanayake and J. Armstrong. Comparison of ACO-OFDM, DCO-OFDM and ADO-OFDM in IM/DD systems. 31(7):1063–1072, January 2013.
142. J. Carruthers and J. Kahn. Multiple-subcarrier modulation for wireless infrared communication. 14(3):538–546, April 1996.
143. O. Gonzalez, R. Perez-Jimenez, S. Rodriguez, J. Rabadan, and A. Ayala. OFDM over indoor wireless optical channel. 152(4):199–204, August 2005.
144. K. Kavehrad. Performance analysis of precoding-based asymmetrically clipped optical orthogonal frequency division multiplexing wireless system in additive white Gaussian noise and indoor multipath channel. *SPIE J. Optical Engineering*, 53(8):1–12, August 2014.
145. J. Carruthers and J. Kahn. Power efficient optical OFDM. *Elect. Lett*, 42(6):370–372, March 2006.
146. J. Guerreiro, R. Dinis, and P. Montezuma. On the impact of the clipping techniques on the performance of optical OFDM. In *Proc. of ICETE'14*, pages 1–5, Austria, Vienna, August.
147. J. Guerreiro, R. Dinis, and P. Montezuma. Analytical characterization and optimum performance of DC-biased optical OFDM signals. *IET Trans. Commun.*, 9(7):969–974, May 2015.
148. J. Guerreiro, R. Dinis, and P. Montezuma. Efficient simulation of nonlinear effects on multicarrier signals. In *Proc. of MIC'14*, pages 1–5, Innsbruck, Austria, February.
149. J. Guerreiro, R. Dinis, and P. Montezuma. An accurate low complexity method for studying quantization effects in base station cooperation systems. In *Proc. of VTC Fall'15*, pages 1–5, Boston, USA, September.
150. P. Fernandes, J. Guerreiro, R. Dinis, and P. Montezuma. On the capacity of nonlinear massive MIMO-OFDM systems. In *Proc. of VTC Fall'16*, pages 1–5, Montreal, Canada, September.
151. M. Akkouchi. On the convolution of Gamma distributions. *Soochow Jornal of Matematics*, 31(2):205–211, April 2005.

Index

Note: *Italic* page numbers refer to *figures* and page numbers followed by 'n' refer to notes.

D

E

F

G

H

I

K

L

M

N

O

For Product Safety Concerns and Information please contact our EU representative GPSR@taylorandfrancis.com
Taylor & Francis Verlag GmbH, Kaufingerstraße 24, 80331 München, Germany

www.ingramcontent.com/pod-product-compliance
Lightning Source LLC
LaVergne TN
LVHW010556110826
845149LV00003B/677

9781032708768